Topics in Applied Physics Volume 8

Topics in Applied Physics Founded by Helmut K. V. Lotsch

1 **Dye Lasers** 2nd Ed. Editor: F. P. Schäfer
2 **Laser Spectroscopy** of Atoms and Molecules. Editor: H. Walther
3 **Numerical and Asymptotic Techniques in Electromagnetics** Editor: R. Mittra
4 **Interactions on Metal Surfaces** Editor: R. Gomer
5 **Mössbauer Spectroscopy** Editor: U. Gonser
6 **Picture Processing and Digital Filtering** 2nd Edition. Editor: T. S. Huang
7 **Integrated Optics** 2nd Ed. Editor: T. Tamir
8 **Light Scattering in Solids** 2nd Ed. Editor: M. Cardona
9 **Laser Speckle** and Related Phenomena Editor: J. C. Dainty
10 **Transient Electromagnetic Fields** Editor: L. B. Felsen
11 **Digital Picture Analysis** Editor: A. Rosenfeld
12 **Turbulence** 2nd Ed. Editor: P. Bradshaw
13 **High-Resolution Laser Spectroscopy** Editor: K. Shimoda
14 **Laser Monitoring of the Atmosphere** Editor: E. D. Hinkley
15 **Radiationless Processes** in Molecules and Condensed Phases. Editor: F. K. Fong
16 **Nonlinear Infrared Generation** Editor: Y.-R. Shen
17 **Electroluminescence** Editor: J. I. Pankove
18 **Ultrashort Light Pulses** Picosecond Techniques and Applications Editor: S. L. Shapiro
19 **Optical and Infrared Detectors** 2nd Ed. Editor: R. J. Keyes
20 **Holographic Recording Materials** Editor: H. M. Smith
21 **Solid Electrolytes** Editor: S. Geller
22 **X-Ray Optics.** Applications to Solids Editor: H.-J. Queisser
23 **Optical Data Processing.** Applications Editor: D. Casasent
24 **Acoustic Surface Waves** Editor: A. A. Oliner
25 **Laser Beam Propagation in the Atmosphere** Editor: J. W. Strohbehn
26 **Photoemission in Solids I.** General Principles Editors: M. Cardona and L. Ley
27 **Photoemission in Solids II.** Case Studies Editors: L. Ley and M. Cardona
28 **Hydrogen in Metals I.** Basic Properties Editors: G. Alefeld and J. Völkl
29 **Hydrogen in Metals II** Application-Oriented Properties Editors: G. Alefeld and J. Völkl
30 **Excimer Lasers** Editor: Ch. K. Rhodes
31 **Solar Energy Conversion.** Solid State Physics Aspects. Editor: B. O. Seraphin
32 **Image Reconstruction from Projections** Implementation and Applications Editor: G. T. Herman
33 **Electrets** Editor: G. M. Sessler
34 **Nonlinear Methods of Spectral Analysis** Editor: S. Haykin
35 **Uranium Enrichment** Editor: S. Villani
36 **Amorphous Semiconductors** Editor: M. H. Brodsky
37 **Thermally Stimulated Relaxation in Solids** Editor: P. Bräunlich
38 **Charge-Coupled Devices** Editor: D. F. Barbe
39 **Semiconductor Devices** for Optical Communication. 2nd Ed. Editor: H. Kressel
40 **Display Devices** Editor: J. I. Pankove
41 **The Computer in Optical Research** Methods and Applications. Editor: B. R. Frieden
42 **Two-Dimensional Digital Signal Processing I** Linear Filters. Editor: T. S. Huang
43 **Two-Dimensional Digital Signal Processing II** Transforms and Median Filters. Editor: T. S. Huang
44 **Turbulent Reacting Flows** Editors: P. A. Libby and F. A. Williams
45 **Hydrodynamic Instabilities and the Transition to Turbulence** Editors: H. L. Swinney and J. P. Gollub
46 **Glassy Metals I** Editors: H.-J. Güntherodt and H. Beck
47 **Sputtering by Particle Bombardment I** Editor: R. Behrisch
48 **Optical Information Processing** Fundamentals. Editor: S. H. Lee
49 **Laser Spectroscopy of Solids** Editors: W. M. Yen and P. M. Selzer
50 **Light Scattering in Solids II.** Basic Concepts and Instrumentation Editors: M. Cardona and G. Güntherodt
51 **Light Scattering in Solids III.** Recent Results Editors: M. Cardona and G. Güntherodt
52 **Sputtering by Particle Bombardment II** Editor: R. Behrisch
53 **Glassy Metals II.** Electronic, Magnetic, and Mechanical Properties, Structure and Applications Editors: H. J. Güntherodt and H. Beck
54 **Light Scattering in Solids IV.** Electronic Scattering, Spin Effects, SERS, and Morphic Effects Editors: M. Cardona and G. Güntherodt

Light Scattering in Solids I

Introductory Concepts

Edited by M. Cardona

With Contributions by
M. H. Brodsky E. Burstein M. Cardona
L. M. Falicov M. V. Klein R. M. Martin
A. Pinczuk A. S. Pine Y.-R. Shen

Second Corrected and Updated Edition

With 111 Figures

Springer-Verlag Berlin Heidelberg New York 1983

Professor Dr. MANUEL CARDONA

Max-Planck-Institut für Festkörperforschung
D-7000 Stuttgart 80, Fed. Rep. of Germany

ISBN 3-540-11913-2 2. Auflage Springer-Verlag Berlin Heidelberg New York
ISBN 0-387-11913-2 2nd edition Springer-Verlag New York Heidelberg Berlin

ISBN 3-540-07354-X 1. Auflage Springer-Verlag Berlin Heidelberg New York
ISBN 0-387-07354-X 1st edition Springer-Verlag New York Heidelberg Berlin

Library of Congress Cataloging in Publication Data. Main entry under title: Light scattering in solids. (Topics in applied physics; v. 8–) Bibliography: v. 1, p. Includes index. 1. Semiconductors – Optical properties. 2. Raman effect. 3. Solids – Optical properties. I. Cardona, Manuel, 1934–. II. Brodsky, M. H. (Marc Herbert), 1938–. III. Series: Topics in applied physics; v. 8, etc. QC611.6.06L53 535'.4 82-17025

Typesetting, printing and bookbinding: Brühlsche Universitätsdruckerei, Gießen 2153/3130-543210

Preface to the Second Edition

The first edition of this book appeared in 1975. Five years later I decided to edit, in collaboration with G. Güntherodt, another volume on light scattering in solids. In the meantime light scattering spectroscopy had become a standard, well established technique with most of the necessary equipment commercially available. Results have been appearing in the literature at an increasing pace and, while the basic principles of the phenomenon were covered in the first volume, we felt that the wealth of new information available warranted issuing a new volume of the Topics in Applied Physics series. The originally planned volume exploded as the authors overran their page allotments and a few more authors were added. Thus, by the time this second edition appears, instead of a new additional volume we shall have three (including this one) and a further volume is still forthcoming.

This paperback edition is essential identical to the original hardcover one, except that Chapter 7 has been supplemented with new results and a new chapter (Chap. 8) has been added. This new chapter outlines the current status of light-scattering spectroscopy applied to the study of solids and provides numerous new references. A few typographical errors in the original edition were corrected.

Because of limitation of space, we can do little more in Chapter 8 than point out to the reader the latest literature as well as to the now completed treatment of *Light Scattering in Solids I–IV*, published in the Topics in Applied Physics series (Vols. 8, 50, 51, and 54). For convenient, cross-referencing we have added the Roman numeral one to the title of this second edition, and supplemented its various chapters with references to related work covered in other volumes of this series: their contents have been listed on pages 335–347. A few relevant references to recent work have also been included. Completeness has not been attempted in the compilation of these references. Instead we have tried to give a sampling of the current works by some of the most representative groups.

Stuttgart, July 1982 MANUEL CARDONA

Preface to the First Edition

This book is devoted to the problem of inelastic light scattering in semiconductors, i.e., to processes in which a photon impinges upon a semiconductor, creating or anihilating one or several quasi-particles, and then emerges with an energy somewhat different from that of the incident photon. In light scattering spectroscopy the incident photons are monochromatic; one measures the energy distribution of the scattered photons with a spectrometer. Because of its monochromaticity, power, and collimation, lasers are ideal sources for light scattering spectroscopy. Consequently, developments in the field of light scattering have followed, in recent years, the developments in laser technology. The scattering efficiencies are usually weak and thus light scattering spectroscopy requires sophisticated double and triple monochromators with high stray light rejection ratio. Both, powerful lasers and good monochromators are specially important for studying the scattering of light to which the samples of interest are opaque, as is the case in most semiconductors. This explains why these materials are relatively latecomers to the field of light scattering.

In spite of these difficulties, the field of light scattering in semiconductors has experienced a boom in recent years, and reached a certain degree of maturity. Because of space limitations, the editor was faced with the necessity of making a choice in the subjects to be included. In spite of the natural bias towards his own research interests he hopes to have gathered a number of articles representative of present-day research in the field.

Chapter 1 contains a historical survey of the field of light scattering in general and of the bibliography in the fields of light scattering by one and two phonons, including resonant Raman scattering, i.e., scattering of phonons whose energy lies in the neighborhood of strong electronic structure in the optical constants.

Chapter 2 discusses the fundamentals of light scattering, its phenomenological description, kinematics and selection rules. It also contains an introduction to the microscopic theory.

Chapter 3 describes in detail the phenomenon of resonant Raman scattering which, in isolating the electronic states which participate in the resonance, yields very detailed information about the scattering mechanism and the electron-phonon interaction. Several models are considered quantitatively in order to give a feeling for the interpretation of observed resonances.

Chapter 4 treats scattering by electronic excitations in semiconductors, a field of considerable technological interest. It includes scattering by free particles, plasmons and excitations between impurity levels. In all cases the specific effects of the intricacies of constant-energy surfaces in semiconductors are taken into account.

Chapter 5 discusses Raman scattering by amorphous semiconductors. This field which has received considerable attention since the discovery by the contributor and his coworkers that the observed first-order spectra often contain information about the density of one-phonon states in alotropic crystalline materials. A discussion of the implications of the observed spectra with respect to short-range order and chemical bonding is included.

Chapter 6 describes the fundamentals of Brillouin scattering in semiconductors, and the spontaneous and stimulated processes which can be observed when phonons are generated in polar materials by means of the acousto-electric effect.

Finally, Chapter 7 is devoted to stimulated Raman scattering, another subject of considerable technological importance because of the tunable spin-flip lasers and the recently developed spectroscopical technique referred to as CARS (coherent antistokes Raman spectroscopy).

The editor, a relatively new comer to the field of light scattering, has profited from his collaboration in this field with a large number of scientists, graduate students, colleagues at Brown University and at the Max Planck Institute and other institutions. It would be cumbersome to name them all here. Most of their names appear in the literature referenced throughout this tract. Last but not least thanks are due to all the contributors for keeping the deadlines as well as they could in spite of numerous other commitments and for their patient consideration of the editor's suggestions.

Stuttgart, May 1975 MANUEL CARDONA

Contents

Contributors

Brodsky, Marc H.

IBM Thomas J. Watson Research Center,
Yorktown Heights, NY 10598, USA

Burstein, Elias

University of Pennsylvania, Department of Physics,
Philadelphia, PA 19174, USA

Cardona, Manuel

Max-Planck-Institut für Festkörperforschung,
D-7000 Stuttgart 80, Fed. Rep. of Germany

Falicov, Leopoldo M.

University of California, Physics Department,
Berkeley, CA 94720, USA

Klein, Miles V.

University of Illinois at Urbana-Champaign, Physics Department,
Urbana, IL 61801, USA

Martin, Richard M.

Xerox Corporation, Palo Alto Research Center,
Palo Alto, CA 94304, USA

Pinczuk, Aron

Bell Laboratories, Holmdel, NJ 07733, USA

Pine, Alan S.

National Bureau of Standarts,
Washington, DC 20234, USA

Shen, Yuen-Ron

University of California, Department of Physics,
Berkeley, CA 94720, USA

1. Introduction

M. CARDONA

With 3 Figures

Licht, mehr Licht ...
J. W. GOETHE

Most of this volume is devoted to light scattering in semiconductors with special emphasis on Raman scattering by phonons, and, in particular, resonant Raman scattering. Among semiconductors, the family of tetrahedrally coordinated materials, including the structures of germanium, zincblende, wurtzite, ternary chalcopyrites, and others, form an excellent laboratory for light scattering experiments. They are simple, their band structures are well understood and offer an enormous variety. Also, considerable information is available about their phonon spectra. Besides, semiconductors muster a number of interesting many-body-type quasi-particle excitations such as excitons, plasmons, polarons, polaritons, exciton drops, excitons bound to impurities, etc. These quasi-particles can sometimes be excited in a light scattering process. They can also be studied indirectly by observing resonances in other scattering processes, when the frequency of the scattered light is in the neighborhood of the quasi-particle frequency. Recently, Cu_2O also became a one-material laboratory to test and analyze the mechanisms for Raman scattering processes. This material has, in contrast to zincblende, a large number of optical phonons at $\boldsymbol{k}=0$ and a forbidden direct edge with sharp excitons. This edge and excitons occur in a wavelength region very convenient for studies with tunable dye lasers.

1.1. Historical Remarks

The foundations of the light scattering process were established long ago. In 1922 BRILLOUIN [1.1] predicted the scattering of light by long wavelength elastic sound waves. SMEKAL [1.2] developed in 1923 the theory of light scattering by a system with two quantized energy levels; this theory contained the essential characteristics of the phenomena discovered by RAMAN [1.3] and, independently, by LANDSBERG and MANDELSTAM [1.4] in 1928. It was soon realized that the newly discovered effect constituted an excellent tool to study excitations of molecules and molecular structure. Such studies dominated the field until about 1940. In the 1940's emphasis shifted to systematic investigations of single crystals in order to obtain information for the semi-empirical treatment of their lattice dynamics. Because of the small scattering cross sections, however, experiments were difficult and the field remained in the hands of relatively few groups. The advent of the laser in 1960 was to change this situation rather drastically. Its monochromaticity, coherence, collimation and power quickly made the old mercury arcs obsolete as sources for light scattering spectroscopy. The first reports on Raman scattering using the 6943 Å line of the pulsed ruby laser and photographic recording were soon to appear [1.5]. The appearance of cw lasers, which made possible photoelectric recording with photon counting, was to make photographic recording quickly obsolete. Ever since the discovery of the laser, progress in light scattering has followed within a short distance the main developments in laser technology, including the cw He–Ne laser, the Ar^+ and Kr^+ ion lasers, and, most recently, the tunable pulsed and cw dye lasers.

The range of applicability of light scattering is at present so wide that an exhaustive review is nearly impossible. I would like therefore to give at the outset a few general references bearing on the field of solids. Excellent references are the Proceedings of the first [1.6] and the second [1.7] International Conferences on light scattering in solids. The proceedings of the third conference to be held in Campinas, Brazil in July 1975 [1.8] are expected to continue the same tradition of excellence. The recent two volumes edited by ANDERSON [1.9], the review of Raman Scattering in Semiconductors by MOORADIAN [1.1] and the earlier review by BRANDMÜLLER [1.11] should also be mentioned. The book by SUSHCHINSKII [1.12], discussing both molecules and crystals, is also recommended. An exhaustive review of the theory of Raman scattering by phonons, especially of its group theoretical aspects has been recently written by BIRMAN [1.13]. In addition, a number of articles on Raman scattering in semiconductors can be found in the proceedings of the biyearly International Semiconductors Conferences since the one held in Kyoto in 1966. A listing of all the proceedings with appropriate sources can be found in [1.14].

1.2. Scattering by Phonons in Semiconductors

In the early days of Raman scattering, work was only possible with materials transparent to the scattering radiation. The scattering volume, limited by the absorption length, was too small in opaque samples to make observation possible. The early observations of first-order Raman scattering by optical phonons in diamond [1.15] and CdS [1.16] belong to the days of the mercury arc. In the same spirit the early measurements with lasers were limited to transparent semiconductors; PORTO and coworkers reported the first-order phonon spectrum of ZnO [1.17] and CdS [1.18]. In this work, a number of overtones of LO phonons was seen and resonance enhancement, observed by using two laser lines, was hinted at. In order to be able to study the III–V semiconductors, mostly opaque to the visible, MOORADIAN and WRIGHT [1.19] used the 1.06 μm infrared radiation of the Nd-YAG laser. They reported the first-order phonon spectra of GaAs, InP, AlSb and GaP. A major breakthrough occurred with the observation by PARKER et al. [1.20] of first- and second-order scattering in silicon and first-order scattering in germanium with the 4880 Å line of the Ar^+ laser to which these materials are opaque. First order back-scattering had already been observed for Si by RUSSELL with a He–Ne laser [1.21]. The measurements were performed in the back-scattering configuration which has later become standard for studies of resonace effects. It is only rather recently that Brillouin measurements have become possible for opaque samples thanks to the use of multiple pass Fabry-Perot interferometers [1.22]. These measurements have been so far limited to Ge and Si. In sharp contrast to the back-scattering experiments, small-angle scattering in transparent polar materials has been used to obtain polariton (i.e. coupled phonon-photon modes) dispersion relations since the pioneer work of HENRY and HOPFIELD [1.23].

The second-order spectra have been investigated for transparent materials from the early days of mercury arcs (see [1.24] for NaCl and [1.25] for diamond) and interpreted in terms of overtone and combination modes [1.26]. These spectra are usually very weak and thus very difficult to study with the old sources. The advent of the laser gave impetus to the investigation of these often broad but very highly structured spectra [1.27]. The phonon assignments, aided by observations of the temperature dependence of the scattered intensities, often remained conjectural. Recently, however, considerable progress has been made by systematic separation of the irreducible tensor components of the second-order spectra, mostly in the back-scattering configuration [1.28, 29]. These studies led to the realization that the completely symmetric component of the spectrum (Γ_1) is composed mostly of two-phonon overtones and represents rather faithfully the density of *one-phonon* states with a change by a factor of two in the energy scale. The Γ_{12} spectrum is nearly negligible. This fortunate simplicity of the two-phonon spectra is not common to other solids (e.g. the alkali halides) and does not apply to the two-phonon infrared absorption of germanium and silicon either.

Crystalline semiconductors exhibit first-order scattering only by phonons with $\boldsymbol{k} \simeq 0$. Phonons with large $\boldsymbol{k}$ must be observed in second order. Many of the tetrahedral semiconductors, however, may be prepared in amorphous form for which $\boldsymbol{k}$ conservation no longer holds. Thus, in the amorphous materials the first-order Raman spectrum is broad and corresponds to the density of *one-phonon* states [1.30]. Another fortunate circumstance is the fact that this density of states is often a broadened version of that of the corresponding crystalline material [1.30]. This implies a large amount of conservation of short-range order. Chapter 4 of this book is devoted to light scattering by amorphous semiconductors (see also [Ref. 8.1, Chap. 2]).

The frequency shift of the sharp lines observed in first-order scattering can be determined with an accuracy far superior to that obtained in neutron scattering. This also applies to some of the features (critical points) in the second-order spectra and makes Raman scattering an ideal technique to study the effect of perturbations on the phonon spectra. The method has been applied to the study of the temperature depedence of phonon frequencies and widths [1.31], the effect of doping, including self-energy shifts and broadenings [1.32] and impurity induced local vibrational modes [1.33], alloying [1.34], uniaxial [1.35], and hydrostatic stress [1.36], and electric-field-induced scattering [1.37]. Scattering by vibrational surface modes (surface polaritons) has been recently observed by EVANS et al. [1.38].

1.3. Resonances in the Scattering by Phonons

The scattering by phonons occurs mainly through virtual intermediate electronic transitions. Thus, for incident and scattered photons near the energy of interband transitions structure in the Raman cross section must appear. This structure can be used for studying electronic transitions in a way similar to that used in modulation spectroscopy [1.39]. The modulation in this case is not applied externally but produced by the appropriate Raman phonon. The most important application of resonance Raman scattering, however, is to elucidate the scattering mechanism (always some form of electron-phonon interaction) and to extract electron-phonon interaction constants or deformation potentials. Resonant Raman scattering by optical phonons is discussed in detail in Chapters 2 and 3 and resonant Brillouin scattering in Chapter 6 (see also [Ref. 8.1, Chap. 2] and [Ref. 8.2, Chaps. 6 and 7]).

The main difficulty is to find a tunable source operating in the region around the critical point under study. Fortunately, the tetrahedral materials have a number of critical points (gaps) which vary drastically in energy from one member of the family to the other. It is therefore often possible to choose the materials so as to have the critical point to be studied in the region where tunable sources are available. Also, the gaps of semiconductors have large temperature and pressure coefficients and can be changed continuously by alloying. These properties can be used advantageously for studies of resonant Raman scattering: one can tune the gap instead of tuning the laser.

The first resonant measurements in semiconductors were performed for CdS using the discrete lines of the Ar^+ laser which cover the range between 2,38 and 2.73 eV [1.40]. Later, measurements for CdS [1.41] confirmed the observed resonance behavior and established the existence of an antiresonance (zero scattering) in the cross section for TO scattering (not for LO) immediately below the absorption edge. Such structure is now understood as a cancellation between the contributions of the edge and of higher transitions: these contributions must thus have opposite signs. Such cancellations do not appear in the Raman cross section of zincblende-type materials, they are common, however, for their stress-induced birefringence constants [1.42] and lead to zeros in the Brillouin scattering cross section [1.43] as discussed in Chapter 6. Early measurements using discrete lines showed that this antiresonance is absent for the E_0 gap of zincblende-type materials such as GaP [1.44].

Measurements taken with a few discrete lines, however, may sometimes miss sharp structure between discrete points. This can be avoided by using a discrete laser line and tuning the gap with an external perturbation, e.g., temperature. By this method, PINCZUK and BURSTEIN [1.45] were able to detect a resonance near the E_1 gap of InSb. This resonance was studied in greater detail by LEITE and SCOTT for InAs using the lines of the Ar^+ laser [1.46]. The shape of the E_1 resonance in Ge was elucidated by RENUCCI et al., shifting the gap by alloying with 22% of Si so as to bring it to the region where the discrete Ar^+ laser lines are found [1.47].

A breakthrough occurred in the field of resonance scattering with the advent of the tunable dye laser [1.48]. The dye laser with the widest tunability range is the pulsed laser. A N_2-laser-pumped pulsed dye laser has been used by BELL et al. [1.49] to study the direct edge (E_0) of GaP and its spin-orbit splitting ($E_0 + \Delta_0$). A similar laser was used for Raman measurements by OKA and KUSHIDA [1.50]. A flash-lamp pumped dye laser was used by DAMEN and SHAH [1.51] to study resonances in the LO scattering near the absorption due to excitons bound to impurities.

The pulsed laser has, when compared with the cw laser, the disadvantage of not permitting conventional use of photon counting [1.52]. Thus, its use for Raman scattering is limited and most resonance measurements have been performed with the cw laser [1.53]. The range of tunability was initially limited to that of the dye Rhodamine 6G

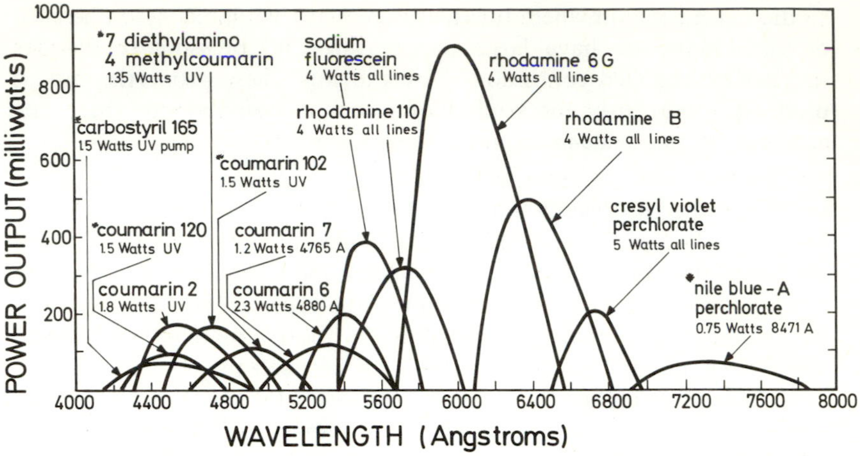

Fig. 1.1. Power obtained for various dyes, pump powers and pump wavelengths as a function of dye laser wavelength in cw operation. (From [1.53].) For a still more up-to-date version see [Ref. 8.1, Fig. 2.7]

(1.85–2.20 eV). The recent work of YARBOROUGH [1.53] has extended the range of the cw dye laser to the region 1.55–2.90 eV (see Fig. 1).

The cw laser has been used to study two typical E_0 resonances (GaP [1.49] and ZnTe [1.54]) and two typical $E_1 - E_1 + \Delta_1$ resonances [1.55, 56].

The appearance near resonance of *LO phonon* scattering in forbidden polarization configurations was discussed by MARTIN and DAMEN [1.57]. This phenomenon, which was attributed to the electric field of the LO phonons (Fröhlich interaction), has received considerable theoretical and experimental attention, as discussed in Chapters 2 and 3 of this book. It is usually only present for parallel polarized incident and scattered fields. Being a forbidden effect, it resonates more strongly than the allowed LO and TO scattering and sometimes overcomes the allowed LO scattering near the resonance. Sharply resonant 2 LO scattering involving two phonons near Γ and due also to Fröhlich interaction, has been observed in a number of materials [1.58, 59]. With the exception of 2 LO structure, the rest of the second-order spectrum resonates in a manner similar to the allowed first-order peaks [1.58]. This realization led to the conclusion that the structure of both first- and second-order scattering is produced by the electron two-phonon interaction [1.58]. The ratio of second- to first-order scattering thus yields the ratio of electron two-phonon to electron one-phonon interaction constants. The electron two-phonon interaction is also of importance in a number of other seemingly unrelated phenomena [1.59]. Non-polar materials, like Ge and Si, also show in the neighborhood of gaps a resonance in the scattering by two optical phonons near Γ [1.60]. This resonance is stronger than that of the remaining second-order spectrum and has been attributed to two iterated electron one-phonon processes [1.60]. A similar iterated resonance has been recently observed near the indirect gap of semiconductors for phonons with the wavevector of the indirect gap [1.61].

In spite of the emphasis on tetrahedral semiconductors, we should mention here some of the recent work on Cu_2O. The lowest exciton in this material is dipole forbidden since the corresponding one-electron bands have the same parity. It occurs at 2.11 eV at liquid He, a region ideal for the cw dye laser. This exciton is very sharp and therefore enables the observation of separate resonances when the incident and scattered frequencies are swept across it. The forbidden nature of the exciton allows only resonances in which one of the participating photons (e.g., the incident one), produces quadrupole transitions. In this case odd parity (usually forbidden) phonons are seen in the Raman scattering [1.62]. An interesting scattering process involving two phonons with iterated electron one-phonon interaction has also been reported for this material [1.63]. YU and SHEN have also observed a shift in the frequencies of the Raman peaks of multiphonon processes when the laser wavelength is swept through resonance in Cu_2O [1.64]. They were able to relate these shifts to the dispersion relations of the phonons near $\boldsymbol{k}=0$.

The resonant Raman experiments enable us to elucidate the nature of the scattering mechanism. Once this is done, the corresponding electron-phonon interaction constants can be obtained, provided one has determined absolute scattering cross sections. This is seldom done; it is particularly difficult in back-scattering experiments where cross sections can be drastically altered as a result of surface conditions. "Absolute" measurements have been performed in solids (e.g. GaAs [1.65]) using transparent radiation and comparing the scattered intensities to those of a standard scatterer (e.g. benzene [1.66]). Because of the difficulties involved in doing absolute measurements, the resonance work has been mostly used for the determination of electron one-phonon coupling constants (deformation potentials) relative to each other, in particular when *two* coupling constants determine *one* given resonance, and of ratios of two-phonon to one-phonon constants. Usually, only one gap (or a spin-orbit split multiplet belonging to a given gap) has been studied so far for a given material. As the spectral range of tunable lasers, parametric oscillators, and frequency doubled tunable sources available is extended, the measurement of resonances around several gaps of a given material, and therefore, of their relative electron-phonon interaction constants, should become possible. We point out that for zincblende-type semiconductors only E_0 and E_1 resonances have been studied. There is a large additional number of relatively well characterized electronic gaps (E_0', E_1', E_2 ...) [1.67] which should exhibit Raman resonances and will be studied when appropriate sources are available. Relative measurements of scattering cross sections of the various materials of the family with respect to each other should also yield the relative values of the coupling constants and thus contribute to the understanding of the systematics of the electron-phonon interaction. (For recent progress see [Ref. 8.1, Chap. 2].)

A method to determine the absolute value of the Raman tensor by measuring two-photon absorption of two laser beams with frequencies differing by the Raman frequency has been proposed [1.68]. Also, GRIMSDITCH and RAMDAS have recently measured the ratio of Brillouin to Raman cross sections in diamond [1.69]. The Brillouin cross section can be calculated from the well known piezo-optical constants. One can therefore determine from these measurements the absolute value of the Raman tensor but not its sign. The sign of the Raman tensor for a given scattering process can be determined when this process interferes with another one of known sign. In this way, it has been possible [1.70] to determine the sign of the one-phonon Raman tensor of silicon by observing its interference with the electronic intra-valence-band scattering for which the sign of the Raman tensor can be easily derived.

1.4. Theory of Scattering by Phonons

From the days of the original theoretical work mentioned in [1.1] it became clear that the quantum mechanical problem of calculating cross sections from first principles could be solved only in special cases. This is true, in particular, for solids. Consequently, phenomenological and semi-phenomenological theories have been used profusely to interpret the experimental results. Central to these theories is always the group theoretical use of the symmetry of the excitation involved in the scattering process to reduce the number of independent parameters of the phenomenological theory and to determine the correlation between the polarization and propagation vectors of the incident and the scattered radiation.

At the core of the phenomenological theory is the concept of Raman tensor, usually a second-rank tensor—in Chapter 2 this second-rank Raman tensor is called transition susceptibility—which connects the incident with the scattered electric field for a given channel of excitation (i.e., the transfer function for a given scattering process). Sometimes, however, one defines as the Raman tensor a fourth-rank tensor which, when contracted twice with the unit vector of the incident field and also twice with that of the scattered field, it gives the scattering probability for these polarizations [1.71]. This fourth-rank tensor is related to products of components of the second-rank Raman tensor in a simple way. Our statements, so far, may suggest that the scattering is a function of the direction of incident and scattered fields but not of the direction of propagation of the radiation. This is indeed true in many cases involving Raman scattering, but not always: like the dielectric tensor, the Raman tensor can have a dependence on the wavevector of the incident and scattered radiation ("spatial dispersion"). The absence of "spatial dispersion" in the usual Raman scattering in solids can be justified as due to the fact that the wavelength of light is large compared with the characteristic wavelength of microscopic processes involved (typically the lattice constant). The situation is fully different for Brillouin scattering since the energy of the excitation depends very strongly on the angle between the incident and the scattered wavevectors, on account of the linearity of the dispersion relation of acoustical phonons near $\boldsymbol{k}=0$. The Brillouin tensor vanishes for $\boldsymbol{k}=0$ and one can say that the observed effect is only a consequence of the spatial dispersion in the "Brillouin tensor".

The second-order Raman tensor is usually decomposed into irreducible symmetry components which therefore correspond to excitations of a well defined group theoretical symmetry. Tables of the form of these irreducible tensors and the number of independent elements have been given by several authors [1.72, 73]. It is interesting to point out that the Raman tensor for phonon scattering is symmetric away from any resonant electronic energies. Antisymmetric components can, however, appear near resonance. They have been listed in [1.72] but so far they have escaped experimental observation in connection with scattering by phonons. Antisymmetric Raman tensors do, however, play an important role in the scattering by magnetic excitations (magnons) [1.74] (see [Ref. 8.1, Table 2.1]).

The concept of polarizability has played a very important historical role in the theory of Raman scattering [1.75, 76]. The Raman polarizability is the quantum mechanical analog of the second-rank Raman tensor and describes the change in the electric polarizability produced by an excitation of the crystal. The polarizability tensor can be expanded in power series of the normal coordinates so as to obtain the Raman tensors of first, second, third order, etc. This expansion leads to a number of phenomenological treatments which have been quite fruitful in the past. One of them is the so-called bond polarizability theory of WOLKENSTEIN [1.77].

WOLKENSTEIN assumed that each bond has a polarizability which can be expanded in power series of its length. The total crystal or molecular polarizability is then obtained as the sum of bond polarizabilities. This theory has been used rather recently to interpret the first- and second-order Raman spectra and the elastic constants of group four semiconductors [1.78, 79]. The concept of bond polarizability is also very helpful for the theoretical analysis of the spectra of amorphous semiconductors (see Chapter 5).

The expansion of the polarizability or susceptibility in terms of the phonon normal coordinates enables us to obtain Raman tensors of solids by using the sophisticated methods developed in recent years for the calculation of band structures and dielectric constants of solids. A zone center phonon produces a lattice deformation which preserves the translational symmetry. One can therefore calculate the dielectric constant as a function of phonon coordinates u. By expanding this dielectric constant (or the susceptibility) as a function of u, we find:

$$\overleftrightarrow{\varepsilon}(u,\omega)=\overleftrightarrow{\varepsilon}(\omega)+\frac{d\overleftrightarrow{\varepsilon}(\omega)}{du}u+\frac{1}{2}\frac{d^2\overleftrightarrow{\varepsilon}}{du^2}u^2+\cdots. \tag{1.1}$$

The various derivatives in (1.1) define the first- and second-order Raman tensor and can be simply obtained from band structure dielectric constant calculation [1.80, 81].

1.4.1. Scattering by One Phonon

For an incident laser field $E_L \exp(-i\omega_L t)$ and a phonon $u=u_0\exp(\pm i\Omega t)$ Eq. (1.1) gives the induced dipole moment (to first order in u_0):

$$P(\omega_L\pm\Omega)=\frac{d\overleftrightarrow{\varepsilon}}{du}u_0E_L\exp(-i(\Omega_L\pm\Omega)t). \tag{1.2}$$

The $-$ and $+$ signs in front of Ω correspond to Stokes and anti-Stokes radiation, respectively. Thus, the intensity of the total (Stokes plus anti-Stokes) scattered dipole radiation is proportional to

$$I\propto\omega_L^4\left|\frac{d\overleftrightarrow{\varepsilon}}{du}\right|^2\langle u_0^2\rangle. \tag{1.3}$$

Equation (1.3) provides for the simplest description of the phenomenon of resonant Raman scattering: in the vicinity of sharp structure in $\overleftrightarrow{\varepsilon}$, sharp structure in $d\overleftrightarrow{\varepsilon}/du$ and thus in the scattered intensity I occurs. This equation contains the thermal average of the phonon displacement (or normal coordinate) $\langle u_0^2\rangle$. This average yields the sum of the Stokes and the anti-Stokes scattered intensities, as corresponds to the assumption of a negligibly small phonon frequency. The Stokes and anti-Stokes components can be separated by using the relation

$$
\begin{aligned}
\langle n|u_0^2|n\rangle &= |\langle n+1|u_0|n\rangle|^2 + |\langle n-1|u_0|n\rangle|^2 \\
&= \frac{\hbar}{2MN\Omega}(n+1) + \frac{\hbar}{2MN\Omega}n \\
&= \quad \text{Stokes} \quad + \text{ anti-Stokes}\,,
\end{aligned}
\tag{1.4}
$$

where M is the reduced mass (for Si half the atomic mass), N the number of unit cells and n is the Bose-Einstein statistical factor.

The foundations of the quantum mechanical theory of Raman and Brillouin scattering in solids were given by BORN and HUANG [1.76] and in a series of articles by LOUDON [1.73, 82, 83]. The Brillouin scattering can be phenomenologically described in terms of the elasto-optic constants [1.76, 84] since the dielectric constant modulation required for scattering is produced by the macroscopic strains associated with long wavelength acoustical phonons. A comparison of the phenomenological cross section for Brillouin scattering [1.76] with the quantum mechanical ones yields microscopic expressions for the elasto-optic constants [1.82].

In homopolar cubic semiconductors (e.g. Ge) the scattering intensities for the first-order optical LO and TO degenerate phonons are determined by a single parameter since at $\boldsymbol{k}=0$ they belong to one irreducible representation ($\Gamma_{25'}$). In polar semiconductors (e.g. GaAs) the LO phonons are accompanied by a longitudinal electric field which produces a modulation of the dielectric constant through the so-called first-order electro-optic effect [1.83]. Consequently, the intensities of LO and TO scattering become unrelated for finite $\boldsymbol{k}$; they are now determined by two independent parameters: the deformation potential Raman tensor and the electro-optic effect. This fact has been recognized by GRECHKO and OVANDER [1.85] as well as by LOUDON [1.83] although the functional dependences of the electro-optic contributions to the Raman tensor near a resonance given in [1.83 and 85] are in disagreement with each other [Ref. 8.1, Sect. 2.1.12].

A recent development of the phenomenological theory helps in the calculation of scattering tensors belonging to several coupled excitations such as phonons, photons, plasmons, etc. [1.86, 87]. In this method, each scattering excitation is described as the fluctuation of some variable (e.g. phonon coordinate or atomic displacement, electric field, electric charge, etc.). These fluctuations are coupled to each other and to the polarization which produce the scattering. This coupling can be described by a matrix formalism. The temperature dependent amplitude of the fluctuations is calculated by means of the fluctuation-dissipation theorem. This method is particularly appropriate to calculate the intensities of the near forward scattering by polaritons, a mixture of phonon-like atomic displacements and accompanying infrared electromagnetic fields which is obtained for small values of $\boldsymbol{k}$ (forward scattering) in polar materials [1.23]. For a more detailed discussion of this method see Chapter 2 and [Ref. 8.1, Sect. 2.1.11].

As we have seen, light scattering by phonons in solids involves the interaction of two rather complicated systems: The system of electrons in their energy bands, with possibly many-body effects such as excitons, and the system of phonons with their dispersion relations and anharmonic interactions. The anharmonic interactions, which are usually described phenomenologically by introducing terms of third and fourth order in the normal coordinates are responsible for the line widths of the phonon states. The electronic states have also line widths, determined in part by electron-phonon interaction. The line widths of resonant electronic states is of great importance in second-order iterated resonant Raman scattering, as discussed in Chapter 3. It also determines the time lag between the incident and the scattered photon at resonance (see also Chapter 3); away from resonance this time lag is negligible. In the region of high and sharp electronic absorption i.e. absorption by excitons, the electronic excitations couple very strongly to the photons to form quasi-particles known as polaritons. As intermediate states of the scattering processes one must take these polaritons instead of bare electronic or excitonic excitations. Similarly, if one deals with first-order scattering by infrared active phonons (non-centrosymmetric crystals), the phonons can couple strongly to infrared photons to form polaritons. These dressed phonons are of importance for configurations close to forward scattering [1.23] (see [Ref. 8.1, Sect. 2.1.12] and [Ref. 8.2, Chap. 7]).

Obviously, it is not possible to include the full complication of *both* the electron and the phonon systems in a calculation of the Raman tensor. Calculations which include a large degree of sophistication for the phonon system, such as anharmonicity, are usually performed within the framework of the phenomenological polarizability theory and neglect electronic resonance effects. In the formulation developed by COWLEY [1.88, 89] the fourth-rank Raman tensor $I_{\alpha\beta\gamma\delta}$ is obtained as the Fourier transform of the correlation function of the polarizability operators $P_{\alpha\beta}$ expressed in the Heisenberg representation [Ref. 8.1, Eq. (2.56)].

$$I_{\alpha\beta\gamma\delta}(\Omega) = \frac{1}{2\pi} \int_{-\infty}^{+\infty} \langle P_{\alpha\beta}(t)^* \, P_{\gamma\delta}(0) \rangle \, \mathrm{e}^{-i\Omega t} \, dt \, . \tag{1.5}$$

This correlation function can be obtained from the corresponding time-ordered thermodynamic Green's function which, in turn, is obtained by means of diagramatic expansions which enable the inclusion of anharmonicity. Calculations of the second-order spectrum of the alkali halides have been performed by this method using the shell model of the lattice dynamics [1.89]. The present volume with its emphasis on electronic resonance effects, does not treat this powerful technique any further.

We have mentioned the important role played by group theory in determining the allowed forms for the Raman tensor, i.e., its irreducible components. Only excitations having the symmetry of an irreducible component of the second-rank Raman tensor are allowed in first-order scattering. For higher-order processes in which several excitations participate, it is the product of the space group symmetries of the different excitations which must contain the symmetry of an irreducible component of the Raman tensor. This fact considerably relaxes the selection rules, since the product of two space group representations of wavevectors $\boldsymbol{k}_1$ and $\boldsymbol{k}_2$ (with $\boldsymbol{k}_1+\boldsymbol{k}_2 \simeq 0$ but $\boldsymbol{k}_1 \neq 0$) usually contains most representations of zero wavevector. Tables of products of space group representations were listed by Birman in [1.13]. The reader should look into this exhaustive work for references to the original literature.

Group theory is also useful in disentangling the *forbidden* effects induced by various agents such as the finite $\boldsymbol{k}$ vector of the photons (an intrinsic effect), defects, or extrinsic external perturbations (stress, electric fields). If the inducing agent can be represented by a tensor of rank r (e.g., $r=1$ for an electric field, $r=2$ for a stress) one can define a "forbidden" Raman tensor of rank $2+r$ for the first order forbidden effect, $2+nr$ for the forbidden effect of n-th order. Scattering induced by the simultaneous presence of two perturbations (e.g. stress and electric field) is also possible. The symmetry properties of these "forbidden" Raman tensors, and their reduction into irreducible components, is also discussed exhaustively in [1.13] where references to the original literature are given.

We have mentioned that in polar materials the scattering cross section for LO scattering contains, beside the deformation potential term which determines the TO scattering, a term produced by the Fröhlich interaction through the linear electro-optic effect. The linear electro-optic coefficient can be described as an *interband* effect of a perturbing electric field (i.e., that associated with the LO phonons). *Intraband* effects of electric fields on the excited electron-hole pairs are also important in determining the LO-scattering probability near resonance. In the case of the Fröhlich interaction, however, they only contribute to forbidden scattering in the form of a term in the Raman tensor proportional to the wavevector of the scattering phonon. The allowed intraband terms vanish because the effect of an electric field on the virtually excited electron exactly cancels the corresponding effect on the hole left behind. This forbidden term has been discussed by MARTIN [1.57, 90]. It appears strong for parallel polarizations of the incident and scattered photon. Whenever the crystal is so oriented that for parallel polarizations allowed LO scattering is also possible, the forbidden terms can dominate under strongly resonant conditions [1.58]. This forbidden scattering is particularly strong for resonances near strong excitons; the electric fields ionize the excitons and thus produce a strong modulation of the dielectric constant [1.90]. Actually, forbidden LO scattering can also be induced by electric fields at surface barriers [1.91, 92] combined with the Fröhlich interaction. It is often difficult to separate the pure Fröhlich effect from that of surface fields although the contribution of surface fields has been conclusively established [1.54, 92]. The dependence of the Fröhlich interaction effect on scattering $\boldsymbol{k}$ vector could be also used to separate both contributions; it has not been done so far because all resonance experiments have been performed in the back-scattering configuration [Ref. 8.1, Sect. 2.2.8].

The theory of resonant first-order Raman scattering has been formulated by GANGULY and BIRMAN [1.93] for arbitrary electronic excitations which may include strongly correlated electron-hole pairs (excitons). By means of a canonical transformation these authors were also able to treat dressed electronic excitations and thus to include polariton effects in the resonant energies. Detailed calculations of polariton scattering were performed by BENDOW and BIRMAN [1.94]. The delicate question of polariton boundary conditions and their effect on light scattering was discussed by ZEYHER et al. [1.95], while the effect of spatial dispersion in the excitonic polaritons was discussed by BRENIG et al. [1.96] (see [Ref. 8.2, Chap. 7]).

1.4.2. Scattering by Two Phonons

The microscopic theory of Raman scattering by two phonons represents an even higher degree of complication. Away from resonance, for photon energies much lower than the lowest absorption edge, considerable use has been made of phenomenological theories. In their most general form [1.76] these theories expand the contribution of each atom to the polarizability to second order in the atomic displacement and then compute the Raman tensor by adding the effects of the displacements of two phonons of equal and opposite $\boldsymbol{k}$ as obtained from some lattice dynamical model. The number of parameters involved in this second-order expansion of the atomic polarizabilities is too large, even after reducing it by using group theory, to be theoretically meaningful. It is often reduced by means of more or less justifiable devices such as that of keeping the mathematically simplest terms. Calculations of this type can be seen in [1.26]. A generalization of the bond polarizability, model [1.77] was used by COWLEY [1.97] to calculate the second-order Raman spectrum of Si. He made the assumption of a *longitudinal* bond polarizability which is, to second order, a function only of the change in bond length. The results obtained do not represent well the experimental spectra [1.28] (see, however, [Ref. 8.1, Fig. 2.13]).

This difficulty has been recently lifted for Ge, Si, and diamond [1.79] by introducing a *transverse* bond polarizability which is also, to second order, a function of the phonon induced changes in bond length. These parameters are obtained, in part, from the first-order Raman tensor and from the elasto-optic constants. A similar model, with a more complete "Ansatz" for the polarizability, has been also successful for interpreting the *first-order* Raman spectra of amorphous tetrahedral semiconductors as discussed in Chapter 5 [1.98]. These spectra arc produced by *second-order polarizabilities*, one set of atomic displacements being the result of the disorder and the other produced by the scattering phonon. As mentioned in Section 1.2 the spectra of tetrahedral semiconductors, once decomposed into irreducible components, separate into overtone and combination spectra which reflect the corresponding combined densities of states. The scattering probabilities are smooth functions of photon energy. The densities of overtones, and of sum and difference combination states of a large number of tetrahedral semiconductors have been recently computed by KUNC et al. [1.99]. They are very useful for the interpretation of two-phonon spectra.

The formal microscopic theory of the second-order scattering is also contained in the paper of GANGULY and BIRMAN [1.93]. We show in Fig. 1.2 typical diagrams for one- and two-phonon Stokes scattering processes with the intermediate state assumed to be correlated electron-hole pairs. These are usually the diagrams which mostly contribute to resonances. Other diagrams can be obtained from these by permutation of the interaction vertices [1.58, 93] and by placing one or two-phonon interaction vertices on the hole side of the diagrams (see Fig. 2.2b). The diagrams in Fig. 1.2d represent processes of higher order (6 interaction vertices) and can, in general, be neglected. Figure 1.2c has the same structure as diagram a of this figure with the electron one-phonon interaction replaced by an electron two-phonon vertex. The electron one-phonon interaction, iterated twice, determines Fig. 1.2b. Resonances for processes of type a and c as shown in Fig. 1.2, with the intermediate states i_1 and i_2 belonging to the same band are called two-band processes (see Fig. 2.4b). If i_1 and i_2 belong to different bands one speaks of three band processes (see Fig. 2.5). Resonances of the iterated two-phonon type (b of Fig. 1.2) can be two-, three-, or four-band processes. It is of

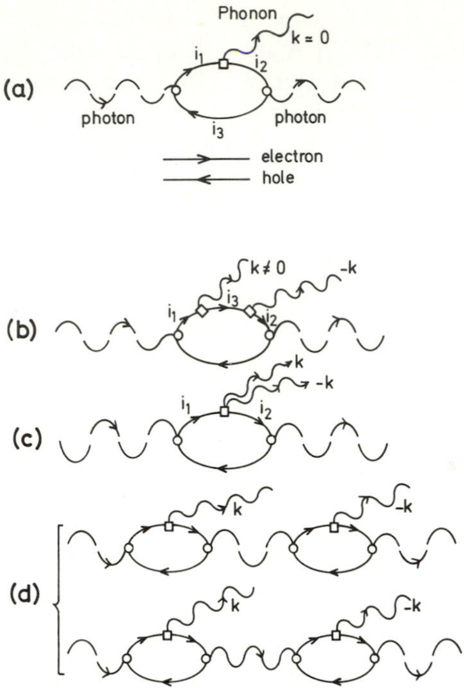

Fig. 1.2a–d. Typical processes involved in first- (a) and second-order (b–d) Raman scattering by phonons

considerable interest to discern whether a given two-phonon process is of the type b or c of the same figure in order to extract the corresponding electron one- phonon or electron two-phonon deformation potentials. Examples of both types can be found even for one given material. Operationally this distinction is possible, crudely speaking, by observing whether the two-phonon process has a resonance similar to a one-phonon process or a much stronger one. In tetrahedral semiconductors the former is true for most of the two-phonon spectra [1.58, 60] except for some sharply resonant features at the energy of two or more phonons each with $\boldsymbol{k} \simeq 0$. These features involve at least one LO phonon and are related to Fröhlich interaction in polar materials [1.58, 99, 100]. They are very strong near resonance. In non-polar materials they are produced by two iterated electron one-phonon deformation potential vertices [1.60]; they are weaker than their Fröhlich-interaction-induced counterparts. These features are due to processes of type b of Fig. 1.2 with the three intermediate states i_1. i_3, and i_2 resonating simultaneously. For scattering frequencies in the continuum of electronic excitations a phonon can be usually found for which these processes diverge. Thus a strong deformation of the two-phonon Raman spectrum is expected for processes Fig. 1.2b as the scattering frequency is changed near a resonance unless one deals with Einstein-model (i.e. non-dispersive) phonons. Since the optical branches are rather flat for $\boldsymbol{k} \simeq 0$, the Einstein model is often a good approximation; the scattering cross section diverges for scattering frequencies in the continuum near a resonance unless the finite width (lifetime broadening) of the intermediate states is taken into account. Once this is done, one finds that the cross section is proportional to the lifetime τ_3 of the intermediate state i_3 [Ref. 8.1, Eq. 2.246]).

It is also interesting to discuss the case in which i_3 is very far away from resonance. A typical two-band process of type Fig. 1.2c has an interaction vertex determined by the matrix element [1.59]

$$\langle i_1 | \nabla^2 V | i_1 \rangle , \tag{1.6}$$

where V is the crystal potential. The equivalent ($i_1 \equiv i_2$) process Fig. 1.2b with i_3 non resonant has the same structures as Fig. 1.2c with (1.6) replaced by the approximate expression

$$\frac{\langle i_1 | \nabla V | i_3 \rangle \langle i_3 | \nabla V | i_1 \rangle}{E_1 - E_3} , \tag{1.7}$$

where E_3 are the energies of the intermediate states. The sum of (1.6) and (1.7) represents the quadratic term in the power series expansion of the change of the energy E_1 as a function of atomic displacement. For an isolated atom this sum must vanish due to translational invariance. Equation (1.6), however, can reach very large values ($\sim 10^6$ eV for a displacement of the order of the lattice constant) [1.58, 59, 101] which must be exactly cancelled by (1.7). This exact cancellation is lifted in solids by the bonding and one is left with residues of the order of 10^3 eV. These rather large "effective" electron-phonon coupling constants are to be understood as representing a "dressed" electron two-phonon interaction vertex, as shown in Fig. 1.3. Thus, operationally, the processes of type Fig. 1.2c and those of type Fig. 1.2c with i_3 are indistinguishable away from resonance and must be lumped together as shown in Fig. 1.3.

Accurate quantum mechanical calculations of the matrix element of (1.6) using the OPW method for the s-like states of the conduction band of tetrahedral semiconductors [1.102] yield values of the order of 10^6 eV, similar to those found for the corresponding free atoms [1.58]. A pseudopotential calculation, however, yields completely different

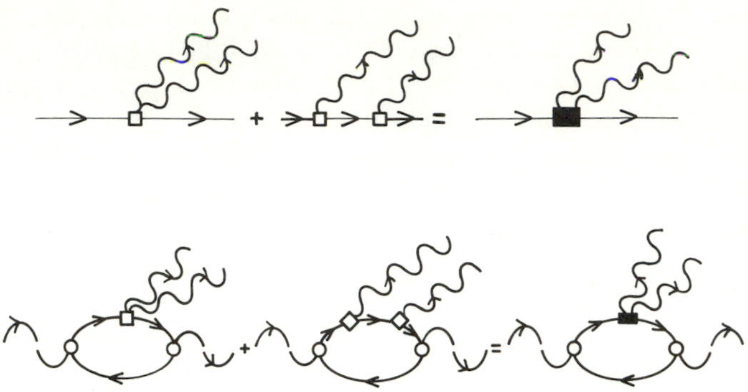

Fig. 1.3. Renormalization of the electron two-phonon interaction vertex, so as to include non resonant iterated electron one-phonon processes. The summands correspond to (1.6) and (1.7), respectively

values, of the order of 100 eV [1.58]. The inability of pseudopotential-pseudofunction theory to calculate matrix elements of ∇^2 has been discussed by LIN-CHUNG [1.103]. For *s*-like states, for instance, it originates in the strong divergence of $\nabla^2 V$ near the atomic core, which yields the main contribution to (1.6) and thus is not well represented by pseudopotential theory. Pseudopotential theory, however, adequately represents the energies of conduction and valence states and the effects of perturbations on these energies. One can therefore expect that the sum of (1.6) and (1.7), i.e., the renormalized electron two-phonon interaction constant, should be correctly given by a pseudopotential calculation. This expectation has been recently computationally confirmed [1.104].

We should also point out that SWENSON and MARADUDIN [1.80] in order to compute the $\Gamma_{25'}$ component of the first-order Raman tensor of Si calculated the dielectric tensor ε for several values of the phonon amplitude u [see (1.2)]. These calculations show a remarkably strong nonlinearity of the components of ε as a function of u, a fact remindful of the large values of the electron two-phonon interaction constant. A fit of SWENSON and MARADUDIN'S calculations with a polynomial quadratic in u yields ratios of the electron two-phonon to the electron one-phonon coupling constant in reasonable agreement with experiment [1.105].

In spite of the recent progress and interest in the two-phonon spectra of solids we have decided not to include a chapter exclusively dedicated to this subject in this tract. For one thing space limitations prevent it. Also, developments both in the theory and measurements of two-phonon spectra, are still so numerous as to make at this point a review article rather perishable [Ref. 8.1, Sects. 2.2.10 and 2.3.3].

As already mentioned in Subsection 1.4.1, the scattering due to intraband Fröhlich interaction is very strong near resonance. Though "forbidden" for one LO phonon, it can even be stronger than the corresponding allowed one-phonon process in polar materials. We have mentioned earlier in this section that Fröhlich interaction *allowed* scattering involving two LO phonons can also be very strong [1.100]. Multiphonon processes, most likely due to intraband Fröhlich interaction, are often seen very near resonance. Scattering up to 9 phonons have been reported for CdS [1.106]. The theory of these processes, induced by intraband Fröhlich interaction, has been recently given by ZEYHER [1.107] (see [Ref. 8.1, Sect. 2.3.5]).

1.4.3. Hot Luminescence

The frequency region in which most of the resonant Raman experiments are performed usually straddles the resonant energy. Therefore in part of this region *real* (not virtual) electronic transitions take place as the incident photons are absorbed. These transitions may be accompanied by the absorption or emission of phonons. The question is then often raised as to whether the observed scattered light is indeed due to Raman scattering (RRS) or to so-called hot luminescence (HL) of the *really* excited carriers spontaneously emitting aided by the emission or absorption of one or more phonons. The more fundamental question may be whether there is any intrinsic difference between those two processes and whether an operational criterion to distinguish them can be given. KLEIN [1.108] reached the conclusion, using the golden rule, that Raman scattering and absorption followed by phonon-aided luminescence were indeed indistinguishable processes. SHEN [1.109], however, performed a calculation based on the density matrix formalism [1.110] which provides for a more subtle treatment of relaxation or energy broadening (due to the effect of random fields, i.e., of unspecified scattering processes). Within this formalism two terms appear in the scattering probability. One of them is proportional to the number of electrons in the excited state n, $\varrho_{nn}(0)$, while the other term is proportional to an off-diagonal element of the density matrix ϱ_{if} between the initial and the final state. This fact provides us with a natural, formal distinction between HL and RRS: the terms proportional to the electron concentration in the excited state are called HL. Those proportional to ϱ_{if}, which represent a direct process independent of how many excited electrons are there, are called RRS, This *formal* distinction does not always lead to an operational distinction between HL and RRS, especially in the steady state. For a sharp temporal discontinuity in the incident beam, however, RRS yields always an instananeous response while the response due to HL is exponential, representing the buildup or decay of electrons in the intermediate state. The spectral width of the HL contains the widths of initial and final states: hence the standard occurrence of broad background due to HL with a superimposed peak due to RRS. The luminescence background is often a handicap to Raman work, especially near resonance. This problem can be eliminated by using the technique of coherent anti-Stokes Raman scattering (CARS). In this method three strong beams are mixed, two of them coming from a pulsed laser (frequency ω_L) and one being the Stokes beam stimulated by a second laser of frequency ω_S (see Chapter 7). The resulting beam has the anti-Stokes frequency [Ref. 8.1, Chap. 4].

$$\omega_A = 2\omega_L - \omega_S = \omega_L - \Omega\,, \tag{1.8}$$

where Ω is the frequency of the elementary excitation generated in the Stokes process. Because of the coherence of the beams involved and the resulting wavevector conservation, the emission takes place only along certain directions and is thus easy to distinguish from the nearly isotropic HL. In some cases the time delay involved in the HL may also help in the separation because of the use of short pulses.

References

1.1. L. BRILLOUIN: Ann. Phys. (Paris) **17**, 88 (1922).
1.2. A. SMEKAL: Naturwiss. **11**, 873 (1923).
1.3. C.V. RAMAN: Ind. J. Phys. **2**, 387 (1928).
1.4. G. LANDSBERG, L. MANDELSTAM: Naturwiss. **16**, 57 (1928).
1.5. S.P.S. PORTO, D.L. WOOD: J. Opt. Soc. Am. **52**, 251 (1962).
1.6. *Light Scattering Spectra of Solids*, ed. by G.B. WRIGHT (Springer, New York, 1969).
1.7. *Light Scattering in Solids*, ed. by M. BALKANSKI (Flammarion Sciences, Paris, 1971).
1.8. *Light Scattering in Solids*, ed. by M. BALKANSKI, R.C.C. LEITE, S.P.S. PORTO (Flammarion, Paris, 1975).
1.9. *The Raman Effect*, ed. by A. ANDERSON (M. Dekker, Inc., New York, 1973), vols. I and II.
1.10. A. MOORADIAN: In *Advances in Solid State Physics*, Vol. 9 (Pergamon-Vieweg, Oxford, Braunschweig, 1969), p. 74.
1.11. J. BRANDMÜLLER: Naturwiss. **54**, 3 (1967).
1.12. M.M. SUSHCHINSKII: *Raman Spectra of Molecules and Crystals* (Israel Program for Scientific Translations, New York, Jerusalem, London, 1972).
1.13. J.L. BIRMAN: In *Encyclopedia of Physics* (ed. by S. FLÜGGE), Vol. XXV/2b, ed. by L. GENZEL (Springer, Berlin, Heidelberg, New York, 1974).
1.14. M. CARDONA: In Proceedings of the 12th Intern. Conf. Physics of Semiconductors, Stuttgart 1974, ed. by M.H. PILKHUN (B. G. Teubner, Stuttgart, 1974), p. 1351.
1.15. C.V. RAMAN: Proc. Ind. Acad. Sci. Sec. A **44**, 99 (1956).
1.16. H. POULET, J.P. MATHIEU: Ann. Phys. (Paris) **9**, 543 (1964).
1.17. T.C. DAMEN, S.P.S. PORTO, B. TELL: Phys. Rev. **142**, 570 (1966).
1.18. B. TELL, T.C. DAMEN, S.P.S. PORTO: Phys. Rev. **144**, 771 (1966).
1.19. A. MOORADIAN, G.B. WRIGHT: Solid State Commun. **4**, 431 (1966).
1.20. J.H. PARKER, JR., D.W. FELDMAN, M. ASHKIN: Phys. Rev. **155**, 712 (1967).

1.21. J. P. Russell: Appl. Phys. Letters **6**, 223 (1965).
1.22. J. R. Sandercock: Phys. Rev. Letters **28**, 237 (1972).
1.23. C. M. Henry, J. J. Hopfield: Phys. Rev. Letters **15**, 964 (1965).
1.24. F. Rasetti: Nature **127**, 626 (1931).
1.25. R. S. Krishnan: Proc. Ind. Acad. Sci. **19**, 216 (1944).
1.26. H. M. J. Smith: Phil. Trans. Roy. Soc. A**241**, 14 (1948);
M. Born, M. Bradburn: Proc. Roy. Soc. A**188**, 161 (1947).
1.27. S. S. Mitra: Proc. Phys. Soc. Japan S**21**, 61 (1966).
1.28. P. A. Temple, C. E. Hathaway: Phys. Rev. B**7**, 3685 (1973).
1.29. B. A. Weinstein, M. Cardona: Phys. Rev. B**7**, 2545 (1973).
1.30. J. E. Smith, Jr., M. H. Brodsky, B. L. Crowder, M. I. Nathan, A. Pinczuk: Phys. Rev. Letters **26**, 642 (1971).
1.31. T. R. Hart, R. L. Aggrawal, B. Lax: Phys. Rev. **1**, 638 (1970).
1.32. F. Cerdeira, M. Cardona: Phys. Rev. B**5**, 1440 (1972).
1.33. F. Cerdeira, T. A. Fjeldly, M. Cardona: Phys. Rev. **9**, 4344 (1974).
1.34. D. W. Feldman, M. Ashkin, J. H. Parker: Phys. Rev. Letters **17**, 1209 (1966).
1.35. F. Cerdeira, C. J. Buchenauer, F. H. Pollak, M. Cardona: Phys. Rev. **5**, 580 (1972).
1.36. O. Brafman, S. S. Mitra: [1.7], p. 284.
1.37. E. Anastassakis, A. Filler, E. Burstein: [1.6], p. 421.
1.38. D. J. Evans, S. Ushioda, J. D. McMullen: Phys. Rev. Letters **31**, 369 (1973).
1.39. M. Cardona: Surface Sci. **37**, 100 (1973).
1.40. R. C. C. Leite, S. P. S. Porto: Phys. Rev. Letters **17**, 10 (1966).
1.41. J. M. Ralston, R. L. Wadsack, R. K. Chang: Phys. Rev. Letters **25**, 814 (1970).
1.42. P. Y. Yu, M. Cardona: J. Phys. Chem. Sol. **34**, 29 (1973).
1.43. D. K. Garrod, R. Bray: Phys. Rev. B**6**, 1314 (1972).
1.44. J. F. Scott, T. C. Damen, R. C. C. Leite, W. T. Silfvast: Solid State Comm. **7**, 953 (1969).
1.45. A. Pinczuk, E. Burstein: Phys. Rev. Letters **21**, 1073 (1968).
1.46. R. C. C. Leite, J. F. Scott: Phys. Rev. Letters **22**, 130 (1969).
1.47. M. A. Renucci, J. B. Renucci, M. Cardona: [1.7], p. 326.
1.48. For a review, see *Dye Lasers*, ed. by F. P. Schäfer, in: Topics in Applied Physics, Vol. 1 (Springer, Berlin, Heidelberg, New York, 1973).
1.49. M. I. Bell, R. N. Tyte, M. Cardona: Solid State Commun. **13**, 1833 (1973).
1.50. Y. Oka, T. Kushida: J. Phys. Soc. Japan **33**, 372 (1972).
1.51. T. C. Damen, J. Shah: Phys. Rev. Letters **27**, 1506 (1971).
1.52. M. I. Bell, R. N. Tyte: Appl. Opt. **13**, 1610 (1974).
1.53. Y. M. Yarborough: Appl. Phys. Letter **24**, 629 (1974).
1.54. R. L. Schmidt, B. D. McCombe, M. Cardona: Phys. Rev. B**11**, 746 (1975).
1.55. F. Cerdeira, W. Dreybrodt, M. Cardona: Solid State Commun. **10**, 591 (1972).
1.56. P. Y. Yu, Y. R. Shen: Phys. Rev. Letters **29**, 468 (1972).
1.57. R. M. Martin, T. C. Damen: Phys. Rev. Letters **26**, 86 (1971).
1.58. B. A. Weinstein, M. Cardona: Phys. Rev. **8**, 2795 (1973).
1.59. K. L. Ngai, E. J. Johnson: Phys. Rev. Letters **29**, 1607 (1972).
1.60. M. A. Renucci, J. B. Renucci, R. Zeyher, M. Cardona: Phys. Rev. B**10**, 4309 (1974).
1.61. J. S. Kline, M. Masui, J. J. Song, R. K. Chang: Solid State Commun. **14**, 1163 (1974).
1.62. A. Compaan, H. Z. Cummins: Phys. Rev. Letters **31**, 41 (1973).
1.63. P. Y. Yu, Y. R. Shen, Y. Petroff, L. Falicov: Phys. Rev. Letters **30**, 283 (1973);
P. Y. Yu, Y. R. Shen: Phys. Rev. Letters **32**, 373 (1974).
1.64. P. Y. Yu, Y. R. Shen: Phys. Rev. Letters **32**, 939 (1974).

1.65. W.D.JOHNSON, JR., I.P.KAMINOV: Phys. Rev. **188**, 1209 (1969).
1.66. J.G.SKINNER, W.G.NIELSEN: J. Opt. Soc. Am. **58**, 113 (1968).
1.67. M.CARDONA: *Modulation Spectroscopy* (Academic Press, New York, 1969).
1.68. E.ANASTASSAKIS, P.N.ARGYRES: Phys. Letters A**49**, 457 (1974).
1.69. M.H.GRIMSDITCH, A.K.RAMDAS: Phys. Rev. B**11**, 3139 (1975).
1.70. M.CARDONA, F.CERDEIRA, T.A.FJELDLY: Phys. Rev. **10**, 3433 (1974).
1.71. R.A.COWLEY: p. 99 in [1.8], Vol. 1.
1.72. L.N.OVANDER: Opt. Spectrosc. **9**, 302 (1960).
1.73. R.LOUDON: Advan. Phys. **13**, 423 (1964).
1.74. P.A.FLEURY, R.LOUDON: Phys. Rev. **166**, 514 (1968).
1.75. G.PLACZEK: In *Handbuch der Radiologie*, ed. by E.MARX (Akademische Verlagsgesellschaft, Leipzig, 1934), vol. VI, p. 205.
1.76. M.BORN, K.HUANG: *Dynamical Theory of Crystal Lattices* (Clarendon Press, Oxford, 1956).
1.77. M.WOLKENSTEIN: Compt. Rend. Acad. Sci. URSS **32**, 185 (1941).
1.78. A.A.MARADUDIN, E.BURSTEIN: Phys. Rev. **164**, 1081 (1967).
1.79. S.GO, H.BILZ, M.CARDONA: Phys. Rev. Letters **34**, 580 (1975).
1.80. L.R.SWENSON, A.A.MARADUDIN: Solid State Commun. **8**, 859 (1970).
1.81. M.CARDONA: Solid State Commun. **9**, 819 (1971).
1.82. R.LOUDON: Proc. Roy. Soc. A**275**, 218 (1963).
1.83. R.LOUDON: J. Phys. **26**, 677 (1965).
1.84. G.B.BENEDEK, K.FRITSCH: Phys. Rev. **149**, 647 (1966).
1.85. L.G.GRETCHKO, L.N.OVANDER: Sov. Phys.–Solid State **4**, 112 (1962).
1.86. A.S.BARKER, JR., R.LOUDON: Rev. Mod. Phys. **44**, 18 (1972).
1.87. D.T.HON, W.L.FAUST: Appl. Phys. **1**, 241 (1973).
1.88. R.A.COWLEY: Advan. Phys. **12**, 421 (1963).
1.89. R.A.COWLEY: Proc. Phys. Soc. **84**, 281 (1964).
1.90. R.M.MARTIN: Phys. Rev. B**4**, 3676 (1971).
1.91. J.G.GAY, J.D.DOW, E.BURSTEIN, A.PINCZUK: [1.7], p. 33.
1.92. A.PINCZUK, E.BURSTEIN: Proc. 10th Intern. Conf. Physics of Semiconductors, Cambridge, Mass., 1970 (US Atomic Energy Commission, 1970), p. 727.
1.93. A.K.GANGULY, J.L.BIRMAN: Phys. Rev. **162**, 806 (1967).
1.94. B.BENDOW, J.L.BIRMAN: Phys. Rev. B**1**, 1678 (1970).
1.95. R.ZEYHER, C.S.TING, J.L.BIRMAN: Phys. Rev. B**10**, 1725 (1974).
1.96. W.BRENIG, R.ZEYHER, J.L.BIRMAN: Phys. Rev. B**6**, 4617 (1972).
1.97. R.A.COWLEY: J. Phys. **26**, 659 (1965).
1.98. R.ALBEN, J.E.SMITH, JR., M.H.BRODSKY, D.WEAIRE: Phys. Rev. Letters **30**, 1141 (1973).
1.99. K.KUNC: Ann. Phys. (Paris) **8**, 319 (1973).
1.100. R.ZEYHER: Phys. Rev. B**9**, 4439 (1974).
1.101. P.J.LIN-CHUNG, K.L.NGAI: Phys. Rev. Letters **29**, 1610 (1972).
1.102. J.IVEY: Phys. Rev. B**10**, 2480 (1974).
1.103. P.J.LIN-CHUNG: Phys. Rev. B**8**, 4043 (1973).
1.104. R.ZEYHER: In [1.8].
1.105. R.ZEYHER, S.GO, M.CARDONA: In [1.8].
1.106. J.F.SCOTT: Phys. Rev. **2**, 1209 (1970).
1.107. R.ZEYHER: Solid State Commun. **16**, 49 (1975), and references therein.
1.108. M.V.KLEIN: Phys. Rev. **8**, 919 (1973).
1.109. Y.R.SHEN: Phys. Rev. **9**, 622 (1974).
1.110. N.BLOEMBERGEN: *Nonlinear Optics* (Benjamin, New York, 1965), Chapter II.
1.111. R.F.BEGLEY, A.B.HARVEY, R.L.BYER: Appl. Phys. Letters **25**, 387 (1972).
1.112. B.S.HUDSON: J. Chem. Phys. **61**, 5461 (1974).

2. Fundamentals of Inelastic Light Scattering in Semiconductors and Insulators

A. PINCZUK and E. BURSTEIN

With 12 Figures

During the past ten years inelastic light scattering spectroscopy has developed into one of the most powerful and most widely used optical techniques for the study of the properties of the low frequency elementary excitations of solids. This has been, to a great extent, a consequence of the availability of lasers as excitation sources and of improved spectrometers and associated electronics for recording weak light scattering spectra. Scattering by elementary excitations has been observed in opaque solids, including metals and narrow gap semiconductors, as well as in transparent crystals. Very recently, with the help of computerized spectra recording techniques it has become possible to observe extremely weak scattering intensities.

A large group of crystal elementary excitations have been observed by inelastic light scattering spectroscopy. This group includes acoustical and optical phonons, surface and bulk polaritons, magnons, electron-gas excitations (plasmons and single-particle excitations) as well as electronic and vibrational excitations of isolated ions in crystals.

One of the most important selection rules for light scattering processes in crystals is the conservation of wave vector. Because the wave vector of light is much smaller than the Brillouin-zone boundary wave vector, first-order inelastic light scattering (in which scattering by one quantum of low frequency crystal excitation is involved) only allows one to study the excitations near the center of the Brillouin zone. This is one of the major limitations of the technique. For the zone center low frequency excitations, light scattering spectroscopy enables us to determine in a relatively simple and straightforward way the energy, lifetime and symmetry properties of these excitations. When the objective is the determination of the dispersion relation of the excitations, the data obtained by light scattering spectroscopy has to be considered as complementary to those obtained by inelastic neutron scattering. In regard to this problem we should mention that second-order scattering by magnons and optical phonons has been used to determine the two-magnon and two-phonon densities of states. In addition, one-phonon density of states has been obtained in defect-induced scattering experiments. These and similar data may be used, after considerable analytical work, to determine the properties of the excitations away from the Brillouin-zone center.

As in the case of the emission of light it is possible to have *spontaneous* inelastic light scattering, in which no scattered photons exist before the scattering event, and *stimulated* scattering in which the process takes place in the presence of already existing scattered photons. It is now well established that light is scattered inelastically by the spatial and temporal fluctuations in the *electronic* contributions to the electric susceptibility, which are associated with the elementary excitations of the crystal. In the case of light scattering by collective excitations such as lattice vibrations, the fluctuations result from the modulation of the electric susceptibility.

The space and time dependent fluctuations in the electric susceptibility are given by the *transition* electric susceptibility. In a microscopic description, the *transition* electric susceptibility is proportional to the matrix elements which characterize the transitions which accompany the annihilation of the incident photon and creation of the scattered photon and the creation or annihilation of quanta of crystal elementary excitations. In the case of the scattering of light by collective excitations, the *transition* electric susceptibility can be written as an expansion in powers of the normal coordinates of the modes. The coefficients of this expansion are the *Raman tensors* of the collective excitations, which describe the mixing of the incident and scattered fields with crystal normal modes. In the case of polar modes, the Raman tensors are also related to the tensor coefficients which describe the nonlinear electric susceptibilities of the crystal. In addition, when the energy of the incident photons is in resonance with the electronic interband transitions, the frequency dependence of the Raman scattering intensity displays features similar to those found in optical modulation spectroscopy of electronic interband transitions. These aspects of Raman scattering add another dimension to the interest in the field in the sense that inelastic light scattering may also be used to obtain data on the nonlinear optical susceptibilities as well as on crystal energy band structure.

This chapter has the dual purpose of giving a description of the basic mechanisms of inelastic light scattering processes as well as that of discussing the properties of the first-order Raman tensors of the collective excitations of non-magnetic semiconductors (lattice vibrations and plasmons). We give a general discussion of the properties of the scattering cross-section, kinematics and selection rules for inelastic light scattering processes. The scattering cross-section is written in terms of the *transition* electric susceptibility. We also discuss the conditions under which conservation of wave vector occurs. This is an aspect of the kinematics of the light scattering process which has to be dealt with carefully in opaque crystals and in media where translational symmetry is destroyed. We then consider first-order Raman scattering by lattice vibrations and plasmons. Among other things we show the relations that exist between the first-order Raman tensors and the tensor coefficients which describe such nonlinear phenomena as three-wave mixing and the linear electro-optic effect.

Finally, we discuss the frequency dependence of the first-order Raman tensors of optical phonons. It is shown that, close to resonance with the interband electronic transitions, the frequency dependence of the Raman tensors is related to the frequency dependence of the electric susceptibility. In the discussion of resonance effects we devote particular attention to the contributions to the Raman tensors which are wave-vector dependent as well as to the contributions which are induced by external or built-in electric fields or stresses. Such contributions have been observed in resonant Raman scattering in semiconductors. They are of interest for two reasons. First, they allow the observation of scattering by modes which are Raman-inactive according to the conventional selection rules. In addition, they give the researcher a powerful tool for elucidating the mechanisms of Raman scattering in crystals. The equations are given in cgs units.

2.1. The Inelastic Light Scattering Process

2.1.1. The Scattering Cross-Section

In inelastic light scattering processes a quantum of the incident radiation is annihilated and a quantum of the scattered radiation is created. This occurs with creation (in the Stokes process) or annihilation (in the anti-Stokes process) of a crystal excitation. The anti-Stokes process can occur only if the crystal is initially in an excited state. Light scattering may result from the coupling of the light waves to the electric or magnetic moments of the crystal. Theoretical estimates, as well as experimental results, have shown that, even in the case of magnetic excitations, the light scattering processes associated with the magnetic moment of the crystal electronic transitions are very weak. The matrix element which enters in the expression for the scattering cross-section can be written in terms of a *transition* electric susceptibility [2.1, 2]. Under the conditions of typical experiments, the most important contribution to the *transition* electric susceptibility arises from the coupling of light to the electric moment of the electronic excitations. Recent calculations [2.3] indicate that Raman scattering processes associated with the coupling to the electric moment of the ionic excitations (the ionic Raman effect) would be observable only when the incident light is resonant with lattice vibration excitations.

In order to calculate the *transition* electric susceptibility we consider first the Hamiltonian which describes the coupling of electrons, with charge $-e$ and mass m, to a radiation field described by a vector potential $\boldsymbol{A}$. This Hamiltonian is written as [2.4, 5]

$$\mathscr{H}_{\mathrm{ER}} = \mathscr{H}_{\mathrm{A}} + \mathscr{H}_{\mathrm{AA}}\,, \tag{2.1}$$

where

$$\mathscr{H}_{\mathrm{AA}} = (e^2/2mc^2) \sum_{\boldsymbol{k}_1,\omega_1} \sum_{\boldsymbol{k}_2,\omega_2} N(-\boldsymbol{k}_1 + \boldsymbol{k}_2)\, \boldsymbol{A}(\boldsymbol{k}_1,\omega_1) \cdot \boldsymbol{A}^*(\boldsymbol{k}_2,\omega_2) \tag{2.2}$$

and

$$\mathscr{H}_{\mathrm{A}} = (e/mc) \sum_{\boldsymbol{k}_1,\omega_2} \boldsymbol{p}(-\boldsymbol{k}_1) \cdot \boldsymbol{A}(\boldsymbol{k}_1,\omega_1)\,, \tag{2.3}$$

where $\boldsymbol{A}(\boldsymbol{k}_1,\omega_1)$ and $\boldsymbol{A}(\boldsymbol{k}_2,\omega_2)$ are the amplitudes of the Fourier components of the vector potential of the incident and scattered photons with wave-vectors and frequencies $(\boldsymbol{k}_1,\omega_1)$ and $(\boldsymbol{k}_2,\omega_2)$, respectively. In (2.2) and (2.3) we have defined

$$N(-\boldsymbol{k}_1 + \boldsymbol{k}_2) = \sum_j \exp \mathrm{i}(\boldsymbol{k}_1 - \boldsymbol{k}_2) \cdot \boldsymbol{r}_j\,, \tag{2.4}$$

$$\boldsymbol{p}(-\boldsymbol{k}_1 + \boldsymbol{k}_2) = \sum_j [\exp \mathrm{i}(\boldsymbol{k}_1 - \boldsymbol{k}_2) \cdot \boldsymbol{r}_j]\, \boldsymbol{p}_j\,, \tag{2.5}$$

where $\boldsymbol{r}_j$ and $\boldsymbol{p}_j$ are the position and momentum operators of the jth electron. $N(-\boldsymbol{k}_1 + \boldsymbol{k}_2)$ and $\boldsymbol{p}(-\boldsymbol{k}_1 + \boldsymbol{k}_2)$ are the Fourier transform of the many-particle number and momentum operators of the electrons. In the second-quantization formalism the $\boldsymbol{A}(\boldsymbol{k},\omega)$ are linear in the photon creation and annihilation operators.

The contribution of the electrons to the *transition* electric susceptibility involves a two-photon process in which the scattering medium, through the electron-radiation interactions, destroys the incident photon and creates the scattered photon. It follows from (2.2) and (2.3) that the time-dependent perturbation calculation of the *transition* electric susceptibility has to be carried out to first-order in $\mathscr{H}_{\mathrm{AA}}$ and to second-order in $\mathscr{H}_{\mathrm{A}}$. The matrix elements of the $\mu\nu$ component of the *transition* electric susceptibility tensor operator can be written as [2.5]

$$\begin{aligned} \langle f|\delta\chi_{\mu\nu}|i\rangle = (e^2/m^2\omega_2^2 V)\{&-m\langle f|N(-\boldsymbol{k}_1+\boldsymbol{k}_2)|i\rangle \\ &+ \sum_b [\langle f|p_\mu(\boldsymbol{k}_2)|b\rangle \langle b|p_\nu(-\boldsymbol{k}_1)|i\rangle/(E_b - E_i - \hbar\omega_1) \\ &+ \langle f|p_\nu(-\boldsymbol{k}_1)\, b\rangle \langle b|p_\mu(\boldsymbol{k}_2)|i\rangle/(E_b - E_i + \hbar\omega_2)]\}\,, \end{aligned} \tag{2.6}$$

where $|i\rangle$, $|f\rangle$, and $|b\rangle$ are the initial, final and intermediate many-particle states of the crystal with energies E_i, E_f, and E_b, and V is the scattering volume. In order to calculate the matrix elements in (2.6) it is necessary to separate, in the many-particle crystal states, the ionic and electronic motions as well as the electronic single-particle and collective coordinates. This is done by using the adiabatic and random phase (RPA) approximations, respectively. $\langle f|\delta\chi_{\mu\nu}|i\rangle$ is a second-rank tensor. Its non-zero components are determined by the symmetry properties of the eigenstates $|i\rangle$, $|f\rangle$, and $|b\rangle$. When $\omega_1=\omega_2$ the initial and final crystal states are necessarily the same. In this case $\langle i|\delta\chi_{\mu\nu}|i\rangle$, as given by (2.6), is identical to the electric susceptibility of the crystal [2.6]. This susceptibility describes the *linear optics* of the crystal for light frequencies much larger than the optical phonon frequencies. Only elastic (Rayleigh) scattering of light is possible in this limit. This will occur in imperfect crystals which are composed of random distribution of scattering units [2.1].

The differential cross-section for spontaneous light scattering is given in terms of the matrix elements of the *transition* susceptibility operator by [2.5] (see [Ref. 8.1, Sect. 2.1.8])

$$(d^2\sigma/d\Omega\, d\omega_2)=(\omega_2/c)^4\,(\omega_1/\omega_2)\,V^2\sum_{i,f}P(E_i)\,\delta\{[(E_i-E_f)/\hbar]-\omega\}\cdot|\hat{e}_2\cdot\overleftrightarrow{\delta\chi}\cdot\hat{e}_1|^2\,, \tag{2.7}$$

where $P(E_i)$ is the (temperature dependent) probability of having the crystal in the initial state; $\overleftrightarrow{\delta\chi}$ is the second-rank tensor with components

$$(\overleftrightarrow{\delta\chi})_{\mu\nu}=\langle f|\delta\chi_{\mu\nu}|i\rangle\,; \tag{2.8}$$

$\hat{e}_1$ and $\hat{e}_2$ are the unit polarization vectors for the incident and scattered photons; and

$$\omega=\omega_1-\omega_2 \tag{2.9}$$

is the *scattering frequency*. In addition, it is convenient to introduce the *scattering wave-vector*

$$\boldsymbol{k}=\boldsymbol{k}_1-\boldsymbol{k}_2\,. \tag{2.10}$$

It follows from (2.6) that in a system with translational symmetry the *transition* susceptibility is a function of the scattering wave vector $\boldsymbol{k}$.

LOUDON [2.7] has described light scattering phenomena in terms of the light scattering efficiency. Its relation to the differential cross-section is given by [(for a clarification see [Ref. 8.1, Sect. 2.1.8])

$$S=(\omega_2/A\,\omega_1)\,[\int(d^2\sigma/d\Omega\,d\omega_2)\,d\omega_2]\,\Delta\Omega\,, \tag{2.11}$$

where $\Delta\Omega$ is the solid angle subtended by the spectrometer, and A is the illuminated area of the crystal. A and V define the *scattering length* as

$$L=(V/A)\,. \tag{2.12}$$

In many important cases a satisfactory description of the inelastic light scattering cross-section can be given without making use of the detailed properties of the crystal intermediate states introduced in (2.6). For these cases it is convenient to put (2.7) into a form which does not involve explicitly the intermediate states. Such an expression, obtained from (2.7) by a standard mathematical procedure is [2.5]

$$\begin{aligned}(d^2\sigma/d\Omega\,d\omega_2)=(\omega_2/c)^4\,(\omega_1/\omega_2)\,(V^2/2\pi)\int dt\,\exp(\mathrm{i}\omega t)\\ \cdot\langle i|\delta\chi_{12}(\boldsymbol{k},t)\,\delta\chi_{12}(k,0)|i\rangle_{\mathrm{T}}\,,\end{aligned} \tag{2.13}$$

where $\delta\chi_{12}(\boldsymbol{k},t)=\hat{e}_2\cdot\delta\chi(\boldsymbol{k},t)\cdot\hat{e}_1$ is the *transition* susceptibility operator of the crystal in the Heisenberg representation and $\langle\ldots\rangle_{\mathrm{T}}$ denotes the thermal ensemble average over the crystal initial states.

Equation (2.13) can be used to assign a macroscopic meaning to the *transition* susceptibility. It was mentioned above that for $|i\rangle=|f\rangle$ the *transition* susceptibility becomes identical to the electric susceptibility of the medium. In order to find the macroscopic meaning of $\delta\chi_{12}(\boldsymbol{k},t)$ for $|i\rangle\neq|f\rangle$, it is useful to compare (2.13) with the expression of the light scattering cross-section by fluctuations in the electric susceptibility of a dielectric medium, as given by classical electromagnetic theory [2.8, 9]. From such a procedure it is found that $\delta\chi_{12}(\boldsymbol{k},t)$ can be related to the spatial and temporal fluctuations in the electronic contribution to the electric susceptibility. $\delta\chi_{12}(\boldsymbol{k},t)$ can in fact be written as

$$\delta\chi_{12}(\boldsymbol{k},t)=(1/V)\int\exp(\mathrm{i}\boldsymbol{k}\cdot\boldsymbol{r})\,\Delta\chi_{12}(\boldsymbol{r},t)\,d^3r\,, \tag{2.14}$$

where $\Delta\chi_{12}(\boldsymbol{r},t)$ represents the fluctuations in the electric susceptibility. In addition, from Maxwell's equations, we deduce that the source of the scattered field is an electric moment per unit volume given by

$$\boldsymbol{M}_2(\boldsymbol{r},t)=\overleftrightarrow{\Delta\boldsymbol{\chi}}_{12}(\boldsymbol{r},t)\,\boldsymbol{E}_1(\boldsymbol{r},t)\,, \tag{2.15}$$

where $\boldsymbol{E}_1(\boldsymbol{r},t)$ is the field of the incident wave inside the crystal.

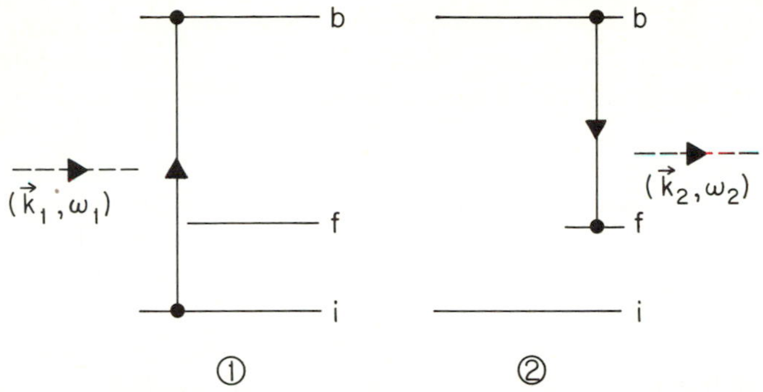

Fig. 2.1. Schematic description of the transitions involved in the contribution of $\mathscr{H}_A$ to the *transition* susceptibility. *1* and *2* indicate the time order of the transitions. The dotted arrows indicate photons. In the process shown, the first transition is accompanied by the annihilation of the incident photon and the second transition is accompanied by the creation of the scattered photon

Let us consider now the first term on the right side of (2.6) involving the $\mathscr{H}_{AA}$ interaction to first order. This term applies to scattering by the elementary excitations which result in fluctuations of the electronic charge density. Using (2.13), its contribution to the cross section may be written as [2.5, 10]

$$(d^2\sigma/d\Omega\, d\omega_2) = (\omega_2/\omega_1)\, r_0^2 (\hat{e}_2 \cdot \hat{e}_1)^2 (1/2\pi)\, V^2 \cdot \int d\omega \exp(i\omega t) \langle i|\varrho(-\boldsymbol{k}, t)\, \varrho(-\boldsymbol{k}, 0)|i\rangle_T , \tag{2.16}$$

where $r_0 = (e^2/mc^2)$ is the classical radius of the electron and $\varrho(-\boldsymbol{k}, t) = N(-\boldsymbol{k}, t)/V$ is the electronic charge density fluctuation associated with the excitation. The charge density correlation function is given by [2.10, 11]

$$\int d\omega \exp(i\omega t) \langle i|\varrho(-\boldsymbol{k}, t)\, \varrho(-\boldsymbol{k}, 0)|i\rangle_T = [\bar{n}(\omega)+1]\, V\mathscr{S}(\boldsymbol{k}, \omega) = [\bar{n}(\omega)+1]\, V(\hbar k^2/4\pi e^2)\, \mathrm{Im}\{1/\varepsilon(\boldsymbol{k}, \omega)\} , \tag{2.17}$$

where $\mathscr{S}(\boldsymbol{k}, \omega)$ is the dynamical structure factor of the electrons, $\varepsilon(\boldsymbol{k}, \omega)$ is their dielectric response function and

$$\bar{n}(\omega) = 1/[\exp(\hbar\omega/k_B T) - 1] . \tag{2.18}$$

where k_B is the Boltzman constant, and T is the absolute temperature. $\bar{n}(\omega)$ has the following important property [2.12],

$$[\bar{n}(-\omega)+1] = -\bar{n}(\omega) , \tag{2.18a}$$

indicating that (2.16) and (2.17) describe both Stokes and anti-Stokes processes [2.12]. The inelastic light scattering processes associated with electronic charge density fluctuations are discussed in detail by KLEIN in this volume [2.11].

Inelastic light scattering by electronic excitations and by lattice vibrations are described by the last term in (2.6), involving the $\mathscr{H}_A$ electron-radiation interaction in second-order. These two-step processes are shown, diagrammatically, in Fig. 2.1. Light scattering by electronic excitations is given by this term when the states $|i\rangle$ and $|f\rangle$, described within the Born-Oppenheimer approximation, have the same phonon occupation numbers and differ only in their electronic states. This mechanism accounts for light scattering by all types of electronic excitations, including spin-density fluctuations [2.11] and magnons [2.2], provided that the effects associated with spin-orbit interactions are taken into consideration.

Light scattering by lattice vibrations i.e. Raman and Brillouin scattering, are represented by the last term in (2.6) when the states $|i\rangle$ and $|f\rangle$, described with the Born-Oppenheimer approximation, differ only in their phonon occupation numbers. LOUDON [2.2] has calculated the matrix elements of the *transition* susceptibility operator under the assumption that the electron wave functions, as well as the energies of the intermediate states, which enter in the matrix elements and denominators of the last term in (2.6), are only weakly perturbed by the time-dependent electron-lattice interactions. In the case of first-order (one-phonon) scattering processes, $\langle f|\delta\chi_{\mu\nu}|i\rangle$ is given by 3rd order time-dependent perturbation theory. For this case the scattering process, which can be viewed as the mixing of three waves, is considered to take place in three steps of photon or phonon induced electronic transitions. In the scattering process described by the particular Feynman diagram of Fig. 2.2 the sequence is as follows:

1) The first electronic transition to an intermediate electron-hole pair state takes place with the annihilation of the incident photon.

2) The second electronic transition to another electron-hole pair state occurs with the creation of a phonon.

3) The third transition to the electronic ground states takes place with the creation of the scattered photon.

Steps 1) and 3) involve the $\mathscr{H}_A$ interaction and step 2) the electron-phonon interaction $\mathscr{H}_{EL}$. Energy is conserved only in the total process, whereas wave vector is conserved in each step. The diagrams of Fig. 2.2 also describe other processes which involve the mixing of three waves via the second-order non-linear electric susceptibility of crystals [2.13]. An example is second harmonic generation (SHG). The calculation of the SHG coefficient is described by the diagram of Fig. 2.2

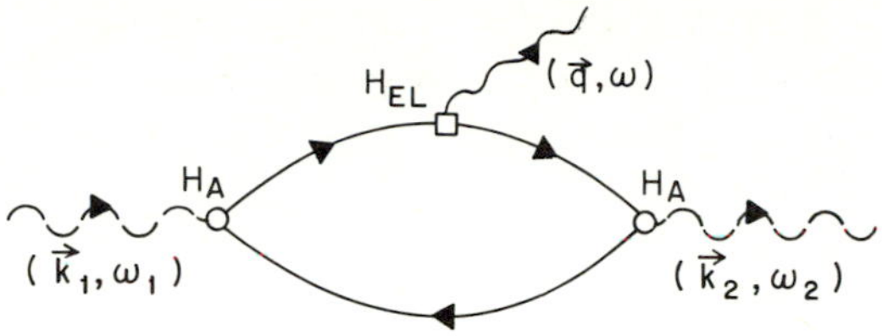

(a)

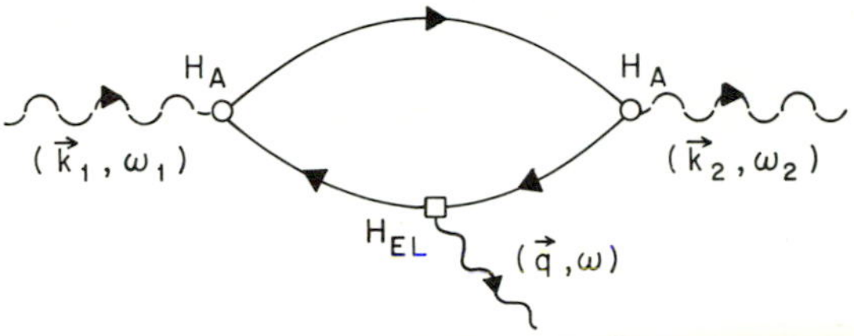

(b)

Fig. 2.2a and b. Feynman diagrams which describe the 3rd order perturbation calculation of first-order Raman scattering by lattice vibrations involving intermediate electron-hole pair states. $\mathscr{H}_A$ is the electron-radiation interaction, and $\mathscr{H}_{EL}$ is the electron-lattice interaction. (a) shows the electron contribution and (b) the hole contribution

provided that $\mathscr{H}_{EL}$ is replaced by $\mathscr{H}_A$. The Feynman diagrams leading to second-order Raman processes (involving two phonons) have been discussed in detail by LOUDON [2.14].

2.1.2. Kinematics of Inelastic Light Scattering Processes

The kinematics of inelastic light scattering processes are determined by conservation of energy and momentum. It follows from the properties of the matrix elements in (2.6), that in the case of media with *translational symmetry*, the conservation conditions can be written in terms of the wave vectors and frequencies of the photons and crystal excitations involved in the process. For crystals transparent to the incident and scattered light, with elementary excitations having an infinite lifetime, the conservation of energy and momentum can be written as

$$\omega = \omega_1 - \omega_2 = \pm \omega_j \tag{2.19}$$

$$\boldsymbol{k} = \boldsymbol{k}_1 - \boldsymbol{k}_2 = \pm \boldsymbol{q}_j \,, \tag{2.20}$$

where ω_j and $\boldsymbol{q}_j$ are the frequency and wave vector of the type j crystal excitation. The plus and minus signs correspond to the Stokes and anti-Stokes processes, respectively.

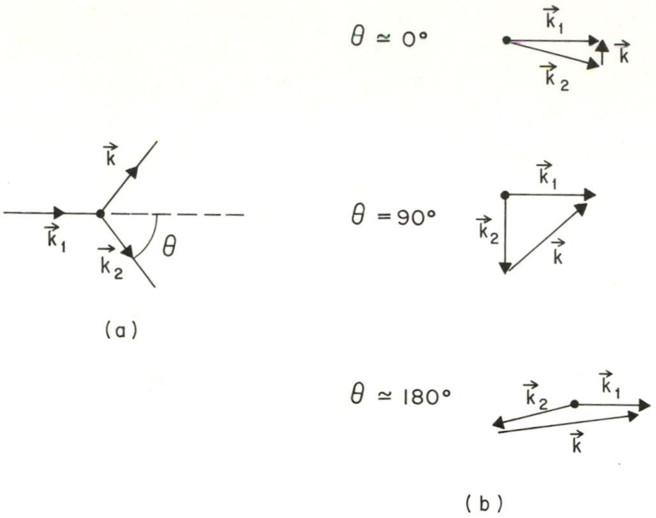

Fig. 2.3a and b. Kinematics of Stokes inelastic light scattering. (a) defines the scattering angle θ. (b) Shows the magnitudes of $\boldsymbol{k}$ for forward, right angle and backward scattering geometries

The magnitude of the scattering wave vector is determined by the scattering geometry. As indicated by Fig. 2.3, the minimum value of $|\boldsymbol{k}|=k$ is obtained in forward scattering, when $\theta=0°$, and in isotropic media is given by

$$k_{\min}=[\eta(\omega_1)\,\omega_1-\eta(\omega_2)\,\omega_2]/c\,, \tag{2.21}$$

where $\eta(\omega_1)$ and $\eta(\omega_2)$ are the refractive indices of the crystal for the incident and scattered light. When $\omega\to 0$, as in the case of phonon-polaritons, $k_{\min}$ may be zero [2.15]. The maximum value of $|\boldsymbol{k}|$ is obtained in back-scattering, when $\theta=180°$, and is given by

$$k_{\max}=[\eta(\omega_1)\,\omega_1+\eta(\omega_2)\,\omega_2]/c\,. \tag{2.22}$$

It follows from (2.21) and (2.22) that for typical light scattering experiments (in or near the visible) the range of scattering wave vectors is

$$0\leqq k\lesssim 10^6\ \mathrm{cm}^{-1}\,. \tag{2.23}$$

This implies that for first-order scattering processes the accessible range of $|\boldsymbol{q}_j|=q_j$, under conditions of wave vector conservation, is small compared to a reciprocal lattice wave vector. However, this is an important range for many excitations which may not be otherwise readily accessible. In higher-order processes the individual wave vectors of the excitations, $\boldsymbol{q}_j$, can range from zero to a reciprocal lattice vector since

$$\boldsymbol{k} = \sum_j \boldsymbol{q}_j \,. \tag{2.24}$$

Equations (2.19) and (2.20) have to be modified when the lifetime of the crystal excitations are strongly limited by their decay into other crystal excitations. In addition, (2.20) breaks down in imperfect crystals, in solids lacking translational symmetry (like solid solutions and amorphous solids) and in crystals which are opaque to the incident and scattered light. The cases in which non-conservation of wave vector may occur are

(i) *The Scattering Medium has no Translational Symmetry.* The absence of translational symmetry allows scattering by modes with $\boldsymbol{q}_j \neq \boldsymbol{k}$. This occurs in crystals with defects, in solid solutions, and in amorphous solids. A detailed discussion of Raman scattering in amorphous solids is given in the Chapter 5 by Brodsky.

(ii) *The Scattering Volume is Small.* CORDEN et al. [2.16] have pointed out that in this case light scattering is due to excitations with wave vectors in a range $\Delta q \sim 2\pi/d$ (where d is a characteristic length in the scattering volume) (see [Ref. 8.1, Eq. (2.117)]).

(iii) *The Incident and Scattered Waves are Damped Inside the Scattering Volume.* Under these conditions, which occur in metals and small gap semiconductors that are opaque to the light, $\boldsymbol{k}_1$ and $\boldsymbol{k}_2$ are complex. This case was discussed by MILLS et al. [2.17] and more recently by ZEYHER et al. [2.18]. The inelastic scattering is due to excitations having a range of wave vectors

$$\Delta q = \mathrm{Im}\{\boldsymbol{k}_1\} + \mathrm{Im}\{\boldsymbol{k}_2\} \tag{2.25}$$

about $\boldsymbol{q} = \mathrm{Re}\{\boldsymbol{k}_1 - \boldsymbol{k}_2\}$. Effects associated with the wave vector uncertainty given by (2.25) have been reported recently in Raman scattering spectra of III–V semiconductor compounds [2.19].

The last case merits some further consideration because of the growing interest in light scattering experiments on substances which are opaque to the incident light. In order to discuss this case we notice that (2.7) and (2.13) give the scattering cross-section *inside* the crystal. To obtain this cross-section let us consider the fluctuation in the electric susceptibility of the crystal, due to an elementary excitation with well defined wave vector $\boldsymbol{q}$, which may be written as

$$\delta\chi_{12}(\boldsymbol{r},t) = \delta\chi_{12}(\boldsymbol{q},t)\exp(\mathrm{i}\boldsymbol{q}\cdot\boldsymbol{r}) + \delta\chi^*_{12}(\boldsymbol{q},t)\exp(-\mathrm{i}\,\boldsymbol{q},\boldsymbol{r})\,, \tag{2.26}$$

where the first term on the right side of (2.26) describes the Stokes process and the other term the anti-Stokes process. For the Stokes term we have

$$\delta\chi_{12}(\boldsymbol{k},t)=\delta\chi_{12}(\boldsymbol{q},t)\,(1/V)\int d^3r\,\exp[\mathrm{i}(\boldsymbol{q}-\boldsymbol{k})\cdot\boldsymbol{r}]\,. \tag{2.27}$$

When $\boldsymbol{k}$ is real we obtain

$$\delta\chi_{12}(\boldsymbol{k},t)=\delta(\boldsymbol{q}-\boldsymbol{k})\cdot\delta\chi_{12}(\boldsymbol{q},t)\,, \tag{2.28}$$

which leads to wave vector conservation. In the case which interests us here, that of an opaque crystal, $\boldsymbol{k}$ is complex and (2.28) becomes

$$\begin{aligned}&\delta\chi_{12}(\boldsymbol{k},t)\\&=\mathrm{i}\,\delta\chi_{12}(\boldsymbol{q},t)\,\{L[(\boldsymbol{q}-\mathrm{Re}\{\boldsymbol{k}\})-\mathrm{i}(\mathrm{Im}\{\boldsymbol{k}_1\}+\mathrm{Im}\{\boldsymbol{k}_2\})]\}^{-1}\,.\end{aligned} \tag{2.29}$$

Equation (2.29) shows that $(\mathrm{Im}\{\boldsymbol{k}_1\}+\mathrm{Im}\{\boldsymbol{k}_2\})$ is a measure of the extent to which wave vector is not conserved. Inserting (2.29) into (2.13) we obtain the following expression for the scattering cross-section *inside* the crystal

$$\begin{aligned}(d^2\sigma/d\Omega\,d\omega_2)_{\mathrm{in}}&=(\omega_2/c)^4\,(\omega_1/\omega_2)\,(V^2/2\pi)\\&\quad\cdot\sum_q\{L^2[(\boldsymbol{q}-\mathrm{Re}\{\boldsymbol{k}\})^2+(\mathrm{Im}\{\boldsymbol{k}_1\}+\mathrm{Im}\{\boldsymbol{k}_2)^2\}^{-1}\\&\quad\cdot\int d\omega\,\exp(\mathrm{i}\omega t)\,\langle i|\,\delta\chi_{12}(\boldsymbol{q},t)\,\delta\chi_{12}(\boldsymbol{q},0)\,|i\rangle_{\mathrm{T}}\,.\end{aligned} \tag{2.30}$$

In (6.20) the reader will find a slightly modified version of the Lorentzian factor of (2.30) which takes into account the presence of a boundary. When the *transition* susceptibility is independent of $\boldsymbol{q}$, as in conventional Raman scattering by optical phonons, (2.30) becomes

$$\begin{aligned}&(d^2\sigma/d\Omega\,d\omega_2)_{\mathrm{in}}\\&=(\omega_2/\mathrm{c})^4\,(\omega_1/\omega_2)\,(V^2/2\pi)\,[L(\mathrm{Im}\{\boldsymbol{k}_1\}+\mathrm{Im}\{\boldsymbol{k}_2\})]^{-1}\\&\quad\cdot\int d\omega\,\exp(\mathrm{i}\omega t)\,\langle i|\,\delta\chi_{12}(0,t)\,\delta\chi_{12}(0,0)\,|i\rangle_{\mathrm{T}}\,.\end{aligned} \tag{2.30a}$$

Equation (2.30a) shows that, when $\boldsymbol{k}_1$ and $\boldsymbol{k}_2$ are complex, wave vector is not conserved and, moreover, the effective length of crystal that contributes to inelastic light scattering is $(\mathrm{Im}\{\boldsymbol{k}_1\}+\mathrm{Im}\{\boldsymbol{k}_2\})^{-1}$.

In order to obtain the light scattering cross-section *outside* the crystal it is necessary to consider the boundary conditions at the crystal surfaces. The general case is treated in [2.17]. For a backscattering geometry, and in the case in which the polarizations of the incident and scattered radiation are along crystal principal axes, the cross-section outside the crystal can be obtained from (2.30a) by multiplying by the transmission coefficients for the incident and scattered light. In this case the expression for the outside cross-section is equivalent to that given by LOUDON [2.20].

2.1.3. Light Scattering Selection Rules

In addition to the conditions imposed by the kinematics of the scattering process, there are other requirements given by

$$(\overleftrightarrow{\delta\chi})_{\mu\nu} = \langle f|\delta\chi_{\mu\nu}|i\rangle \neq 0 \tag{2.31}$$

which establishes conditions on the transformation properties of the states $|i\rangle$ and $|f\rangle$ under the symmetry operations of the crystal space group. Furthermore, for a given pair of states $|i\rangle$ and $|f\rangle$ the condition

$$\hat{e}_1 \cdot \overleftrightarrow{\delta\chi} \cdot \hat{e}_2 \neq 0 \tag{2.32}$$

determines the selection rules on the polarizations of the incident and scattered light $\hat{e}_1$ and $\hat{e}_2$, respectively.

The selection rules that are associated with the contribution of the $\mathscr{H}_{\mathrm{AA}}$ electron-radiation interaction were discussed in detail by KLEIN [2.11]. We shall consider here the selection rules of the contributions associated with $\mathscr{H}_{\mathrm{A}}$, given by the last two terms in the right side of (2.6). These terms may be separated into symmetric and antisymmetric components. When using the electric dipole approximation, the symmetric part is

$$\begin{aligned}&\langle f|\delta\chi_{\mu\nu}|i\rangle_{\mathrm{S}}\\ &= (e^2/2m^2\omega_2 V)\sum_b \{[(E_b - E_i - \hbar\omega_1)^{-1} + (E_b - E_i + \hbar\omega_2)^{-1}]\\ &\quad\cdot [\langle f|p_\mu(\boldsymbol{k}_2)|b\rangle\langle b|p_\nu(-\boldsymbol{k}_1)|i\rangle + \langle f|p_\nu(-\boldsymbol{k}_1)|b\rangle\langle b|p_\mu(\boldsymbol{k}_2)|i\rangle]\}\end{aligned} \tag{2.33}$$

and the antisymmetric part is

$$\begin{aligned}&\langle f|\delta\chi_{\mu\nu}|i\rangle_{\mathrm{A}}\\ &= (e^2/2m^2\omega_2 V)\sum_b \{[(E_b - E_i - \hbar\omega_1)^{-1} - (E_b - E_i + \hbar\omega_2)^{-1}]\\ &\quad\cdot [\langle f|p_\mu(\boldsymbol{k}_2)|b\rangle\langle b|p_\nu(-\boldsymbol{k}_1)|i\rangle - \langle f|p_\nu(-\boldsymbol{k}_1)|b\rangle\langle b|p_\mu(\boldsymbol{k}_2)|i\rangle]\}\,.\end{aligned} \tag{2.34}$$

We note from (2.34) that the antisymmetric part of the *transition* susceptibility tends to zero when the photon energies are far from resonance with the energies of the transitions to intermediate crystal states ($E_b - E_i \gg \hbar\omega_1$).

The conventional selection rules for Raman scattering (or *allowed* scattering) by optical phonons are obtained when the contributions of these modes to the *transition* electric susceptibility are independent of $\boldsymbol{q}$ (or $\boldsymbol{k}$). For *allowed* scattering, the conditions under which (2.31) is satisfied will be determined only by the transformation properties of the initial and final states under the symmetry operations of the crystal point group. In the language of group theory the condition for an *allowed* scattering process may be written as [2.22].

$$\Gamma_l \times \Gamma_{\mu\nu} \supset A_1 \,, \tag{2.35}$$

where Γ_l is the irreducible representation of the crystal point group which describes the symmetry of the phonon-wave-functions, A_1 is the totally symmetric representation of the point group and $\Gamma_{\mu\nu}$ is the irreducible representation which describes the symmetry properties of the $\mu\nu$ component of a 2nd rank tensor. Some general properties of *allowed* scattering are straightforwardly deduced from (2.35). For example, in a crystal with inversion symmetry $\Gamma_{\mu\nu}$ has even parity while Γ_l, for $q \simeq 0$ optical phonons may have even or odd parity depending on the crystal structure and on the phonon branch being considered. From (2.35) it follows that, in crystals with inversion symmetry, only phonons with even parity will be Raman active in allowed scattering.

It has been shown that in the case of optical phonons with normal coordinates which transform only like polar vectors or polar tensors, $\langle f|\delta\chi_{\mu\nu}|i\rangle_A$ will be small and may exist only as a result of the small difference in the frequencies of the incident and scattered light [2.14]. The form of the symmetric part of the *transition* susceptibility tensors for *allowed* scattering by $q \sim 0$ optical phonons have been tabulated for all crystal classes by LOUDON [2.14]. OVANDER [2.22], MARADUDIN and WALLIS [2.3] have considered both the symmetric and the antisymmetric parts. It should be noted that some crystals have $q \sim 0$ optical phonons which transform like axial vectors for which the *transition* susceptibility tensor is purely antisymmetric. Examples are the A_2 modes in Se, Te, and quartz.

The q-dependent terms in the contribution of optical phonons to the *transition* susceptibility give rise to selection rules which are different than those of *allowed* scattering. For this reason the spectra due to these terms are often called *forbidden* scattering. The q-dependent terms in the *transition* susceptibility originate in the wave vector dependence of the matrix elements and resonant denominators in (2.33) and (2.34). Selection rules which are different from those for *allowed* scattering are also found for Raman scattering processes in the presence of an external force (usually an external electric field or mechanical stress). External-force induced Raman scattering falls in the category of *morphic effects*. q-dependent and morphic effects in Raman scattering [2.23, 24] are discussed further in Section 2.2 and 2.3.

2.2. Raman Scattering by Collective Excitations of Semiconductors and Insulators

The non-magnetic collective excitations of semiconductors and insulators which participate in inelastic light scattering are lattice vibrations (optical and acoustical phonons) and plasma waves (plasmons) of the electron-gas of small gap or doped semiconductors. They also include coupled photon-TO phonon modes (polaritons) and coupled phonon-plasmon modes. In this section we shall discuss first-order inelastic light scattering processes involving this group of elementary excitations.

2.2.1. Microscopic Formulation

The quantum-mechanical expression of the *transition* susceptibility for first-order Brillouin and Raman scattering is obtained by 3rd order time-dependent perturbation calculations of processes of the type described by the diagrams of Fig. 2.2. Plasmons have an additional contribution due to the $\mathscr{H}_{\mathrm{AA}}$ electron-radiation interaction, given by (2.16) and (2.17). This contribution is discussed in further detail in Chapter 4 by KLEIN [2.11].

The intermediate electronic excitations of the crystal involved in the calculation of the *transition* susceptibility are electron-hole pair states. When the electron and hole motions in the intermediate electronic excitations are uncorrelated by their Coulomb interaction (when exciton effects are absent), the electron and hole contributions to the *transition* susceptibility may be separated, as was done in Fig. 2.2a and b. In the processes described in these diagrams, the creation and annihilation of the three quanta involved may occur in any time order. Thus, each diagram of Fig. 2.2 contributes six terms to the first-order *transition* susceptibility which are written as [2.7].

$$
\begin{aligned}
(\delta\chi_{\mu\nu}) = &\langle \bar{n}\pm 1|\delta\chi_{\mu\nu}|\bar{n}\rangle \\
= &(e^2/m^2\omega_2^2 V)\sum_{\alpha\beta}\{\langle 0|p_\mu(\boldsymbol{k}_2)|\beta\rangle\langle\beta,\bar{n}\pm 1|\mathscr{H}_{\mathrm{EL}}|\alpha,\bar{n}\rangle \\
&\cdot\langle\alpha|p_\nu(-\boldsymbol{k}_1)|0\rangle/(E_\beta-\hbar\omega_2)(E_\alpha-\hbar\omega_1) \\
&+\langle 0|p_\nu(-\boldsymbol{k}_1|\beta\rangle\langle\beta,\bar{n}\pm 1|\mathscr{H}_{\mathrm{EL}}|\alpha,\bar{n}\rangle \\
&\cdot\langle\alpha|p_\mu(\boldsymbol{k}_2)|0\rangle/(E_\beta+\hbar\omega_1)(E_\alpha+\hbar\omega_2) \\
&+\langle 0|p_\mu(\boldsymbol{k}_2)|\beta\rangle\langle\beta|p_\nu(-\boldsymbol{k}_1)|\alpha\rangle \\
&\cdot\langle\alpha,\bar{n}\pm 1|\mathscr{H}_{\mathrm{EL}}|0,\bar{n}\rangle/(E_\beta-\hbar\omega_2)(E_\alpha+\hbar\omega) \\
&+\langle 0|p_\nu(-\boldsymbol{k}_1)|\beta\rangle\langle\beta|p_\mu(\boldsymbol{k}_2)|\alpha\rangle \\
&\cdot\langle\alpha,\bar{n}\pm 1|\mathscr{H}_{\mathrm{EL}}|0,\bar{n}\rangle/(E_\beta+\hbar\omega_1)(E_\alpha+\hbar\omega) \\
&+\langle 0,\bar{n}\pm 1|\mathscr{H}_{\mathrm{EL}}|\beta,\bar{n}\rangle \\
&\cdot\langle\beta|p_\mu(\boldsymbol{k}_2)|\alpha\rangle\langle\alpha|p_\nu(-\boldsymbol{k}_1)|0\rangle/(E_\beta-\hbar\omega)(E_\alpha-\hbar\omega_1) \\
&+\langle 0,\bar{n}\pm 1|\mathscr{H}_{\mathrm{EL}}|\beta,\bar{n}\rangle \\
&\cdot\langle\beta|p_\nu(-\boldsymbol{k}_1)|\alpha\rangle\langle\alpha|p_\mu(\boldsymbol{k}_2)|0\rangle/(E_\beta-\hbar\omega)(E_\alpha+\hbar\omega_2)\,,
\end{aligned}
\tag{2.36}
$$

where $|0\rangle$, $|\alpha\rangle$, and $|\beta\rangle$ are the wave functions of the single-particle electronic ground and intermediate states. $|\alpha,\bar{n}\rangle$ are crystal wave functions described within the adiabatic (for phonons) and RPA (for plasmons) approximations, in which the collective excitation has an occupation number $\bar{n}$. The six terms which result from the diagram of Fig. 2.2a correspond to processes in which the electron-hole pair states $|\alpha\rangle$ and $|\beta\rangle$ differ in the state of an electron in the conduction bands. The six terms which result from the diagram of Fig. 2.2b correspond to processes in which $|\alpha\rangle$ and $|\beta\rangle$ differ in the holes in the valence bands. The $\langle\alpha,\bar{n}\pm 1|\mathscr{H}_{\mathrm{EL}}|\beta,\bar{n}\rangle$ are the matrix elements of the electron-collective excitation interaction $\mathscr{H}_{\mathrm{EL}}$. The plus sign corresponds to the Stokes processes and the minus sign to the anti-Stokes processes. The particular time order for the annihilation and creation of the quanta shown in Fig. 2.2 corresponds to the first term in (2.36).

The form of the first-order *transition* susceptibility for the case in which the intermediate electron-hole pair states are Wannier excitons was given by GANGULY and BIRMAN [2.26]. It was shown that for Raman scattering by optical phonons the expression for the *transition* susceptibility is similar to that of (2.36) except that in this case $|\alpha\rangle$ and $|\beta\rangle$ are exciton wave functions.

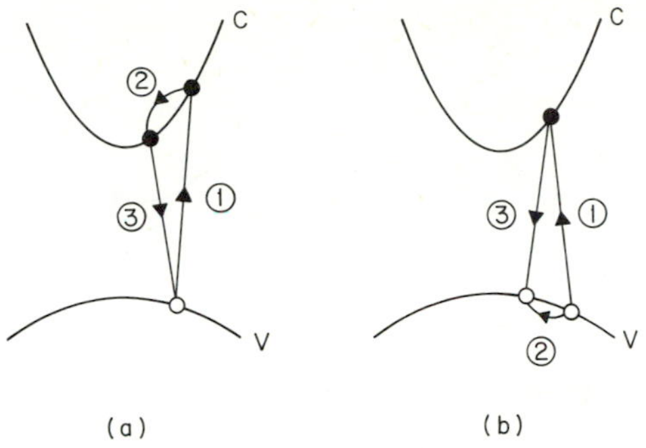

Fig. 2.4a and b. Schematic diagram of the interband and intraband electronic transitions which play a role in *two-band* Raman processes. The numbers indicate the order of the electronic transitions. (a) and (b) correspond to the Feynman diagrams of Fig. 2.2a and b, respectively. v is a filled valence band and c is an empty conduction band. The numbers indicate the order of the transitions

It has been pointed out (see [2.15] and references therein) that one should treat the incident and scattered radiation inside the crystal as coupled photon-electronic excitation modes (polaritons). This polariton theory of inelastic light scattering is, potentially, advantageous when the frequency of radiation is close to resonance with the electronic transitions (the resonance Brillouin and Raman effects). However, it has been shown [2.12] that the results obtained with the polariton picture, for first-order Raman scattering processes, are equivalent to those obtained by 3rd order time-dependent perturbation theory.

The electronic transitions involved in the calculation of the first-order *transition* susceptibility may be conveniently described by diagrams of the type shown in Figs. 2.4 and 2.5 [2.25]. In these diagrams we show the processes of Fig. 2.2a and b for crystals with empty conduction bands and full valence bands. Similar diagrams can be drawn for processes involving intermediate exciton states. The time order of the three electronic transitions in Figs. 2.4 and 2.5 is always the same. The diagrams of Fig. 2.4 describe the *two-band* contributions to $(\overleftrightarrow{\delta\chi})_{\mu\nu}$. Similarly, the diagrams of Fig. 2.5 correspond to the *three-band* terms. This separation of the Raman process into two-band and three-band is convenient for the discussion of the resonant Raman effect in which the dominant contributions to $(\overleftrightarrow{\delta\chi})_{\mu\nu}$ correspond to the time order of the diagrams of Fig. 2.2 and given by the first term in the right side of (2.36). In this case, the two-band processes involve *intraband* matrix elements of $\mathscr{H}_{EL}$, while the three-band processes involve the *interband* matrix elements of $\mathscr{H}_{EL}$.

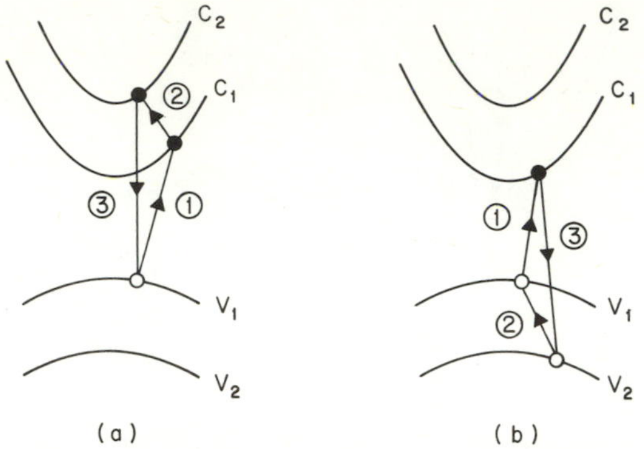

Fig. 2.5a and b. Schematic diagram of the interband and intraband electronic transitions which play a role in *three-band* Raman processes. The numbers indicate the order of the electronic transitions. (a) and (b) correspond to the Feynman diagrams of Fig. 2.2a and b. v_1 and v_1 are filled valence bands and c_1 and c_2 are empty conduction bands. The numbers indicate the order of the transitions

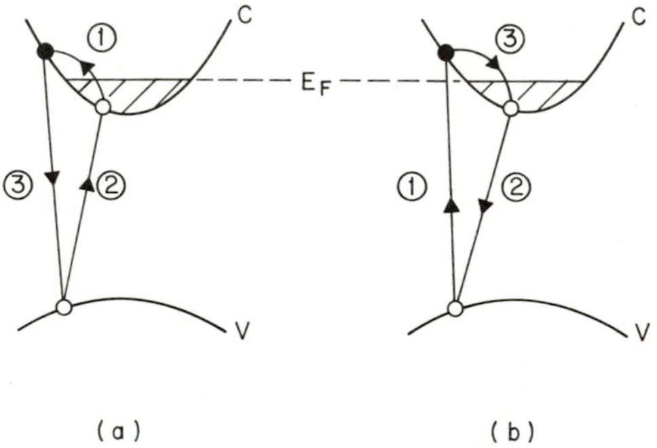

Fig. 2.6a and b. Schematic diagram of the additional interband and intraband electronic transitions which play a role in scattering processes which take place in an *n*-type semiconductor. E_F shows the position ot the Fermi level. The numbers indicate the order of the transitions

The diagrams of Figs. 2.4 and 2.5 also show that in the case of crystals with empty conduction bands and full valence bands, the initial and final electronic transitions are necessarily *interband*. In the case of crystals having free carriers, like semiconductors or metals, it is also possible to have initial or final electronic transitions which are *intraband* [2.25]. This is shown in the diagrams of Fig. 2.6 for the case of a degenerate *n*-type semiconductor. Consequently, crystals with free carriers have contributions to the *transition* susceptibility which do not exist for crystals with full valence bands and empty conduction bands. These additional contributions, which are associated with the charge density fluctuations of the electron-gas, are important light scattering mechanisms for semiconductor plasmons [2.25] and for phonons in metals [2.17] (see [Ref. 8.7, Chap. 2]).

Before entering into the detailed discussion of the light scattering mechanisms of lattice vibrations and plasmons it is useful to comment briefly on the possible occurrence of wave vector dependences in (2.36). We have seen in Subsection 2.1.3 that the *conventional* selection rules for first-order Raman scattering by optical phonons are due to the wave vector *independent* terms in $(\overleftrightarrow{\delta\chi})_{\mu\nu}$. Thus, an understanding of the origin of the wave vector *dependent* terms in (2.36) will, in addition to giving further insight into the properties of the *transition* susceptibility, define the conditions under which the conventional selection rules will hold. In Subsection 2.1.3 it was pointed out that the wave vector dependent terms in the transition susceptibility originate in the wave vector dependence of the electron momentum matrix elements and of the resonant denominators. In the case of light scattering by collective excitations it also arises from the wave vector dependence of the matrix elements of $\mathscr{H}_{\mathrm{EL}}$. The role played by wave vector dependent terms in resonance Raman scattering will be discussed in Section 2.3 and in further detail in Chapter 3 by Martin and Falicov [2.27]. (See [Ref. 8.1, Sect. 2.2.8].)

Brillouin Scattering by Acoustical Phonons

In order to discuss the first-order *transition* susceptibility for Brillouin scattering, we have to consider the electron-acoustical phonon interaction. For a $q \simeq 0$ acoustical vibration, electron-lattice interaction will result from the strain associated with the mode, which is given, in component from, by [Ref. 8.1, Sect. 2.1.14].

$$S_{\lambda\gamma} = \nabla_\lambda w_\gamma = \mathrm{i} q_\lambda w_\gamma \,, \tag{2.37}$$

where q_λ are the components of the phonon wave vector, and the w_γ are the components of the elastic displacement normal coordinate. In crystals without an inversion center, and in centro-symmetric crystals having atoms which are not at centers of symmetry (like the semiconductors with the diamond structure), $\overleftrightarrow{S}$ will be accompanied by an optical or relative displacement $\boldsymbol{u}$ of the atoms, and in piezoelectric crystals it will also produce a macroscopic electric field $\mathscr{E}$ [2.28]. These give rise to electron-lattice interactions which are typical of $q \simeq 0$ optical phonons. We consider here the electron-acoustical phonon interaction due to $(\overleftrightarrow{S})_{\lambda\gamma}$ at zero $\boldsymbol{u}$ and $\mathscr{E}$, whose matrix elements may be expressed in terms of deformation potentials. For first-order scattering processes the matrix elements are given by [2.7]

$$\langle \alpha, \bar{n} \pm 1 | \mathscr{H}_{\mathrm{EL}}^{\mathrm{D}} | \beta, \bar{n} \rangle = (\bar{n}+1)^{\frac{1}{2}} \sum_{\lambda(\gamma)} D_{\lambda(\gamma)}^{(\mathrm{ac})}(\alpha, \beta) \bar{S}_{\lambda\gamma}, \tag{2.38}$$

where $D_{\lambda\gamma}^{(\mathrm{ac})}(\alpha, \beta)$ are elastic strain deformation potentials, and

$$\bar{S}_{\lambda\gamma} = \mathrm{i} q_\lambda \hat{\xi}_\gamma (\hbar / 2\varrho V \omega_q)^{\frac{1}{2}} \tag{2.39}$$

is the amplitude of the strain associated with an acoustical oscillation of frequency ω_q in a medium with a density ϱ. $\hat{\xi}_\lambda$ are the components of the unit polarization vector of the elastic displacement.

The first-order *transition* susceptibility, due to $S_{\lambda\gamma}$ of an acoustical phonon, may be obtained by inserting (2.38) in (2.36). It is clear that we may write

$$(\delta \chi_{\mu\nu})_{\mathrm{ac}} = (\bar{n}+1)^{\frac{1}{2}} \sum_{\lambda\gamma} f_{\mu\nu\lambda\gamma}(\omega_1, \omega_q) \bar{S}_{\lambda\gamma}, \tag{2.40}$$

where $f_{\mu\nu\lambda\gamma}$, as we will show, is related to the elasto-optic tensor. In order to find the physical meaning of $f_{\mu\nu\lambda\gamma}$ and $(\delta\chi_{\mu\nu})_{\mathrm{ac}}$ which results from (2.36)–(2.40) we notice that, because interband electronic excitations enter as intermediate states in the Raman process, $(\delta\chi_{\mu\nu})_{\mathrm{ac}}$ can be expected to represent a change (induced by $S_{\lambda\gamma}$) in the interband susceptibility of the crystal. The expression of the interband susceptibility of the crystal is [2.6]

$$\begin{aligned} \chi_{\mu\nu}(\omega_1) = (e^2/m^2 \omega_1^2 V) \sum_\alpha \{ & \langle 0 | p_\mu(0) | \alpha \rangle \langle \alpha | p_\nu(0) | 0 \rangle / (E_\alpha - \hbar\omega_1) \\ & + \langle 0 | p_\nu(0) | \alpha \rangle \langle \alpha | p_\mu(0) | 0 \rangle / (E_\alpha + \hbar\omega_1) \} . \end{aligned} \tag{2.41}$$

In the *quasi-static* approximation the phonons are assumed to act (through the electron-phonon interactions) like static perturbations of the electronic band structure of the crystal. This perturbation causes the following first-order changes in the interband transition energies and momentum matrix elements which enter in (2.41) [2.7]

$$\delta E_\alpha = - \sum_{\lambda\gamma} D_{\lambda\gamma}^{(\mathrm{ac})}(\alpha, \alpha) \bar{S}_{\lambda\gamma} \tag{2.42}$$

$$\begin{aligned} \delta[\langle 0 | p_\mu(0) | \alpha \rangle] = \sum_{\lambda\gamma} \Bigg\{ & \sum_{\beta \neq \alpha} \langle 0 | p_\mu(0) | \beta \rangle D_{\lambda\gamma}^{(\mathrm{ac})}(\alpha, \beta) / (E_\beta - E_\alpha) \\ & + \sum_{\beta \neq 0} \langle 0 | p_\mu(0) | \beta \rangle D_{\lambda\gamma}^{(\mathrm{ac})}(0, \beta) / E_\beta \Bigg\} \bar{S}_{\lambda\gamma} . \end{aligned} \tag{2.43}$$

δE_α corresponds to the *intraband* matrix elements of the electron-phonon interaction which only occur in *two-band* processes. $\delta[\langle 0|p_\mu|\alpha\rangle]$ involves interband matrix elements of the electron-phonon interaction and may occur in either *two-band* or *three-band* processes. The properties of the deformation potentials, which enter in (2.42) and (2.43) for diamond- and zinc blende-type semiconductors, have been discussed by CARDONA and coworkers [2.29, 30].

From (2.36), (2.41), and (2.43) it follows that in the quasi-static approximation we have

$$f_{\mu\nu\lambda\gamma}(\omega_1, \omega_q) \simeq f_{\mu\nu\lambda\gamma}(\omega_1, 0) = (\partial \chi_{\mu\nu}(\omega_1)/\partial S_{\lambda\gamma})_{u,\mathscr{E}}\,, \tag{2.44}$$

where $(\partial\chi_{\mu\nu}/\partial S_{\lambda\gamma})_{u,\mathscr{E}}$ is the first derivative of the interband susceptibility with respect to strain at zero $\boldsymbol{u}$ and $\mathscr{E}$. In crystals in which the optical atomic displacement associated with $S_{\lambda\gamma}$ is zero, $f_{\mu\nu\lambda\gamma}(\omega_1, 0)$ can also be written as [2.28]

$$f_{\mu\nu\lambda\gamma}(\omega_1, 0) = -(\eta_\mu^2 \eta_\nu^2/4\pi)\, p^{\mathscr{E}}_{\mu\nu\lambda\gamma}\,, \tag{2.45}$$

where $p^{\mathscr{E}}_{\mu\nu\lambda\gamma}$ is the elasto-optic coefficient at constant electric field, and η_μ and η_ν are the principal values of the refractive index along the μ and ν axes, respectively.

Equation (2.44) indicates, as could be expected, that the quasi-static approximation is equivalent to neglecting the phonon-frequency in (2.36). The conditions under which this approximation is valid are found by direct inspection of (2.36). For three-band processes it is necessary that

$$\begin{aligned} |E_\alpha - \hbar\omega_1| &\gg \hbar\omega \\ |E_\beta - \hbar\omega_1| &\gg \hbar\omega\,. \end{aligned} \tag{2.46}$$

For two-band processes it is required that

$$|E_\alpha - \hbar\omega_1| \gg \hbar\omega\,. \tag{2.47}$$

The condition expressed by (2.46) and (2.47) are satisfied in most semiconductor and insulator crystals away from resonance. These inequalities may not hold under resonance conditions. However, they will hold even at resonance provided the damping (lifetime broadening) of the electronic transitions is much larger than the phonon energy [2.31]. In the particular case of $q \simeq 0$ acoustical phonons, $\omega = \omega_q$ is very small and, consequently, the inequalities in (2.46) and (2.47) will always be satisfied [Ref. 8.2, Chap. 7].

Raman Scattering by Optical Phonons

LOUDON [2.7] has pointed out that two types of electron-lattice interactions have to be considered in the calculation of the *transition* susceptibility of $q \simeq 0$ optical phonons. One is the *deformation potential interaction* due to the modulation of the crystal periodic potential by the relative atomic displacement of the phonons. The other is the *Fröhlich interaction* of the electrons with the *longitudinal* macroscopic electric fields associated with the infrared active lattice vibrations. Coupled TO phonon-photon modes (polaritons) also have *transverse* macroscopic electric fields which are associated with the phonon part of the modes. BURSTEIN and PINCZUK [2.25] have pointed out that these fields give rise to a $\mathscr{H}_A$ electron-polariton interaction.

The macroscopic electric fields associated with optical phonons and polaritons have been discussed recently by MILLS and BURSTEIN [2.15]. Neglecting the damping of the modes, the expression for the electric field components along symmetry directions is [2.15]

$$\mathscr{E}_\lambda = (4\pi N e^*/\varepsilon_\infty)\,[(\omega_T^2 - \omega^2)/(\omega_L^2 - \omega_T^2)]\, u_\lambda\,, \tag{2.48}$$

where u_λ is the λ component of the optical vibration normal coordinate; ω_T and ω_L are the TO and LO $q \simeq 0$ phonon frequencies; N is the density of unit cells; e^* is the Born or transverse effective charge; and ε_∞ is the high frequency dielectric constant. Equation (2.48) shows that macroscopic electric fields exist for LO phonons and polaritons provided that they have $\omega \neq \omega_T$. The macroscopic fields in more complicated cases are discussed in [2.12, 15].

The matrix elements of the deformation potential interaction, for first order Stokes processes, may be written as

$$\langle \alpha, \bar{n} \pm 1 | \mathscr{H}_{EL}^{D} | \beta, \bar{n} \rangle = \sum_l D_l^{(\mathrm{opt})}(\alpha, \beta)\, \langle \bar{n} \pm 1 | u_l | \bar{n} \rangle\,, \tag{2.49}$$

where $D_l^{(\mathrm{opt})}(\alpha, \beta)$ are the atomic displacement deformation potentials. Equation (2.49) can be used either when $|\alpha\rangle$ and $|\beta\rangle$ are uncorrelated electron-hole pair states (2.7) or when they are exciton-like states described within the effective-mass approximation [2.26]. In addition, we have [2.15]

$$\begin{aligned} \langle \bar{n} \pm 1 | u_l | \bar{n} \rangle &= (\bar{n}+1)^{\frac{1}{2}}\, \hat{\xi}_l (\hbar/2N\bar{M}\omega_l)^{\frac{1}{2}} \\ &= (\bar{n}+1)^{\frac{1}{2}}\, \bar{u}_l(\omega_l)\,, \end{aligned} \tag{2.50}$$

where $\bar{M}$ is the reduced mass of the optical phonon with frequency ω_l. $\hat{\xi}_l$ and $\bar{u}_l(\omega_l)$ are the eigenvectors and amplitudes of the atomic displacement normal coordinate.

The *elastic strain* deformation potentials are second-rank tensors and, consequently, they have non-zero components in *all* crystal classes. The tensor rank of the *atomic displacement* deformation potentials is determined by the transformation properties of the $q \sim 0$ optical phonon normal coordinate. In centrosymmetric crystals, the $D_l^{(\mathrm{opt})}(\alpha, \beta)$ will have non-vanishing components only for even-parity phonons. In the case of simple diatomic crystals, like those with the rock salt or diamond structures, the components of the relative atomic displacement do not always transform like a vector under the symmetry operations of the crystal point group [2.32]. Consequently, the $q \sim 0$ (infrared inactive) even-parity optical phonons of crystals with the diamond structure are Raman-active. On the other hand, the (infrared active) odd-parity optical phonons of crystals with the rock salt structure have no linear deformation potentials and, consequently, are Raman-inactive. Similar arguments lead to the conclusion that the $q \sim 0$ optical phonons in the non-centrosymmetric crystals with the zinc blende structure are infrared- *and* Raman-active [Ref. 8.1, Fig. 2.9].

The first-order *transition* susceptibility of $q \sim 0$ optical phonons, involving the optical deformation potential interaction, is obtained by using (2.50) in (2.36). We find

$$(\delta\chi_{\mu\nu})_{\mathrm{u}} = (\bar{n}+1)^{\frac{1}{2}} \sum_l a_{\mu\nu l}(\omega_1, \omega, \boldsymbol{q})\, \bar{u}_l(\omega_j)\,, \tag{2.51}$$

where $a_{\mu\nu l}(\omega_1, \omega, \boldsymbol{q})$ is the *atomic displacement* Raman tensor. The perturbation of the electronic band structure of the crystal by the optical phonons is given in the quasi-static approximation, by expressions similar to (2.42) and (2.43). Consequently, in the $q=0$ quasi-static approximation it is possible to write

$$a_{\mu\nu l}(\omega_1, \omega, \boldsymbol{q}) \simeq a_{\mu\nu l}(\omega_1, 0, 0) = (\partial\chi_{\mu\nu}/\partial u_l)_{\mathscr{E}}\,, \tag{2.52}$$

where $(\partial\chi_{\mu\nu}/\partial u_l)_{\mathscr{E}}$ is the first derivative of the interband susceptibility with respect to the optical phonon displacement at zero macroscopic electric field.

In the calculation of the first-order *transition* susceptibility associated with the macroscopic electric fields of optical lattice vibrations, it is necessary to consider the contributions due to the Fröhlich interaction with the longitudinal field of LO phonons and the $\mathscr{H}_{\mathrm{A}}$ interaction with the transverse fields of polaritons [2.25]. In addition, we have to consider the cases in which the intermediate electronic excitations are either uncorrelated electron-hole pairs, or discrete or continuum exciton states.

Let us consider first the case in which the intermediate states are uncorrelated electron-hole pairs. The matrix element of the Fröhlich interaction for Stokes processes may be written as [2.25] (see [Ref. 8.1, Sect. 2.0.0]).

$$\begin{aligned}&\langle \alpha, \bar{n} \pm 1 | \mathscr{H}_{\mathrm{EL}}^{\mathrm{F}} | \beta, \bar{n} \rangle \\ &= \mp (\mathrm{i}e/q)(\bar{n}+1)^{\frac{1}{2}} \langle \alpha | \exp(\mathrm{i}\boldsymbol{q} \cdot \boldsymbol{r}) | \beta \rangle \bar{\mathscr{E}}(\omega_{\mathrm{L}}) ,\end{aligned} \tag{2.53}$$

where, from (2.48), the electric field amplitude for the LO phonons is

$$\bar{\mathscr{E}}(\omega_{\mathrm{L}}) = -(4\pi N e^*/\varepsilon_\infty)\bar{u}(\omega_{\mathrm{L}}) , \tag{2.54}$$

and $\bar{u}(\omega_{\mathrm{L}})$ is the amplitude of the LO phonon normal coordinate, given by (2.50). The $(-)$ and $(+)$ signs correspond to the electron and hole contributions, respectively. In the case of the transverse fields of polaritons, the matrix elements of the $\mathscr{H}_{\mathrm{A}}$ interaction are [2.25]

$$\begin{aligned}&\langle \alpha, \bar{n} \pm 1 | \mathscr{H}_{\mathrm{EL}}^{\mathrm{A}} | \beta, \bar{n} \rangle \\ &= \mp (\mathrm{i}e/m\omega)(\bar{n}+1)^{\frac{1}{2}} \sum_{\lambda} \langle \alpha | p_\lambda(\boldsymbol{q}) | \beta \rangle \bar{\mathscr{E}}_\lambda(\omega) ,\end{aligned} \tag{2.55}$$

where $\bar{\mathscr{E}}_\lambda(\omega)$, calculated with (2.48) and (2.50), are the components of the amplitude of the polariton macroscopic electric field.

In the $q=0$ approximation the intraband terms of (2.53), which involve states $|\alpha\rangle$ and $|\beta\rangle$ in the same pair of bands, do not contribute to the first-order *transition* susceptibility because the electron contribution exactly cancels that of the hole [2.7]. HAMILTON [2.33] pointed out that these terms do make a finite contribution for $q \neq 0$. This results from the q-dependence of $\langle \alpha | \exp(\mathrm{i}\boldsymbol{q} \cdot \boldsymbol{r}) | \beta \rangle$ in (2.53) and of the resonant denominators in (2.36). It can also be shown that in the $q=0$ approximation the intraband terms of (2.55) do not contribute to the first-order *transition* susceptibility for scattering by polaritons in crystals with the zinc blende structure. This term is identical to the two-band terms of the second-order nonlinear susceptibility, which are known to vanish in crystals with the zinc blende structure [2.34].

The interband terms of (2.53) and (2.55), in which the states $|\alpha\rangle$ and $|\beta\rangle$ have holes in different valence bands or electrons in different conduction bands may be written as [2.7, 25]

$$\begin{aligned}&\langle \alpha, \bar{n} \pm 1 | \mathscr{H}_{\mathrm{EL}}^{\mathrm{F}} | \beta, \bar{n} \rangle \\ &= \mathrm{i}e(\bar{n}+1)^{\frac{1}{2}} \sum_{\lambda} \langle \alpha | p_\lambda(0) | \beta \rangle \bar{\mathscr{E}}_\lambda(\omega_{\mathrm{L}})/(E_\alpha - E_\beta) .\end{aligned} \tag{2.56}$$

The summation in (2.56) runs over the components of the longitudinal electric field of LO phonons, or over components of the transverse electric field in the case of polaritons.

In the case in which the intermediate electronic excitations are correlated electron-hole pairs (either discrete or continium excitons described with the effective-mass approximation) the *interband* matrix element of the Fröhlich interaction between two excitonic states α and β belonging to the same pair of bands is given by [2.25]

$$\begin{aligned}&\langle \psi_\alpha, \bar{n} \pm 1 | \mathscr{H}_{\mathrm{EL}}^{\mathrm{F}} | \psi_\beta, \bar{n} \rangle \\ &= (\mathrm{i}e/q)(\bar{n}+1)^{\frac{1}{2}} \qquad (2.57)\\ &\quad \cdot \langle \psi_\alpha | \exp[\mathrm{i}(m_\mathrm{h}^*/M)\,\boldsymbol{q}\cdot\boldsymbol{r}] - \exp[-\mathrm{i}(m_\mathrm{e}^*/M)\,\boldsymbol{q}\cdot\boldsymbol{r}] | \psi_\beta \rangle \bar{\mathscr{E}}(\omega_\mathrm{L}),\end{aligned}$$

where m_e^* and m_h^* are the conduction and valence band effective masses, and $M = m_\mathrm{e}^* + m_\mathrm{h}^*$; $\boldsymbol{r} = \boldsymbol{r}_\mathrm{e} - \boldsymbol{r}_\mathrm{h}$ is the relative coordinate of the electron-hole pair; and $|\psi_\alpha\rangle$ and $|\psi_\beta\rangle$ are the exciton wave functions. The properties of this matrix element, which like the intraband term in (2.53) vanishes for $q=0$, are discussed further in Chapter 3 and [Ref. 8.1, Sect. 2.2.9].

The contributions to the first-order *transition* susceptibility which result from the macroscopic electric fields of the optical modes can be separated into q-dependent two-band terms and q-independent three-band terms. In the case of LO phonons, the two-band contribution, obtained using the intraband term of (2.53) in (2.36), may be written as

$$(\delta\chi_{\mu\nu})_{\mathscr{E}} = (\bar{n}+1)^{\frac{1}{2}} \sum_\lambda b''_{\mu\nu\lambda}(\omega_1, \omega_\mathrm{L}, \boldsymbol{q})\, \bar{\mathscr{E}}_\lambda(\omega_\mathrm{L}), \qquad (2.58)$$

where $b''_{\mu\nu\lambda}(\omega_1, \omega_\mathrm{L}, \boldsymbol{q})$ is the *two-band* electro-optic Raman tensor. This tensor may be expanded in powers of q_γ, the components of the phonon wave vector. Moreover the wave vector independent term $b''_{\mu\nu\lambda}(\omega_1, \omega, 0)$ is zero and the first non-vanishing term is linear in the q_γ. The latter is important under resonance conditions [2.27].

The three-band contribution, obtained using (2.56) in (2.36), may be written as

$$(\delta\chi'''_{\mu\nu\lambda})_{\mathscr{E}} = (\bar{n}+1)^{\frac{1}{2}} \sum_\lambda b'''_{\mu\nu\lambda}(\omega_1, \omega, 0)\, \bar{\mathscr{E}}_\lambda(\omega_\mathrm{L}), \qquad (2.59)$$

where $b'''_{\mu\nu\lambda}(\omega_1, \omega, 0)$ is the wave vector independent *three-band* electro-optic Raman tensor which exists only in non-centro-symmetric crystals. After some algebra it is possible to show that, in the quasi-static approximation,

$$b'''_{\mu\nu\lambda}(\omega_1, \omega, 0) \simeq b'''_{\mu\nu\lambda}(\omega_1, 0) = (\partial\chi_{\mu\nu}/\partial\mathscr{E}_\lambda)_{\mathrm{u}}\,. \tag{2.60}$$

We note that $b'''_{\mu\nu\lambda}(\omega_1, \omega, \boldsymbol{q})$ is also called the electro-optic Raman tensor. It is of interest to point out that there is no interference between the terms linear in q of $(\delta\chi''_{\mu\nu})$ and $(\delta\chi'''_{\mu\nu})$, at least in the region of transparency. This follows from the fact that (2.56) is real and the terms which are linear in q in the intraband terms of (2.53) and (2.57) are imaginary, namely, from the fact that the two contributions are 90° out of phase. The q-independent contributions to the first-order *transition* susceptibility of infrared active optical vibrations, which determine the intensities of "allowed" scattering, may be written as

$$\begin{aligned}(\delta\chi_{\mu\nu}) &= (\delta\chi_{\mu\nu})_{\mathrm{u}} + (\delta\chi'''_{\mu\nu})_{\mathscr{E}}\\ &= (\bar{n}+1)^{\frac{1}{2}} \sum_\lambda [a_{\mu\nu\lambda}\bar{u}_\lambda(\omega) + b'''_{\mu\nu\lambda}\bar{\mathscr{E}}_\lambda(\omega)]\\ &= (\bar{n}+1)^{\frac{1}{2}} \sum_\lambda [a_{\mu\nu\lambda} + (4\pi N e^*/\varepsilon_\infty)(\omega_{\mathrm{T}}^2-\omega^2)/(\omega_{\mathrm{L}}^2-\omega_{\mathrm{T}}^2)\, b'''_{\mu\nu\lambda}]\,\bar{u}_\lambda(\omega)\\ &= (\bar{n}+1)^{\frac{1}{2}}\, a^*_{\mu\nu\lambda}(\omega_1, \omega, 0)\,\bar{u}_\lambda(\omega)\,.\end{aligned} \tag{2.61}$$

The $a^*_{\mu\nu\lambda}$ may be larger or smaller than $a_{\mu\nu\lambda}$ depending on the magnitude and relative signs of the atomic displacement and electric field contributions. The ω-dependence of $a^*_{\mu\nu\lambda}$ is discussed in detail in [2.12] and [2.15]. In the case of *allowed* scattering by infrared active vibrations, $a^*_{\mu\nu\lambda}$ transforms like a 3rd rank tensor.

In order to show the relation between Raman scattering and electro-optic phenomena it is convenient to rewrite (2.61) in the following form (for $\omega \neq \omega_{\mathrm{T}}$)

$$\begin{aligned}&(\delta\chi_{\mu\nu})\\ &= (\bar{n}+1)^{\frac{1}{2}} \sum_\lambda [(\varepsilon_\infty/4\pi N e^*)(\omega_{\mathrm{L}}^2-\omega_{\mathrm{T}}^2)/(\omega_{\mathrm{T}}^2-\omega^2)\, a_{\mu\nu\lambda} - b'''_{\mu\nu\lambda}]\,\bar{\mathscr{E}}_\lambda(\omega)\\ &= (\bar{n}+1)^{\frac{1}{2}} \sum_\lambda b^*_{\mu\nu\lambda}(\omega_1, \omega, 0)\,\bar{\mathscr{E}}_\lambda(\omega)\,.\end{aligned} \tag{2.62}$$

In addition, from (2.60) we have [2.2]

$$b'''_{\mu\nu\lambda}(\omega_1, 0) = -(\eta_\mu\eta_\nu/4\pi)\, r^{\mathrm{u}}_{\mu\nu\lambda}\,, \tag{2.63}$$

where $r^{\mathrm{u}}_{\mu\nu\lambda}$ is the linear electro-optic tensor *for zero u*.

Consequently, $b^*_{\mu\nu\lambda}(\omega_1, 0)$ is proportional to the total linear electro-optic tensor. The ratio [Ref. 8.1, Eq. (2.93)]

$$C = (e^* a_{\mu\nu\lambda}/\bar{M}\,\omega_T^2\, b'''_{\mu\nu\lambda}) \tag{2.64}$$

is the Faust-Henry coefficient [2.35] which determines the relative magnitude of the atomic displacement and electro-optic contributions to $a^*_{\mu\nu\lambda}(\omega_1, \omega, 0)$. The value of the Faust-Henry coefficient may be obtained from data on the relative intensity of Raman scattering by polaritons, TO phonons and LO phonons [2.36, 37]. The tensor $b^*_{\mu\nu\lambda}(\omega_1, \omega)$ determines the nonlinear interactions, inside the crystal, of two electromagnetic waves with frequencies ω_1 and ω to produce a third wave at a frequency $\omega_2 = \omega_1 \pm \omega$ [2.12, 15].

Raman Scattering by Plasmons

We consider next the case of a crystal with a partially filled conduction band or one with a partially empty valence band. A convenient example is an *n*-type degenerate semiconductor with charge carriers in only one conduction band with isotropic effective mass. Other cases, as well as those which occur in multi-component plasmas, are reviewed by KLEIN [2.11] in Chapter 4. The longitudinal plasma waves (plasmons), together with the single particle excitations, constitute the elementary excitations of the electron-gas. In the long wavelength limit the plasmons correspond to an almost uniform displacement of the free carriers. This displacement produces a longitudinal electric field with components [2.10].

$$\mathscr{E}_\lambda(\omega_p) = -4\pi n_e e x_\lambda(\omega_p), \tag{2.65}$$

where n_e is the density of charge carriers; m^* is their effective mass; $\omega_p = (4\pi n_e^2/m^*\varepsilon_\infty)^{\frac{1}{2}}$ is the plasma frequency; and $x_\lambda(\omega_p)$ are the components of the plasmon displacement coordinate [2.10, 25]. In the case of undamped modes, the amplitude of this normal coordinate is given by [2.10]

$$\bar{x}(\omega_p) = \hat{\xi}_\lambda(\hbar/2m^* n_e V \omega_p)^{\frac{1}{2}}. \tag{2.66}$$

where $\hat{\xi}_\lambda$ are the unit polarization vectors of the charge carrier displacements. In the case of an isotropic plasma, the associated charge density fluctuation can be expressed as,

$$\varrho_e(\omega_p, \boldsymbol{q}) = \mathrm{i} n_e \boldsymbol{\nabla} \boldsymbol{x}(\omega_p, \boldsymbol{q}) = \mathrm{i} n_e e \sum_\lambda q_\lambda x_\lambda(\omega_p, \boldsymbol{q}). \tag{2.67}$$

The longitudinal electric field leads to a Fröhlich electron-plasmon interaction with matrix elements similar to those of $q \sim 0$ LO phonons. Thus, the expressions given in (2.53) and (2.56) for the intraband and interband matrix elements can be carried over to the case of scattering by plasmons provided $\bar{\mathscr{E}}(\omega_L)$ is replaced by $\bar{\mathscr{E}}(\omega_p)$.

We expect three distinct contributions to the first-order *transition* susceptibility of plasmons. One is due to the electro-optic terms associated with $\mathscr{E}_\lambda(\omega_p)$. This contribution is obtained by replacing in (2.58) and and (2.59) $\bar{\mathscr{E}}_\lambda(\omega_L)$ by $\bar{\mathscr{E}}_\lambda(\omega_p)$ Another contribution, involving the $\mathscr{H}_{AA}$ electron-radiation interaction, results in a cross-section given by (2.16) and (2.17). The expression for this contribution to the *transition* susceptibility, due to plasmon charge density fluctuations, is

$$(\delta \chi'_{\mu\nu})_\varrho = -(\bar{n}+1)^{\frac{1}{2}}(e/m\omega_1^2)\bar{\varrho}_e(\omega_p, \boldsymbol{q}) \tag{2.68}$$

where $\bar{\varrho}_e(\omega_p, \boldsymbol{q})$ is the amplitude of the charge density fluctuation in the plasma wave. The third contribution results from the diagrams of Fig. 2.6 which, as pointed out in subsect. 2.2.1, do not exist in insulating crystals. These diagrams contribute two types of terms. One corresponds to the case in which the first transition in Fig. 2.6a or the last transition in Fig. 2.6b is associated with the creation or annihilation of a photon. This type of term involves an *intraband* matrix element of $\mathscr{H}_A$ and an *interband* matrix element of the Fröhlich electron-plasmon interaction. Its contribution will be small [2.25]. The other type of terms correspond to the case in which the first transition in Fig. 2.6a or the last transition in Fig. 2.6b is associated with the creation or annihilation of a plasmon. These terms involve the *intraband* matrix element of the Fröhlich electron-plasmon interaction. Its contribution to the *transition* susceptibility, also associated with the charge density fluctuations of the charge-carriers, may be written as [2.25, 38].

$$\begin{aligned}(\delta \chi''_{\mu\nu})_\varrho &= -(\bar{n}+1)^{\frac{1}{2}}(e/m\omega_1^2)\\ &\cdot\left[2\sum_\alpha \langle 0|p_\mu(0)|\alpha\rangle\langle\alpha|p_\nu(0)|0\rangle E_\alpha/(E_\alpha^2-\hbar^2\omega_1^2)\right]\bar{\varrho}_e(\omega_p, \boldsymbol{q})\,.\end{aligned} \tag{2.69}$$

Equations (2.68) and (2.69) may be combined into one term to give

$$\begin{aligned}(\delta\chi_{\mu\nu})_\varrho &= [(\delta\chi'_{\mu\nu})_\varrho + (\delta\chi''_{\mu\nu})_\varrho]\\ &= (\bar{n}+1)^{\frac{1}{2}}\, C_{\mu\nu}(\omega_1)\,\bar{\varrho}_e(\omega_p, \boldsymbol{q})\,,\end{aligned} \tag{2.70}$$

where $C_{\mu\nu}(\omega_1)$ is the *charge density* Raman tensor. From (2.69)–(2.71) we see that in the region of transparency $C_{\mu\nu}(\omega_1)$ is real and that $(\delta\chi_{\mu\nu})$ is purely imaginary and linear un $\boldsymbol{q}_\lambda$. MCWHORTER and ARGYRES [2.38] have shown that, in the quasi-static approximations $C_{\mu\nu}(\omega_1)$ is the derivative of the electric susceptibility with respect to the charge carrier density.

The first-order *transition* susceptibility of plasmons including the electro-optic and charge density mechanisms may be written as

$$(\delta\chi_{\mu\nu})_{pl} = (\delta\chi_{\mu\nu})_{\mathscr{E}} + (\delta\chi_{\mu\nu})_{\varrho}$$
$$= (\bar{n}+1)^{\frac{1}{2}}[b'''_{\mu\nu\lambda}(\omega_1, 0)\,\bar{\mathscr{E}}_\lambda(\omega_p) \qquad (2.71)$$
$$+ b'''_{\mu\nu\lambda}(\omega_1, \omega, \boldsymbol{q})\,\bar{\mathscr{E}}_\lambda(\omega_p) + C_{\mu\nu}(\omega_1)\bar{\varrho}_e(\omega_p, \boldsymbol{q})]\,.$$

The term linear in $\boldsymbol{q}$ in $(\delta\chi''_{\mu\nu})_\varrho$ and $(\delta\chi_{\mu\nu})_\varrho$ are imaginary. They will interfere among themselves but not with $(\delta\chi'''_{\mu\nu})_{\mathscr{E}}$ which is real in the region of transparency. Interference is possible when the sample becomes opaque.

Raman Spectra Line-Shapes

In order to discuss the shapes of the Raman scattering peaks it is necessary to express the *transition* susceptibility in terms of the time dependent normal mode coordinate operators of the collective excitations. Thus, we write the *transition* susceptibility operator in the following form

$$\delta\chi_{\mu\nu}(\boldsymbol{q}, t) = \sum_m R_{\mu\nu m} Q_m(\boldsymbol{q}, t) + \text{higher-order terms}\,, \qquad (2.72)$$

where $Q_m(\boldsymbol{q}, t)$ is a schematic notation for the normal coordinate operators and the $R_{\mu\nu m}$ represent the first-order Raman tensors associated with the Q_m.

The Raman scattering cross-sections are calculated by using (2.72) in (2.28) and (2.13). We see that the cross-section for first-order scattering is proportional to the correlation function

$$A_{mm'}(\boldsymbol{q}, \omega) = (\tfrac{1}{2}\pi) \int_{-\infty}^{\infty} dt \exp(\mathrm{i}\omega t)\, G_{mm'}(\boldsymbol{q}, t)\,, \qquad (2.73)$$

where

$$G_{mm'}(\boldsymbol{q}, t) = \langle i | Q_m(\boldsymbol{q}, t)\, Q_{m'}(\boldsymbol{q}, 0) | i \rangle_{\mathrm{T}} \qquad (2.74)$$

is a temperature-dependent Green's function of the normal coordinates of the excitations. In the absence of interactions among the different modes of the crystal, the $Q_m(\boldsymbol{q}, t)$ are independent variables and, consequently, the $A_{mm'}(\boldsymbol{q}, \omega)$ and $G_{mm'}(\boldsymbol{q}, t)$ are zero unless $m = m'$. When there are interactions that couple different crystal excitations, it is possible to have correlation and Green's functions which are non-zero for $m \neq m'$. The coupling between different excitations may have an important effect on the shapes of the Raman spectra. In the particular case in which interaction produces correlation between two different Raman active modes, interference effects will occur in the Raman scattering cross-sections of these modes.

For undamped excitations and in the absence of interactions among the modes, the correlation function assumes the simple form

$$A_{mm'}(\boldsymbol{q}, \omega) = [\bar{n}(\omega) + 1]\, |\bar{Q}_m(\boldsymbol{q}, \omega)|^2\, [\delta(\omega - \omega_m) - \delta(\omega + \omega_m)]\,, \quad (2.75)$$

where $\bar{Q}_m(q, \omega)$ are the amplitudes of the normal modes. Thus, as expected, the cross section calculated with (2.75) for undamped lattice vibrations and plasmons are equivalent to those obtained previously.

In the case of optical phonons and polaritons, the expression for the *transition* susceptibility operator given by (2.72) may be written, as a generalization of (2.61), in the following form

$$\delta\chi_{\mu\nu}(\boldsymbol{q}, t) = \sum_l a^*_{\mu\nu l}(\omega_1, \omega, \boldsymbol{q})\, u_l(\boldsymbol{q}, t) + \text{higher-order terms}\,, \quad (2.76)$$

where the normal coordinates $u_l(\boldsymbol{q}, t)$ are related to the fluctuations in the atomic displacements due to optical phonons by

$$u_l(\boldsymbol{r}, t) = (1/V)^{\frac{1}{2}} \sum_{\boldsymbol{q}} u_l(\boldsymbol{q}, t) \exp(\mathrm{i}\boldsymbol{q} \cdot \boldsymbol{r})\,. \quad (2.77)$$

In addition, we have introduced the Raman tensor which may be written as

$$a^*_{\mu\nu l}(\omega_1, \omega, \boldsymbol{q}) = a^*_{\mu\nu l}(\omega_1, \omega, 0) + \delta a^*_{\mu\nu l}(\omega_1, \omega, \boldsymbol{q})\,, \quad (2.78)$$

where $a^*_{\mu\nu l}(\omega_1, \omega, 0)$ is identical to the Raman tensor in (2.61), which describes "allowed" scattering by optical phonons and polaritons, and $\delta a^*_{\mu\nu l}(\omega_1, \omega, \boldsymbol{q})$ is the q-dependent part of the Raman tensor, which describes "forbidden", q-dependent Raman scattering by optical phonons and polaritons. Previously we have shown that one contribution to $\delta a^*_{\mu\nu}(\omega_1, \omega, \boldsymbol{q})$ comes from the *two-band* processes involving the Fröhlich interaction.

The Raman scattering cross-section of optical lattice vibrations is obtained by using (2.76) in (2.27) and (2.13). This leads to an expression for the cross-section in the form of a series in which the first term is proportional to the correlation function

$$A^{\mathrm{u}}_{ll'}(\boldsymbol{q}, \omega) = (\tfrac{1}{2}\pi) \int_{-\infty}^{\infty} \exp(\mathrm{i}\omega t) \langle i|u_l(\boldsymbol{q}, t)\, u_{l'}(\boldsymbol{q}, 0)|i\rangle_{\mathrm{T}}\,. \quad (2.79)$$

This is the term which describes first-order processes. In the case of a perfect harmonic lattice, the phonons have an infinite lifetime and $A^{\mathrm{u}}_{ll'}(\boldsymbol{q}, \omega)$ is of the form given in (2.75) with $|\bar{Q}_m(\boldsymbol{q}, \omega)| \equiv |\bar{u}_l(\boldsymbol{q}, \omega)|$. In the case of an anharmonic lattice, it was shown by COWLEY [2.39] that a complete theory of first-order scattering must also include the terms arising from correlations between one-phonon and higher-order vibrational states of the crystal. The vibrational correlation functions may be calculated by the methods of many-body theory using a lattice Hamiltonian which includes the anharmonic terms in the potential energy. In the case of optical phonons, the diagonal element in (2.79) may be written as [2.40]

$$A^{\mathrm{u}}_l(\boldsymbol{q}, \omega) = (\bar{n}+1)\,|\bar{u}_l(\omega_l)|^2\, 4\omega_l^2 \Pi_{l\mathrm{I}}(\boldsymbol{q}, \omega, T) \cdot (\pi\{[\omega_l^2(\boldsymbol{q}, \omega, T) - \omega^2]^2 + 4\omega_l^2\, \Pi_{l\mathrm{I}}^2(q, \omega, T)\})^{-1}\,, \tag{2.80}$$

where

$$\omega_l^2(\boldsymbol{q}, \omega, T) = \omega_l^2(\boldsymbol{q}) + 2\omega_l(\bar{q})\,\Pi_{l\mathrm{R}}(\boldsymbol{q}, \omega, t) \tag{2.81}$$

is the temperature and ω-dependent renormalized mode frequency. $\omega_l(q)$ is the harmonic frequency; and $\Pi_{l\mathrm{R}}$ and $\Pi_{l\mathrm{I}}$ are the real and imaginary parts of the diagonal elements of the self-energy matrix. $\Pi_{l\mathrm{I}}$ is associated with anharmonic damping of the mode and gives the line-width of the Raman bands. General expressions for $\Pi_{l\mathrm{R}}$ and $\Pi_{l\mathrm{I}}$ for phonons have been given by MARADUDIN and FEIN [2.40] and for polaritons by BENSON and MILLS [2.15, 41, 42]. In lowest-order perturbation theory the contributions to $\Pi_{l\mathrm{I}}$ come from 3rd order anharmonic processes in which the mode at $\omega_l(\boldsymbol{q})$ decays into a two-phonon state with conservation of wave vector.

In many cases the observed first-order Raman line-shapes may be obtained from (2.80) by assuming that $\Pi_{l\mathrm{R}}$ and $\Pi_{l\mathrm{I}}$ are temperature dependent but ω-independent. This appears to be the case in diamond [2.43], silicon [2.44] and germanium [2.45, 46]. The relatively small temperature dependence of the widths of the first-order Raman bands from these crystals has been accounted for by the 3rd order contributions to $\Pi_{l\mathrm{I}}$ [2.39, 47]. Information on the frequency dependence of $\Pi_{l\mathrm{I}}$ has been obtained by analyzing the shapes of the first-order Raman bands of TO phonons [2.48] and polaritons [2.49]. In the case of GaP USHIODA [2.49] has pointed out that the frequency dependence of $\Pi_{l\mathrm{I}}$ is due to contributions from the decay of the optical lattice vibrations into two Brillouin zone boundary acoustical phonons. Raman scattering has also been used intensively to study soft optical phonons. Such modes have frequencies and widths which are strongly temperature dependent. The spectroscopy of soft optical modes has been reviewed by WORLOCK [2.50] and by SCOTT [2.51].

The off-diagonal elements of the correlation function matrix, defined in (2.79), are proportional to the off-diagonal terms of the self-energy matrix [2.40, 52]. These terms may exist when the modes l and l' belong to the same irreducible representation of the crystal space group and are coupled by some interaction. The off-diagonal elements of the correlation functions results in an interference between the contributions of modes l and l' to the *transition* susceptibility. In this case, the line-shapes observed in the Raman spectra show a considerable departure from the damped oscillator line-shape predicted by (2.80). The interaction which couples the two modes may result, for example, from the anharmonic terms in the vibrational potential energy [2.52, 53]. It has been shown that interactions which lead to mode coupling and interference effects may also occur through the macroscopic electric fields of polaritons [2.42], of coupled LO phonon-plasmon modes [2.54] and of acoustic phonons in piezoelectric crystals [2.55]. In anharmonic crystals the one phonon states couple to higher-order vibrational states, involving two or more phonons, via cubic or higher terms in the vibrational potential energy. In this case it is necessary to consider the interference terms between the first-order and higher-order contributions to the transition susceptibility [2.39, 56]. BARKER and LOUDON [2.12] have pointed out that in the cases of optical phonons and polaritons $A_l^{\mathrm{u}}(\boldsymbol{q}, \omega)$ may be calculated from phenomenological response functions. In this approach, the correlation function is written (using Nyquist's theorem) as [Ref. 8.1, Sect. 2.1.11].

$$A_l^{\mathrm{u}}(\boldsymbol{q}, \omega) = -\left(\frac{\hbar}{\pi}\right)[n(\omega)+1]\,\mathrm{Im}\{T_l(\boldsymbol{q}, \omega)\}\,, \tag{2.82}$$

where

$$T_l(\boldsymbol{q}, \omega) = \bar{u}_l(\boldsymbol{q}, \omega)/\bar{F}_{\mathrm{ext}}(\boldsymbol{q}, \omega) \tag{2.83}$$

is the response function which gives the amplitude of the atomic displacements set-up in the crystal by the an "external driving force" with amplitude $\bar{F}_{\mathrm{ext}}(\boldsymbol{q}, \omega)$. In the case of a purely mechanical wave, like nonpolar or TO phonons, $T_l(\boldsymbol{q}, \omega)$ may be obtained from the simple classical equation of motion of a driven damped oscillator. In the case of a $q=0$ mode the correlation function is given by

$$A_l^{\mathrm{u}}(0, \omega) = (\bar{n}+1)\,|\bar{u}_l(\omega)_l|^2\, 2\omega_l \omega \gamma_l \{\pi[(\omega_l^2-\omega^2)^2+\omega^2\gamma_l^2]\}^{-1}\,, \tag{2.84}$$

where γ_l is a phenomenological damping constant. When $\omega_l \gg \gamma_l$ (2.84) is identical to (2.80) with $\gamma_l = 2\Pi_{l1}$. However, in using (2.84) it is necessary to make the "ansatz" that ω_l as well as γ_l are temperature and frequency dependent.

BARKER and LOUDON [2.12] also discussed the case of polar vibrational modes with macroscopic electric fields, (i.e., polaritons and LO phonons) in diatomic and polyatomic crystals. In these cases, the correlation functions are similar to that in (2.84). For LO phonons A_l^{u} is obtained from (2.84) by using ω_{L}, the LO phonon frequency, instead of ω_l. In the case of polariton modes, A_l^{u} is obtained from (2.84) using $q^2c^2/\varepsilon(\omega)$ [where c is the velocity of light and $\varepsilon(\omega)$ is the dielectric constant] in place of ω_l. In addition (2.84) has to be multiplied by a function which gives the *phonon content* of the mode [2.15]. HON and FAUST [2.57] have similarly calculated the classical response functions for coupled LO phonon-plasmon modes.

2.2.2. Phenomenological Formulation

As noted earlier, Raman scattering of light by collective excitations may be viewed as a nonlinear interaction of the incident and scattered electromagnetic waves with the collective excitations. The excitations may be treated, classically, as plane waves characterized by a frequency ω, a wave vector $\boldsymbol{q}$ and a collective excitation coordinate amplitude $Q_m(\boldsymbol{q}, \omega)$ which has the transformation properties of the excitation under the symmetry operations of the crystal point group at $\boldsymbol{q}$. In the case of optical phonons, Q_m corresponds to a relative atomic displacement coordinate μ_l; in the case of acoustical phonons it corresponds to the elastic displacement coordinate $\boldsymbol{w}_\lambda$; and in the case of plasmons it is the free-carrier displacement coordinate u_l. In the phenomenological treatment of Raman scattering the time and space averaged potential energy density of the medium is expanded in powers of the amplitudes of the incident and scattered fields $\mathscr{E}_1(\boldsymbol{k}_1, \omega_1)$ and $\mathscr{E}_2(\boldsymbol{k}_2, \omega_2)$, and of $Q_m(\boldsymbol{q}, \omega)$ [2.23]. The contribution of the nonlinear interactions to the energy density is given by

$$\delta\phi = \sum_m \left(\frac{\partial^3\phi}{\partial\mathscr{E}_1\,\partial\mathscr{E}_2\partial Q_m}\right) Q_m\mathscr{E}_1\mathscr{E}_2 + \text{higher order terms}\,, \tag{2.85}$$

where coefficients like $(\partial^3\phi/\partial\mathscr{E}_1\,\partial\mathscr{E}_2\,\partial Q_m)$ describe the first-order nonlinear optical phenomena associated with the collective excitations. Notice that (2.85) may be rewritten as

$$\delta\phi = (\delta\chi_{12})\,\mathscr{E}_1\mathscr{E}_2 + \text{higher order terms}\,, \tag{2.86}$$

where

$$\delta\chi_{12} = \sum_m \left(\frac{\partial^3 \phi}{\partial \mathscr{E}_1 \partial \mathscr{E}_2 \partial Q_m} \right) Q_m = \Sigma_m R_{12m} Q_m \tag{2.87}$$

is the first-order *transition* susceptibility and R_{12m}, as in (2.72), are the first-order Raman tensors of the excitations.

The Raman tensors R_{12m} are, in the general case, asymmetrical in the polarisation vectors of the incident and scattered electromagnetic waves. When the collective coordinate has transformation properties like those of polar vectors or polar tensors, the first-order Raman tensors have both symmetric and antisymmetric parts. However, the antisymmetric part vanishes in the quasi-static approximation [2.14]. This is because the asymmetry comes from the energy denominators. When the collective excitations transform as an axial vector, as in the case of the A_2 symmetry optical phonons of the tellurium structure [2.58], the R_{12m} is purely antisymmetric in the polarisations of the incident and scattered waves. In this case the Raman tensor does not vanish in the quasi-static approximation. This is because the asymmetric terms are linear in wavevector, see (2.90). As yet, there have been no reports in the literature of the observation of an antisymmetric contribution to the q-independent first-order Raman scattering tensor of optical phonons [Ref. 8.8, Chap. 4].

We note that in the limit $q \to 0$, $k_1 \to 0$ and $k_2 \to 0$ the R_{12m} represent wave-vector-independent scattering tensors and their non-zero components can be obtained by group theoretical methods using the point group symmetry operations of the crystal. For example, the normal coordinates of the $q=0$ Raman active optical phonons of centrosymmetric diatomic crystals with the diamond structure (diamond, silicon, germanium and gray-tin) transform like the F_{2g} irreducible representation of O_h group [2.32], [Ref. 8.1, Fig. 2.9]. This representation has the symmetry of a uniaxial stress along the (111) direction. Consequently, the first-order Raman tensor in the diamond structure transforms like a fourth-rank tensor and is non-vanishing. In the case of the F_{1u} optical phonons of the centrosymmetric diatomic crystals with the rocksalt structure, which transform like a polar vector, the $R_{12\lambda}$ transforms like a 3rd rank tensor and vanishes for the rocksalt structure. Thus, F_{1u} optical phonons are inactive in *allowed* Raman scattering.

Wave Vector Dependence and Morphic Effects

The wave vector dependence of the Raman scattering tensors arises from the spatial variation of the incident and scattered electromagnetic fields and of the normal coordinates of the collective excitation. In order to obtain the explicit dependence of the Raman scattering tensor on wave vector, the energy density is expanded around zero wave vector in powers of $\mathscr{E}_1$, $\mathscr{E}_2$, and $\boldsymbol{Q}$ and their gradients and higher spatial derivatives [2.23]. These gradients, which in this phenomenological approach enter as independent variables, are

$$\boldsymbol{\nabla}\mathscr{E}_1 = -\mathrm{i}\boldsymbol{k}_1\mathscr{E}_1 \tag{2.88a}$$

$$\boldsymbol{\nabla}\mathscr{E}_2 = \mathrm{i}\boldsymbol{k}_2\mathscr{E}_2\,, \tag{2.88b}$$

$$\boldsymbol{\nabla}Q_m = \mathrm{i}\boldsymbol{q}\,Q_m\,, \quad \text{etc.} \tag{2.88c}$$

To first-order in $\boldsymbol{q}$, the contribution to $\delta\phi$ associated with first-order scattering, is given by

$$\begin{aligned}\delta\phi = \Bigg[\left(\frac{\partial^3\phi}{\partial\mathscr{E}_1\,\partial\mathscr{E}_2\,\partial Q_m}\right) + \left(\frac{\mathrm{i}\boldsymbol{q}\,\partial^3\phi}{\partial\mathscr{E}_1\,\partial\mathscr{E}_2\,\partial\boldsymbol{\nabla}Q_m}\right)\\ -\mathrm{i}\boldsymbol{k}_1\left(\frac{\partial^3\phi}{\partial\boldsymbol{\nabla}\mathscr{E}_1\,\partial\mathscr{E}_2\,\partial Q_m}\right) + \mathrm{i}\boldsymbol{k}_2\left(\frac{\partial^3\phi}{\partial\mathscr{E}_1\,\partial\boldsymbol{\nabla}\mathscr{E}_2\,\partial Q_m}\right)\Bigg]\mathscr{E}_1\mathscr{E}_2 Q_m\,.\end{aligned} \tag{2.89}$$

In the quasi-static approximation $(\partial^3\phi/\partial\boldsymbol{\nabla}\mathscr{E}_1\,\partial\mathscr{E}_2\,\partial Q_m) = (\partial^3\phi/\partial\mathscr{E}_1\,\partial\boldsymbol{\nabla}\mathscr{E}_2\,\partial Q_m)$, and the linear q-dependent first-order Raman tensor is given by

$$\begin{aligned}R_{12m} = \left(\frac{\partial^3\phi}{\partial\mathscr{E}_1\,\partial\mathscr{E}_2\,\partial Q_m}\right)\\ +\mathrm{i}\boldsymbol{q}\left[\left(\frac{\partial^3\phi}{\partial\mathscr{E}_1\,\partial\mathscr{E}_2\,\partial\boldsymbol{\nabla}Q_m}\right) - \left(\frac{\partial^3\phi}{\partial\boldsymbol{\nabla}\mathscr{E}_1\,\partial\mathscr{E}_2\,\partial Q_m}\right)\right].\end{aligned} \tag{2.90}$$

We note in (2.90) that the linear q-dependent terms of the first-order Raman tensor are 90° out of phase with the wave vector-independent terms and, as a consequence, the two contributions (always assuming ϕ to be real, i.e., that we are in the region of transparency) will not interfere. We also note that the q-dependent contribution to R_{12m} results either from the spatial variation of the electromagnetic field and/or from the spatial variation of the normal mode coordinate.

We have already encountered in Subsection 2.2.1 two cases in which the *transition* susceptibility has q-dependent contributions associated with the spatial variation of the collective normal coordinate. They are Brillouin scattering by acoustical phonons involving the elastic strain $\boldsymbol{\nabla}\boldsymbol{w}$ deformation potential and Raman scattering by plasmons due to the charge density fluctuations $(n_\mathrm{e}\boldsymbol{\nabla}\boldsymbol{x})$.

The form of the tensor coefficients $(\partial^3 \phi/\partial \nabla \mathscr{E}_1 \, \partial \mathscr{E}_2 \, \partial Q_m)$ and $(\partial^3 \phi/\partial \mathscr{E}_1 \, \partial \mathscr{E}_2 \, \partial \nabla Q_m)$ determines the selection rules for linear q-dependent scattering. These coefficients correspond to zero wave-vector and therefore their transformation properties are determined by the crystal point group. The selection rules for linear q-dependent scattering by optical phonons is discussed at length in [2.23–25]. It is noted, as can be inferred by examination of the second term in the right side of (2.90), that linear q-dependent scattering exists for the polar optical phonons of centrosymmetric crystals, which are normally Raman inactive. The Raman active optical phonons of non-centro-symmetric crystals may also have linear q-dependent scattering. The form of the linear q-dependent terms of the *transition* susceptibility tensor, for cubic and hexagonal crystals, has been given by BURSTEIN and PINCZUK [2.25]. q-dependent scattering, which appears as a breakdown in the *allowed* scattering selection rules, is found to be very weak in ordinary Raman scattering experiments. However, such *forbidden* scattering is often observable in experiments carried out under resonance conditions [2.27].

Morphic Effects [Ref. 8.7, Chap. 8]

The changes in the properties of a crystal which follow from the lowering of its symmetry by an external force are called *morphic effects*. In crystal lattice dynamics these effects appear as changes in the tensors which describe the properties of the crystal and also as splittings of degenerate normal-mode frequencies. Typical external forces are electric and magnetic fields and stresses (or strains). Effects which are similar to morphic phenomena may also be observed in imperfect crystals which have built-in electric or magnetic fields and strains. Morphic effects on Raman scattering by optical phonons may be described phenomenologically by expanding the Raman tensors in powers of the tensor components of the external force [2.23, 24]. Let us consider, as an example, the case in which this force is an electric field. The Raman tensor may be written as follows

$$a^*_{\mu\nu l}(\omega_1, \omega, \mathscr{E}_a) = a^*_{\mu\nu l}(\omega_1, \omega, 0) + \sum_\eta (\partial a^*_{\mu\nu l}/\partial \mathscr{E}_\eta) \, \mathscr{E}_\eta + \text{higher-order terms}\,, \tag{2.91}$$

where $\mathscr{E}_\eta$ are the components of $\boldsymbol{\mathscr{E}}_a$, the external electric field. The forms of the tensor coefficients $(\partial a^*_{\mu\nu l}/\partial \mathscr{E}_\eta)$ are determined by the point symmetry of the crystal in the absence of the electric field. Since $\boldsymbol{\mathscr{E}}$ and $\boldsymbol{q}$ are both vectors, the selection rules for linear $\boldsymbol{\mathscr{E}}$-induced Raman scattering are the same as those for linear $\boldsymbol{q}$-dependent scattering. Thus, linear

$\mathscr{E}$-induced Raman scattering may exist for the Raman inactive optical phonons of centro-symmetric crystals as well as for infrared and Raman active optical modes of non-centro-symmetric crystals. The Raman active and infrared inactive optical phonons of centro-symmetric crystals may exhibit quadratic q-dependent scattering, quadratic $\mathscr{E}$-induced effects, as well as linear q-dependent and $\mathscr{E}$-induced scattering.

A similar phenomenological description can be given for the cases in which the external force is a stress (which causes a homogeneous strain) or a magnetic field [2.24]. Since these forces have even parity they will not induce Raman scattering by Raman inactive modes of centro-symmetric crystals.

Electric field induced Raman scattering has been observed in the paraelectric perovskite crystals, $KTaO_3$ and $SrTiO_3$ [2.50]. In semiconductor crystals morphic effects in first-order Raman spectra have been reported in experiments done under resonance conditions. Electric field induced first-order scattering by LO phonons has been reported in InSb [2.59] and InAs [2.60], in the rock salt-type IV–VI compound semiconductors [2.61] and in CdS [2.62]. Uniaxial stress-induced first-order Raman scattering by TO and LO phonons has been observed in InSb [2.63] and InAs [2.64].

2.3. The Frequency Dependence of the First-Order Raman Tensors of Optical Phonons

The dependence of the Raman tensors of optical phonons on the incident photon frequency comes from the resonant denominators in (2.36). It follows from the discussion in Subsection 2.2.1 that this frequency dependence is determined by the superposition of all the contributions from *two-band* and *three-band* scattering processes throughout the Brillouin zone. The latter may involve direct or phonon-assisted interband electronic excitations. Phonon-assisted interband transitions have been found to play an important role on the intensity of *second-order* Raman scattering in Si, GaP and CuO_2 [2.65, 66]. However, current experimental evidence [2.67] indicates that the terms involving phonon-assisted transitions do not make important contributions for the *first-order* Raman scattering. In the case of direct interband transitions, the frequency dependences of the several terms contributing to the Raman tensors are determined by the energies at which critical points in the combined density of states for direct interband transitions occur. In addition, the contribution from each critical point (or energy gap) may be separated into terms involving discrete exciton and continuum electron-hole pair states [Ref. 8.1, Chap. 2].

2.3.1. Resonance Enhancement

Equation (2.36) indicates that a resonance enhancement in the first-order Raman scattering by optical phonons should occur when the energies of the incident photons are sufficiently close to the energies of the intermediate electronic excitations. This resonant enhancement has been observed at photon energies close to the fundamental and higher-energy gaps of a large group of semiconductor crystals. The observation of resonance enhancement of first-order Raman scattering intensities for photon energies at the electric-dipole forbidden yellow exciton series of CuO_2 has also been reported recently [2.68]. DAMEN and SHAH [2.69] have found that in CdS there is a large resonance enhancement at photon energies which correspond to the excitation of excitons bound to impurities [Ref. 8.1, Chap. 2].

The frequency dependence of the atomic displacement Raman tensor may be obtained experimentally by measuring the frequency dispersion of the first-order scattering by $q \sim 0$ optical phonons which have no macroscopic electric field. The measurement of absolute scattering intensities is avoided by comparing the scattering intensity under investigation to that of a Raman peak in a reference medium, which exhibits negligible dispersion with incident photon energy. The frequency dispersion below the lowest E_0 energy gap has been measured for the TO phonons of semiconductor compounds with the zinc blende and wurtzite structures, including ZnTe [2.70], ZnSe [2.71], ZnS [2.71], AlSb [2.70], CdS [2.72], and GaP [2.67, 73]. The experimental results for A_1 and E symmetry TO phonons CdS are shown in Fig. 2.7. In this figure we see that for incident photon energies close to the E_0 gap there is a large (resonance) enhancement of the scattering intensity. However, for smaller photon energies the scattering intensity goes through a minimum. Similar minima have been found for scattering by F_2 symmetry phonons in ZnS [2.71] and for the scattering by the non-polar E_2 phonons of CdS [2.74]. These minima have been explained as a cancellation between the contributions to the atomic displacement Raman tensor, which are resonant at the E_0 gap and non-resonant contributions. The frequency dependence of the first-order Raman scattering intensities by $q \sim 0$ TO phonons of GaP was recently reported by BELL et al. [2.73]. Using a dye laser operating in the pulsed mode, it was possible to obtain data for photon energies spanning the E_0 and $E_0 + \Delta_0$ energy gaps. These results, which are reproduced in Fig. 2.8, show that the enhancement at the $E_0 + \Delta_0$ spin-orbit gap is smaller than the enhancement at the E_0 gap.

The frequency dispersion of first-order Raman scattering by $q \sim 0$ optical phonons with no macroscopic electric field was also measured for

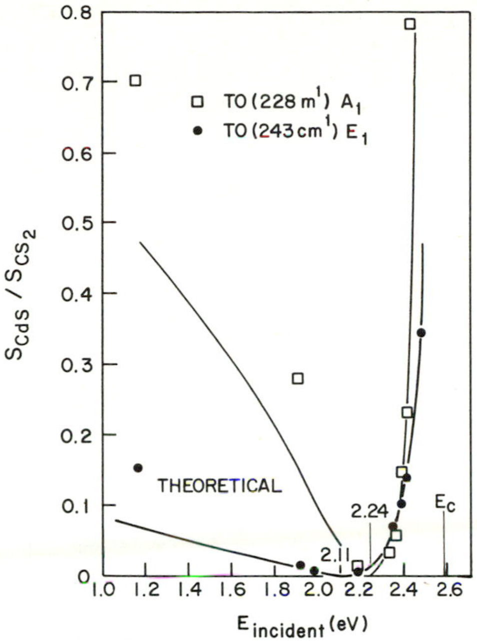

Fig. 2.7. The normalized Raman scattering intensity of A (TO) and E_1 (TO) symmetry phonons of CdS (80 K) as a function of incident photon energy. E_g is the fundamental energy gap. The solid lines are theoretical curves. (After RALSTON et al. [2.72]

photon energies near other critical points. In the case of the E_1 and $E_1 + \Delta_1$ gaps of diamond- and zincblende-type semiconductors, results were reported for the optical phonons of Ge [2.75] as well as for the TO phonons of InSb [2.76, 78], InAs [2.79, 80], $InSb_{1-x}As_x$ [2.79], and GaSb [2.78]. The results for Ge are shown in Fig. 2.9. One of the most important features of these results is that the maxima in the resonance enhancements do not occur at the gap energies, as determined by optical measurements, but are shifted towards higher energies. The frequency dispersion of the first-order Raman scattering intensities by the Raman active, infrared inactive F_{2g} phonons of the II–IV semiconductors with the antifluorite structure was reported by ANASTASSAKIS and PERRY [2.81] at photon energies near the E_1 gap.

The frequency dispersion of *allowed* scattering by LO phonons has been measured for photon energies near the E_0 and for higher energy gaps in several semiconductors with the zinc blende and wurtzite structures (see, for instance, [2.70–72] and [2.77–80] as well as references therein). It is found that the dispersions of the *allowed* LO phonon

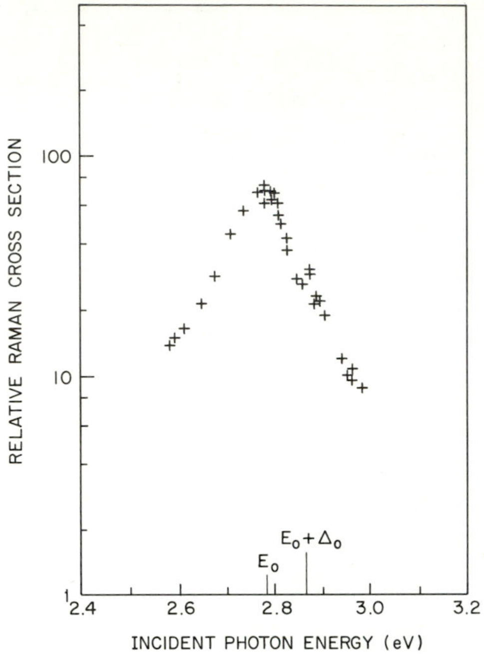

Fig. 2.8. Relative Raman cross-section as a function of incident photon energy for allowed, first-order, TO phonon scattering in GaP at room temperature. The energies of the E_0 and $E_0 + \Delta_0$ gaps are shown in the figure. (After BELL et al. [2.73])

intensities are different from those of the *allowed* TO phonon intensities. This is due to the fact that in the *allowed* scattering cross-section by LO phonons there are contributions due to the *three-band* electro-optic Raman tensor.

We have seen in Section 2.2 that the *two-band* electro-optic Raman tensor arises from *forbidden*, q-dependent, scattering by LO phonons. *Forbidden* scattering may also originate in morphic effects associated with built-in or external electric fields. The available experimental evidence indicates that *forbidden* first-order scattering by LO phonons is, in the case of perfect crystals, observable only under resonance conditions. On the other hand, unambiguous q-dependent scattering by TO phonons has not been observed in "perfect" crystals even under resonance enhanced conditions. The frequency dispersions of the resonant *forbidden* scattering by Raman active and inactive LO phonons were measured in several semiconductor crystals. The *forbidden* scattering by Raman active modes was studied intensively at the E_0 gap of CdS ([2.82–85]

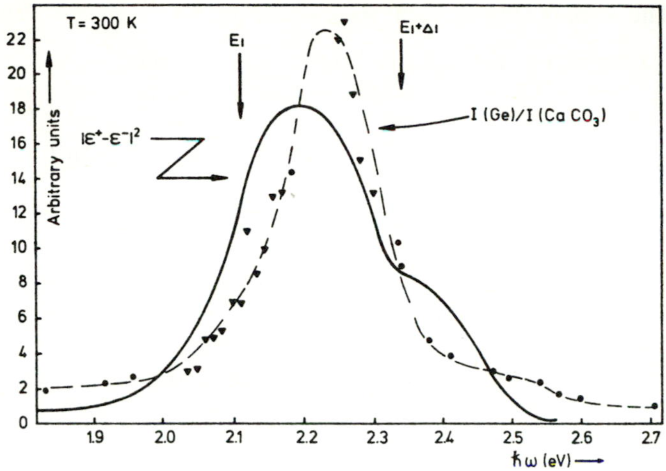

Fig. 2.9. Relative Raman intensity as a function of incident photon energy for allowed first-order scattering by optical phonons in Ge. The solid line was obtained using (2.102) and experimental data for the electric susceptibility. (From CERDEIRA et al. [2.75])

and references therein), and also at the E_1 gap of a group of III–V semiconductor compounds including InSb [2.76–78], InAs [2.80], and GaSb [2.78]. It was found that the intensities of *forbidden* scattering by LO phonons have a much larger resonance enhancement than those of *allowed* scattering. In the case of Raman inactive LO phonons, resonant first-order *forbidden* scattering has been reported in II–IV centrosymmetric semiconductors [2.86], in CuO_2 [2.68, 87] and in TlCl [2.88].

In the following subsections we discuss the expressions for the resonant atomic displacement and electro-optic Raman tensors. The contributions to the first-order *transition* susceptibility which have the most important dependence on the photon frequencies involve the first term on the right side of (2.36). It was pointed out in Subsection 2.2.1 that this term corresponds to the particular time-order shown in the diagrams of Fig. 2.2. For these diagrams, the *two-band* terms involve the intraband matrix element of the electron-phonon interaction, while the *three-band* terms are associated with the interband matrix elements. The intermediate electronic transitions of the crystal which are involved in these terms are shown, schematically, in the diagrams of Figs. 2.4 and 2.5. We shall consider the contributions to the Raman tensors from these terms.

2.3.2. The Atomic Displacement Raman Tensor

The expression for the atomic displacement Raman tensor may be obtained from (2.36) and (2.49). A convenient form is

$$a_{\mu\nu l}(\omega_1, \omega, \boldsymbol{q}) = (e^2/mV\omega_2) \sum_{\alpha,\beta} \Phi_\alpha(0) \langle 0|p_\mu(\boldsymbol{k}_2)|\alpha\rangle \langle\beta|p_\gamma(-\boldsymbol{k},)|0\rangle \cdot \Phi^*_\beta(0)\, D^{(\mathrm{opt})}_l(\alpha,\beta)\,[(E_\alpha - \hbar\omega_1)(E_\beta - \hbar\omega_2)]^{-1} + B(\omega_1)\,, \tag{2.92}$$

where $\Phi_\alpha(\boldsymbol{r})$ and $\Phi_\beta(\boldsymbol{r})$ are envelope functions which describe the dynamics of the intermediate electron hole pair states within the effective-mass approximations [2.6]. $\langle 0|p_\mu|\alpha\rangle$ and $\langle\beta|p_\nu|0\rangle$ are the interband matrix elements of the momentum operator between the electronic ground state of the crystal pair states $|\alpha\rangle$ and $|\beta\rangle$. $B(\omega_1)$ describes the less-resonant or non-resonant terms.

Resonant Two-Band Terms

The wave vector independent two-band contributions are calculated by using the dipole approximations for the momentum matrix elements and by taking into consideration a single pair of conduction and valence bands. We consider first the case of uncorrelated electron-hole pair intermediate states. In this case $\Phi_\alpha(0) = \Phi_\beta(0) = 1$. Thus the resonant contributions to the two-band atomic displacement Raman tensor are given by [2.89].

$$a''_{\mu\nu l}(\omega_1, \omega, 0) = (e^2/mV\omega_2)\langle c|p_\mu|v\rangle\langle v|p_\nu|c\rangle \cdot\Big\{\sum_{\vec{\kappa}} [(E_{\vec{\kappa}} - \hbar\omega_1)^{-1} - (E_{\vec{\kappa}} - \hbar\omega_2)^{-1}] \cdot [(\hbar\omega + \hbar\boldsymbol{q}\cdot\boldsymbol{v}_\mathrm{v})^{-1} d^\mathrm{v}_l + (\hbar\omega + \hbar\boldsymbol{q}\cdot\boldsymbol{v}_\mathrm{c})^{-1} d^\mathrm{c}_l\Big\}\,, \tag{2.93}$$

where $E_{\vec{\kappa}}$ are the energies of the direct interband transactions at electron-wave vector $\vec{\kappa}$, and $\langle c|p_\mu|v\rangle$ is the interband momentum matrix element. d^c_l and d^v_l are intra-band deformation potentials for the conduction and valence bands, respectively. In addition, $\boldsymbol{v}_\mathrm{c}(\vec{\kappa}) = (1/\hbar)\nabla_{\vec{\kappa}} E_\mathrm{c}(\vec{\kappa})$ and $\boldsymbol{v}_\mathrm{v}(\vec{\kappa}) = (1/\hbar)\nabla_{\vec{\kappa}} E_\mathrm{v}(\vec{\kappa})$, where $E_\mathrm{c}(\vec{\kappa})$ and $E_\mathrm{v}(\vec{\kappa})$ are the energy vs. wave vector relations for the conduction and valence bands. It follows from (2.41) and (2.92) that in the case of *cubic crystals* the resonant term of $a''_{\mu\nu l}(\omega_1, \omega, 0)$ can be written in terms of the frequency dependent electric susceptibility $\chi(\omega_1)$ as [2.89]

$$a'' = [\chi(\omega_1) - \chi(\omega_1 - \omega)]\,(d^\mathrm{c} + d^\mathrm{v})/\hbar\omega \tag{2.94}$$

Similar expressions may be written for the components of $a''_{\mu\nu l}(\omega_1, \omega, 0)$ which enter in the diagonal components of the *transition* susceptibility tensor of uniaxial and biaxial crystals. In the quasi-static limit we have

$$\lim_{\omega\to 0} a''(\omega_1, \omega, 0) = [d\chi(\omega_1)/d\omega_1]\,(d_l^{\rm c} + d_l^{\rm v})/\hbar\,. \tag{2.95}$$

Equations (2.94) and (2.95) indicate that the resonant first-order Raman scattering intensities by $q \sim 0$ optical phonons have features which are similar to those encountered in the modulation spectroscopy of interband electronic transitions. In the case of Raman scattering the modulation of the interband optical properties of the crystal is induced by the lattice vibrations. This point of view has been emphasized by CARDONA [2.31]. Equations (2.94) and (2.95) have been used to obtain the frequency dependence of resonance Raman scattering from experimental data on the frequency dispersion of the dielectric susceptibility. This procedure was used in the case of the Raman active $F_{2\rm g}$ phonons in the II–IV semiconductors [2.81] as well as in Se and Te [2.90].

Equation (2.94) may be used to calculate $a''_{\mu\nu l}(\omega_1, \omega, 0)$ from theoretical expressions for $\chi(\omega_1)$. In the case of intermediate uncorrelated electron-hole pairs we may use the $\chi(\omega_1)$ obtained from model density of states calculations [2.29, 30]. For an M_0 critical point (which occurs at the E_0 gap of typical semiconductors) one obtains [Ref. 8.1, Sect. 2.2.6]

$$\begin{aligned}\chi(\omega_1) &= (\tfrac{1}{8}\pi)\, C_0''(1/\omega_{\rm g})\,[2\omega_{\rm g}^{\frac{1}{2}} - (\omega_1 + \omega_{\rm g})^{\frac{1}{2}} - (\omega_{\rm g} - \omega_1)^{\frac{1}{2}}]\\ &= (\tfrac{1}{8}\pi)\, C_0''\,\omega_{\rm g}^{\frac{1}{2}}\, f(\omega_1, \omega_{\rm g})\,,\end{aligned} \tag{2.96}$$

where $\omega_{\rm g} = E_{\rm g}/\hbar$ is the gap frequency and $C_0'' = (4P^2/3)(2\mu^*/\omega_{\rm g})^{\frac{3}{2}}$. P is the magnitude of the interband momentum matrix element, and μ^* is the reduced effective mass at the energy gap. From (2.94) and (2.96) we obtain

$$\begin{aligned}a''(\omega_1, \omega, 0) \simeq{} & (\tfrac{1}{8}\pi)\, C_0''\,\omega_{\rm g}^{\frac{3}{2}}\,\omega_1^{-2}\,[(d^{\rm v,0} + d^{\rm c,0})/\hbar\omega]\\ & \cdot [(\omega_{\rm g} - \omega_1 + \omega)^{\frac{1}{2}} - (\omega_{\rm g} - \omega_1)^{\frac{1}{2}}]\,,\end{aligned} \tag{2.97}$$

where $d^{\rm c,0}$ and $d^{\rm v,0}$ are the deformation potentials at the M_0 critical point. It follows from (2.97) that once broadening of the order of ω is included $a''(\omega_1, \omega, 0)$ exhibits a peak at $\omega_1 = \omega_{\rm g} + \omega/2$. We note further that for $\omega_1 < \omega_{\rm g}$, $a''(\omega_1, \omega, 0)$ is real, indicating that only virtual interband transitions contribute to resonance Raman scattering. When $\omega_2 < \omega_{\rm g} < \omega_1$, $a''(\omega_1, \omega, 0)$ is complex because real as well as virtual interband

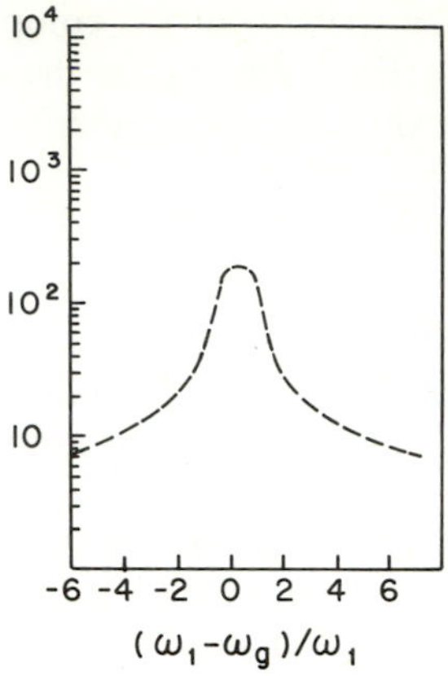

Fig. 2.10. Frequency dependence of the squared modulus of the resonant contribution to the two-band atomic displacement Raman tensor in the vicinity of an M_0 critical point, based on (2.97). The peak occurs at $\omega_1=\omega_g+\omega/2$. (After ZEYHER et al. [2.18])

transitions are involved. Finally when $\omega_2>\omega_g$, the contribution from real transitions dominate those from the virtual transitions and $a''(\omega_1,\omega,0)$ is largely imaginary.

Equation (2.97) was first derived by LOUDON [2.7], who considered only the contributions from virtual interband transitions. The possible role of real intermediate interband transitions in Raman scattering processes was first proposed by HAMILTON [2.35]. The ω_1-dependence of $|a''(\omega_1,\omega,0)|^2$ for $\omega_1\approx\omega_g$ was discussed recently by ZEYHER et al. [2.18]. Their result, which includes the contributions from the real and imaginary terms, is reproduced in Fig. 2.10. In this figure we see that the resonant enhancement in the scattering cross section has a symmetric peak centered about $\omega_1=\omega_g+\omega/2$.

Equations (2.92) and (2.97) were used (after taking the squared modulus) by CHANG and co-workers [2.72, 72] to fit their experimental results for the frequency dispersion of the Raman scattering intensities by TO phonons at the E_0 gap of CdS, ZnS, and ZnSe. Results of this fitting procedure for the case of CdS, can be seen in Fig. 2.7. The minimum in the scattering intensity, for photon energies below the gap, results from the destructive interference between the resonant and non-resonant contributions. BELL et al. [2.73] suggested that (2.97) can account for their experimental results in GaP, shown in Fig. 2.8, when the contributions of both the E_0 and the $E_0+\Delta_0$ energy gaps are included.

The two-band terms of the resonant atomic displacement Raman tensor at a M_1 critical point has been calculated by JAIN and CHOUDHURY

[2.91]. For intermediate uncorrelated electron-hole excitations it was found that

$$a''(\omega_1, \omega, 0) \sim (1/\omega)[(\omega_1 - \omega_g)^{\frac{1}{2}} - (\omega_1 - \omega_g - \omega)^{\frac{1}{2}}] . \tag{2.98}$$

Unfortunately these authors only considered the real part of a'' and thus their results are incomplete.

The frequency dependence of the contributions to the Raman tensors and scattering cross sections which involve, intermediate, continuum and discrete exciton states have been discussed extensively in the literature ([2.18, 25, 92] and references therein). We shall not reproduce these calculations here. However, it is worthwhile to point out that, here also, the two-band term of the atomic displacement Raman tensor may, in the $q=0$ approximation, be calculated by means of (2.94). In the case of the 1S hydrogenic exciton, the dielectric susceptibility associated with this exciton is given by [2.6]

$$\chi_{ex}(\omega_1) \sim [\omega_{ex}^2 - (\omega_1 + i\Gamma)^2]^{-1} , \tag{2.99}$$

where $\hbar\omega_{ex}$ is the energy of the exciton transition, and Γ is the exciton damping constant. The expression for $a''(\omega_1, \omega, 0)$ which is obtained with (2.94) and (2.99) is identical to that obtained with the equations in [2.18] (for $\Gamma = 0$). The cases of non-hydrogenic exciton intermediate states, like those occurring at M_1 critical points, have been discussed by JAIN and CHOUDHURY [2.91].

We should point out that in the calculations of $a''_{\mu\nu l}(\omega_1, \omega, 0)$ given above, we have not included the "double resonance" contributions for which both denominators in (2.92) vanish simultaneously. It has been shown [2.93] that for terms involving free electron-hole pairs, the "double resonance" terms are small for crystals with energy bands like those of diamond- and zinc blende-type semiconductors. "Double resonance" may however be important for terms which involve transitions to discrete exciton states [2.27].

Resonant Three-Band Terms

When the three-bands involved in the Raman process are well separated, resonance exists with only one energy gap. In this case, it follows from (2.36) and (2.41) that the resonant part of the three-band atomic displacement Raman tensor is proportional to the interband dielectric susceptibility. On the other hand, when the energy separation between two of the bands is not much larger than the optical phonon energy, the structure of the resonant Raman tensor is more complicated. In order to discuss this case it is convenient to consider the situation in which the separation of two energy bands, participating in the Raman process, corresponds to their spin-orbit splitting. This important case occurs in the diamond- and zinc blende-type semiconductors. For the two diagrams of Fig. 2.5(b) the energy denominators may be written as

$$[(E_{\vec{\kappa}}^{-}-\hbar\omega_1)(E_{\vec{\kappa}}^{+}-\hbar\omega_2)]^{-1} = [\Delta-\hbar\omega]^{-1}[(E_{\vec{\kappa}}^{+}-\hbar\omega_2)^{-1}-(E_{\vec{\kappa}}^{-}-\hbar\omega_1)^{-1}],$$

and (2.100)

$$[(E_{\vec{\kappa}}^{+}-\hbar\omega_1)(E_{\vec{\kappa}}^{-}-\hbar\omega_2)]^{-1} = [\Delta+\hbar\omega]^{-1}[(E_{\vec{\kappa}}^{+}-\hbar\omega_1)^{-1}-(E_{\vec{\kappa}}^{-}-\hbar\omega_2)^{-1}],$$

where $\Delta=E_{\vec{\kappa}}^{-}-E_{\vec{\kappa}}^{+}$ is the spin-orbit splitting of the valence bands, and we have neglected the q-dependent terms in (2.100). In the case of the $(E_0, E_0+\Delta_0)$ and $(E_1, E_1+\Delta_1)$ doublets, $(\Delta\pm\hbar\omega_0)^{-1}$ may be expanded in powers of $\hbar\omega/\Delta$. Retaining only the term of first-order in $\hbar\omega/\Delta$, one obtains the following expression for $a'''(\omega_1,\omega,0)$ [2.94]:

$$a'''(\omega_1,\omega,0)=(D^{\mathrm{v}}/\Delta)\{[\chi^{+}(\omega_1)+\chi^{+}(\omega_1-\omega)-\chi^{-}(\omega_1)-\chi^{-}(\omega_1-\omega)] -(\hbar\omega/\Delta)[\chi^{-}(\omega_1)-\chi^{-}(\omega_1-\omega)+\chi^{+}(\omega_1)-\chi^{+}(\omega_1-\omega)]\}, \quad (2.101)$$

where $\chi^{\pm}(\omega_1)$ are the contributions of the spin-orbit split gaps to the electric susceptibility. D^{v} is the valence band deformation potential associated with the interband matrix elements of the electron-phonon interaction.

The quasi-static approximation for $a'''(\omega_1,\omega,0)$ is given by [2.75].

$$a'''(\omega_1,0,0)=(2D^{\mathrm{v}}/\Delta)[\chi^{+}(\omega_1)-\chi^{-}(\omega_1)]. \quad (2.102)$$

In the case of the $(E_0, E_0+\Delta_0)$ doublet, using (2.96), $a'''(\omega_1,0,0)$ is given by

$$a'''(\omega_1,0,0)=(2D^{\mathrm{v},0}/\Delta_0)(C_0''/8\pi)\{f(\omega_1,E_0/\hbar) - [E_0/(E_0+\Delta_0)]^{\frac{3}{2}} f(\omega_1,E_0+\Delta_0/\hbar\}. \quad (2.103)$$

Equation (2.103) was first derived by Cardona [2.30]. It was found [2.73] that the terms of the form given by (2.103) are needed to explain the results in GaP (shown in Fig. 2.8) for photon energies larger than $E_0+\Delta_0$. Equation (2.102) together with values of $\chi^{\pm}(\omega_1)$ obtained from optical measurements were used by Cerdeira et al. [2.75] to describe the resonance enhancement at the $(E_1, E_1+\Delta_1)$ doublet of Ge. Their results are shown in Fig. 2.9. We may observe that the peak in the resonance enhancement results from overlapping contributions from the E_1 and $E_1 + \Delta_1$ gaps which are relatively close to one another.

q-Dependent Terms

Because the atomic displacement deformation potentials are essentially q-independent, the q-dependent contributions to the atomic displacement Raman tensor arise either from the q-dependence of the resonant denominators or from the $\boldsymbol{k}_1$- and $\boldsymbol{k}_2$-dependences of the momentum matrix elements in (2.93). The latter occurs when electric quadrupole or magnetic dipole interactions are included in the momentum matrix elements. The contributions which originate in the electric quadrupole or magnetic dipole terms of the momentum matrix element are particularly important when the interband electronic transitions are electric dipole forbidden. The *forbidden* resonance enhanced Raman scattering which was observed at the dipole forbidden 1S yellow exciton series of Cu_2O [2.68] has been explained in terms of such electric quadrupole and magnetic dipole terms [Ref. 8.1, Sect. 2.3.6].

Morphic Effects

The influence of an external uniaxial stress on the resonant Raman scattering by TO phonons has been studied at the E_1 energy gap of InSb [2.63] and InAs [2.64]. It was found that the changes in the scattering intensities occur only for photon frequencies close to resonance. The data for InAs, shown in Fig. 2.11, have been explained in terms of the changes in the two-band and three-band contributions to the atomic displacement Raman tensor caused by the uniaxial stress induced shift and splitting of the four-fold degenerate E_1 gap into triplet and singlet gaps.

There are two distinctive ways in which an internal or built-in electric field will modify the Raman scattering processes. In one, the field produces a relative displacement of the ions which modify the periodic potential of the crystal as well as the eigenvectors and energies of the lattice vibrations. In the other the electric field causes *interband* and/or *intraband* mixing of the one-electron wave function of the crystal. The intraband mixing corresponds to the *Franz-Keldysh effect*. The atomic displacement mechanism, which should be important in crystals in which the electric field produces large atomic displacements, has been invoked to explain the observation of electric field induced Raman scattering by "soft" TO phonons in paraelectric crystals [2.50, 95]. It has been suggested [2.89] that the Franz-Keldysh mechanism should occur in semiconductors, particularly at resonance.

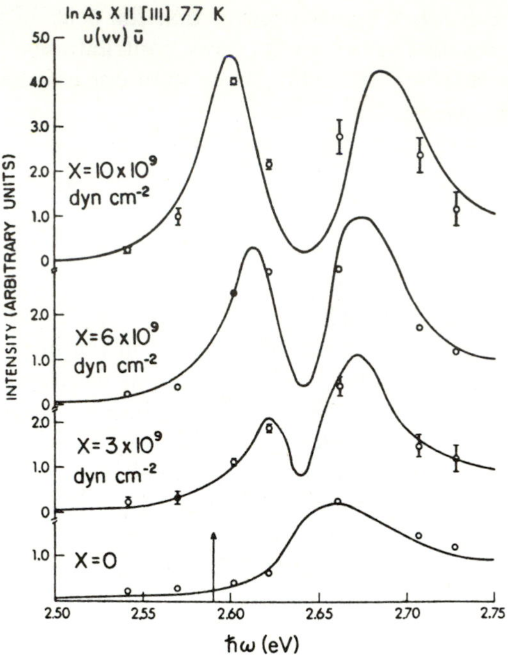

Fig. 2.11. Relative Raman scattering intensities for allowed TO phonon scattering in InAs at 77 K for various stresses $X''[1,1,1]$. The solid lines are the theoretical curves. (From ANASTASSAKIS et al. [2.64])

Electric field induced changes in the first-order scattering by nonpolar or TO phonons has been reported in diamond [2.96] and in CdS [2.62]. The small changes in the atomic displacement Raman tensor, which were quadratic in the external field in both cases, could be measured only because they interfered with field independent scattering. In CdS [2.62] the field-induced change in scattering by Raman active TO phonons was observed only under strong resonance of the photon energy with the exciton transition. This was taken as an indication that the dominant mechanism for electric field induced scattering in CdS is associated with the Franz-Keldysh effect. A more detailed discussion of this mechanism will be given in the next subsection.

2.3.3. The Electro-Optic Raman Tensor

Three-Band Electro-Optic Raman Tensor

The frequency dispersion of the three-band electro-optic Raman tensor may be calculated by means of (2.60), from the frequency dispersion of the *direct* linear *electro-optic* tensor. As noted by BELL [2.97], the frequency dependence of the *direct* electro-optic tensor is essentially the same as the frequency dependence of the second-order susceptibility $[\chi^{(2)}_{\mu\nu\lambda}(\omega_1,\omega)_{\mathrm{u}}]$. This nonlinear susceptibility describes the mixing of two electromagnetic waves at frequencies ω_1 and ω (for $\omega \gg \omega_{\mathrm{T}}$) to generate a third electromagnetic wave at ω_2. Thus, the *three-band* electro-optic Raman tensor is given by

$$b'''_{\mu\nu\lambda}(\omega_1,0) = (\partial\chi_{\mu\nu}/\partial\mathscr{E}_\lambda)_{\mathrm{u}} = [\chi^{(2)}_{\mu\nu\lambda}(\omega_1,0)]_{\mathrm{u}}\,. \tag{2.104}$$

Expressions for $[\chi^{(2)}_{\mu\nu\lambda}(\omega_1,0)]_{\mathrm{u}}$ of zinc blende-type semiconductors have been obtained by BELL [2.34] using a model density of states to describe the interband transitions at the E_0, $E_0+\Delta_0$, E_1 and $E_1+\Delta_1$ energy gaps.

The resonant part of the *three-band* electro-optic Raman tensor may also be calculated using (2.101) and (2.102) and replacing the deformation potential coefficients by factors which are proportional to the interband momentum matrix elements. As a consequence, the frequency dispersion of the resonant part of $b'''_{\mu\nu\lambda}(\omega_1,\omega)$ should be similar to that of the resonant three-band contributions to the atomic displacement Raman tensor.

Two-Band Electro-Optic Raman Tensor: q-Dependent Terms

The q-dependent terms of the resonant *two-band* electro-optic Raman tensor arise from the q-dependence of the resonant denominators and of the intraband matrix elements of the Fröhlich electron-phonon interaction given by (2.53) and (2.57). There are additional contributions associated with the $\boldsymbol{k}_1$- and $\boldsymbol{k}_2$-dependent terms of the momentum matrix elements. However, as in the case of the atomic displacement Raman tensor, the latter will generally be important only for dipole forbidden transitions to the intermediate electronic states [Ref. 8.1, Sect. 2.2.8].

The terms in $b''_{\mu\nu\lambda}(\omega_1,\omega,\boldsymbol{q})$ which are associated with the q-dependence of the *intraband* matrix elements of the Fröhlich interaction can be calculated by expanding these matrix elements, given by (2.55) and (2.59), in powers of $\boldsymbol{q}$. The contributions to the first-order *transition* susceptibility obtained in this way are discussed by MARTIN and FALICOV [2.75] in Chapter 3. It is shown that in this case the leading term of $b''_{\mu\nu\lambda}(\omega_1,\omega,\boldsymbol{q})$ is linear in $\boldsymbol{q}$ [Ref. 8.1, Sect. 2.2.8].

Electric Field Induced Contributions to the Electro-Optic Raman Tensors

We have pointed out in the previous subsection that two basic mechanisms have been proposed to explain the effects of an external (or built-in) electric field on the Raman scattering processes. One mechanism is associated with the field induced displacements of the ions in the crystal. It accounts for the field induced contributions to the atomic displacement Raman tensor of "soft" TO phonons of paraelectric crystals. The other mechanism is associated with field induced *interband* and *intraband* (or Franz-Keldysh) mixing of the one-electron states of the crystal. The mixing of wave functions brings about changes in the matrix elements of the $\mathscr{H}_A$ and the $\mathscr{H}_{EL}$ interactions, as well as in the interband transition energies. Such a mechanism is particularly effective in inducing a $q=0$ contribution to the resonant *two-band* electro-optic Raman tensor of LO phonons [Ref. 8.1, Sect. 2.2.8d].

In the $q=0$ limit the macroscopic electric field of LO phonons is similar to that of a dc electric field. Consequently, the terms in the electro-optic Raman tensors linear in the external or built-in electric field are, in the $q=0$ quasi-static approximation, identical to the frequency-dependent quadratic electro-optic tensor for two dc applied electric fields. The interband and Franz-Keldysh mixing of wave functions induced by electric fields has also been invoked in the calculation of the third-order nonlinear susceptibility of semiconductor crystals [2.98, 99]. It is now well established [2.99] that the dominant contribution to the resonant third-order nonlinear susceptibility comes from the Franz-Keldysh effect. We expect therefore that the largest effects of the electric field on the resonant Raman scattering by LO phonons in semiconductor crystals will also come from Franz-Keldysh (two-band) effects.

The resonant *two-band* electro-optic Raman tensor, in the presence of a homogeneous electric field, may be written as

$$b''_{\mu\nu\lambda}(\omega_1, \omega, \boldsymbol{q}, \mathscr{E}_a) = (e^2/m^2\omega_2^2 V)\langle c|p_\mu|v\rangle\langle v|p_\nu|c\rangle \\ \cdot \sum_{\alpha\beta} \varphi_\alpha(0, \mathscr{E}_a)\,\varphi_\beta^*(0, \mathscr{E}_a)\,\mathscr{H}_{EL}^{F}(\alpha, \beta, \boldsymbol{q}, \mathscr{E}_a)\,[(E_\alpha - \hbar\omega_1)(E_\beta - \hbar\omega_2)]^{-1}, \tag{2.105}$$

where $\varphi_\alpha(\boldsymbol{r}, \mathscr{E}_a)$ are envelope functions which describe the dynamics of the electron-hole pair intermediate states in a homogeneous electric field and $H_{EL}^{F}(\alpha, \beta, \boldsymbol{q}, \mathscr{E}_a)$ is the intraband matrix element of the *Fröhlich* interaction. $H_{EL}^{F}(\alpha, \beta)$ may be calculated with (2.57), provided $\psi_\alpha(\boldsymbol{r})$ and $\psi_\beta(\boldsymbol{r})$ are replaced by $\varphi_\alpha(\boldsymbol{r}, \mathscr{E}_a)$ and $\varphi_\beta(\boldsymbol{r}, \mathscr{E}_a)$. The q-independent term of this matrix element is given by [2.100]

$$H_{EL}^{F}(\alpha, \beta, 0, \mathscr{E}_a) = -\bar{\mathscr{E}}(\omega_L)\,\hat{\xi}\cdot\boldsymbol{d}_{\alpha\beta}(\mathscr{E}_a), \tag{2.106}$$

where

$$\boldsymbol{d}_{\alpha\beta}(\mathscr{E}_a) = e\int d^3r\,\varphi_\alpha(\boldsymbol{r}, \mathscr{E}_a)\,\boldsymbol{r}\,\varphi_\beta^*(\boldsymbol{r}, \mathscr{E}_a) \tag{2.107}$$

are the dipole moments of the electron-hole pair states induced by the electric field and $\hat{\xi}$ is the phonon unit polarization vector. Since the induced dipole moments are parallel to the field, q-independent scattering involves only LO phonons for which $\hat{\xi} \cdot \mathscr{E}_a \neq 0$.

The $\boldsymbol{d}_{\alpha\beta}(\mathscr{E}_a)$ have been calculated in closed form only for the case of 1S hydrogenic exciton intermediate states. In this case the term in $H^{\mathrm{F}}_{\mathrm{EL}}(1\mathrm{S}, 1\mathrm{S}, \boldsymbol{q}, \mathscr{E}_a)$, which is linear in $\boldsymbol{q}$ and $\mathscr{E}_a$, may be written as ([2.100] and references therein)

$$H^{\mathrm{F}}_{\mathrm{EL}}(1\mathrm{S}, 1\mathrm{S}, q, \mathscr{E}_a) \simeq -e\overline{\mathscr{E}}(\omega_{\mathrm{L}}) \{(\tfrac{9}{4})(a_0^3/e^2)\mathscr{E}_a - \mathrm{i}(m_{\mathrm{e}}^2 - m_{\mathrm{h}}^2/M)(q/2a_0^2)\}\,, \tag{2.108}$$

where a_0 is the exciton Bohr radius. In (2.108) $\boldsymbol{q}$, the phonon wave vector, is equal to the scattering wave vector. Wave vector conservation occurs in the Franz-Keldysh mechanism for field induced scattering because a uniform electric field does not affect the motion of the center of mass of the electron-hole pairs. The terms linear in $\mathscr{E}_a$ and $\boldsymbol{q}$ in the resonant *two-band* electro-optic Raman tensor may be obtained by using (2.108) in (2.105). The magnitude of this tensor is given by

$$b''(\omega_1, \omega, \boldsymbol{q}, \mathscr{E}) \simeq -(e/\omega_{\mathrm{L}})\,[(\tfrac{9}{4})(a_0^3/e^2)\mathscr{E}_a - \mathrm{i}(m_{\mathrm{e}}^2 - m_{\mathrm{h}}^2/2Ma_0^2)\,q] \cdot [\chi_{1\mathrm{S}}(\omega_1) - \chi_{1\mathrm{S}}(\omega_1 - \omega_{\mathrm{L}})]\,. \tag{2.109}$$

where $\chi_{1\mathrm{S}}(\omega_1)$ is the contribution of the 1S exciton to the electric susceptibility. In deriving (2.109), only the term independent of $\mathscr{E}_a$ in $\chi_{1\mathrm{S}}(\omega_1)$ was considered. In addition, we neglected the Stark shift of the exciton energy. The fact that in (2.109) the term linear in q is purely imaginary and the q-independent $\mathscr{E}$-induced term is real means that there is no interference term between their contributions to the scattering cross-section.

Gay et al. [2.100] have calculated by means of (2.108) the parameter

$$X_{1\mathrm{S}}(\boldsymbol{q}, \mathscr{E}) = |\varphi_{1\mathrm{S}}(0, \mathscr{E}_a|^2\, H^{\mathrm{F}}_{\mathrm{EL}}(1\mathrm{S}, 1\mathrm{S}, \boldsymbol{q}, \mathscr{E}_a) \tag{2.110}$$

which determines the magnitude of the field induced terms of the *two-band* electro-optic Raman tensor. Figure 2.12 shows the result of this calculation for crystals with properties similar to CdS. In this figure we see that $|X_{1\mathrm{S}}(\boldsymbol{q}, \mathscr{E}_a)|^2$ for small fields increases proportionally to $\mathscr{E}_a^2$ until it reaches a maximum and decreases. This decrease is a consequence of the electric field induced separation of the electron and hole pair in the electric field. The latter causes a decrease in $\varphi_{1\mathrm{S}}(0, \mathscr{E}_a)$ which measures the overlap of the electron and hole wave functions. For large

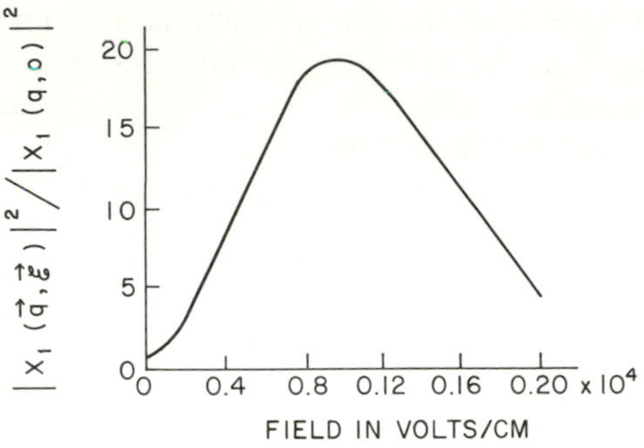

Fig. 2.12. Plot of the ratio $|X_{1S}(\boldsymbol{q}, \mathscr{E})|^2/|X_{1S}(\boldsymbol{q}, 0)|^2$ for field induced backward Raman scattering by LO phonons involving the ground state (1S) excitons of CdS. (After GAY et al. [2.100])

fields, the decrease in $\varphi_{1S}(0, \mathscr{E}_a)$ overrides the increase in the induced dipole moment $\boldsymbol{d}^2_{1S}(\mathscr{E}_a)$. The decreases in the envelope functions of electron-hole pairs in an electric field is not limited to hydrogenic excitons but also occurs for other electronic interband transitions of the crystal. The field induced decrease in the envelope function of the electron-hole pair states will also affect the resonant atomic displacement Raman tensor. SHAND and BURSTEIN [2.62] suggested that this effect may explain the field-induced decrease in the intensities of resonant *allowed* scattering by TO phonons in CdS mentioned in the previous subsection.

The q-independent part of the resonant electric field induced scattering cross-section of optical phonons was also considered by PEUKER et al. [2.101]. These authors discussed the case in which there are no excitonic effects in the intermediate electronic interband transitions. For scattering by LO phonons involving the q-independent, field dependent, *two-band* electro-optic Raman tensor, the cross-section, like in the case of resonance with $1S$ exciton transitions, is quadratic in $\mathscr{E}_a$ for small fields until it reaches a maximum and decreases. This depends on resonance conditions through the parameter $\zeta = (\omega_g - \omega_1)/\Omega$, where $\Omega = (\hbar e^2 \mathscr{E}_a/2\mu^*)^{1/3}$ is the electro-optic frequency and μ^* is the reduced effective mass for the electronic interband transitions. The maximum in the cross-section occurs for $\zeta \approx 1$. This behavior results from the fact that, for the case under consideration, the envelope functions which enter in (2.105) are Airy functions [2.6, 89].

Resonant electric field induced Raman scattering by LO phonons has been observed in CdS [2.62], InSb [2.59], InAs [2.16, 60], PbTe [2.61], and SnTe [2.61]. In the case of CdS the experiments were carried out, for photon energies below the exciton ground state, by application of a homogeneous electric field. The results are in qualitative agreements with the predictions of the Franz-Keldysh mechanism shown in Fig. 2.12. In the case of InSb, InAs, PbTe, and SnTe the scattering was induced by surface space charge electric fields. A comparison of the results of surface electric field induced Raman scattering experiments with the predictions of the Franz-Keldysh mechanism is difficult because the surface electric field is highly inhomogeneous. Nevertheless, this is a subject which deserves further attention because it opens up the possibility of studying the surface properties of crystal by light scattering spectroscopy [2.16] (see [Ref. 8.1, Fig. 2.40]).

2.4. Concluding Remarks

We have attempted in this chapter to present a "unified" discussion of the physics of inelastic light scattering in semiconductors and insulators with particular emphasis on Raman scattering by collective excitation. Toward this end, we have stressed basic concepts and underlying physics rather than formalism or quantitative comparison with experiment. Some aspects, such as the polariton formulation of light scattering, were not covered largely because they are adequately covered by recent papers in the literature—for example the polariton formulation of light scattering was reviewed by MILLS et al. [2.15]—or are discussed in detail in other chapters in this volume. In any case, the particular approach that we have followed, and the subject matter that we have emphasized, reflects, as it always does in any review, our personal viewpoint and that of the editor.

References

2.1. M. BORN, K. HUANG: *Dynamical Theory of Crystal Lattices* (Clarendon Press, Oxford, 1954).
2.2. R. LOUDON: Proc. Intern. School in Physics "E. Fermi", Course XLII (Academic Press, New York, 1969), pp. 297–320.
2.3. A. A. MARADUDIN, R. F. WALLIS: Phys. Rev. B **3**, 2063 (1971).
2.4. W. HEITLER: *The Quantum Theory of Radiation* (Clarendon Press, Oxford, 1954).
2.5. S. S. JHA: Nuovo Cimento B **63**, 331 (1969); also Comm. Sol. State Phys. **4**, 111 (1972).
2.6. M. CARDONA: *Modulation Spectroscopy* (Academic Press, New York, 1969).
2.7. R. LOUDON: Proc. Roy. Soc. A **275**, 218 (1963).
2.8. L. LANDAU, E. LIFSCHITZ: *Electrodynamics of Continuous Media* (Pergamon Press, London, 1960), Chap. 14.
2.9. H. CUMMINS: *Proc. Intern. School in Physics "E. Fermi"*, Course XLII (Academic Press, New York, 1968), pp. 247–296.
2.10. D. PINES, P. NOZIERES: *The Theory of Quantum Liquids I* (W. A. Benjamin, New York, 1966).

2.11. M.V.KLEIN: *Electronic Raman Scattering*, this volume, p. 147.
2.12. S.BARKER, R.LOUDON: Rev. Mod. Phys. **44**, 18 (1972).
2.13. N.BLOEMBERGEN: *Nonlinear Optics* (W.A.Benjamin, New York, 1965).
2.14. R.LOUDON: Advan. Phys. **13**, 423 (1964).
2.15. D.L.MILLS, E.BURSTEIN: Rept. Progr. Phys. **37**, 817 (1974).
2.16. P.CORDEN, A.PINCZUK, E.BURSTEIN: *Proc. 10th Intern. Conf. Physics of Semiconductors* (U.S. Atomic Energy Commission, Washington, D. C., 1970), pp. 739–745.
2.17. D.L.MILLS, A.A.MARADUDIN, E.BURSTEIN: Ann. Phys. (N.Y.) **56**, 504 (1970).
2.18. R.ZEYHER, C.S.TING, J.L.BIRMAN: Phys. Rev. B**10**, 1725 (1974).
2.19. S.BUCHNER, E. BURSTEIN: Phys. Rev. Letters **33**, 908 (1974); also K.MURASE, S.KATAYAMA, Y.ANDO, H.KAWAMURA: Phys. Rev. Letters **33**, 1481 (1974).
2.20. R.LOUDON: J. Phys. (Paris) **26**, 677 (1965).
2.21. A.KIEL: In *Light Scattering Spectra of Solids*, ed. by G.B.WRIGHT (Springer, New York, 1969), pp. 245–253.
2.22. L.N.OVANDER: Opt. Spectrosc. **9**, 302 (1960).
2.23. E.BURSTEIN: In *Atomic Structure and Properties of Solids*, ed. by E.BURSTEIN (Academic Press, New York, 1972), pp. 3–21.
2.24. E.ANASTASSAKIS: *Atomic Structure and Properties of Solids* (Academic Press, New York, 1972), pp. 294–324.
2.25. E.BURSTEIN, A.PINCZUK: In *The Physics of Opto-Electronic Materials*, ed. by W.A.ALBERS (Plenum Press, New York, 1971), pp. 33–79.
2.26. A.K.GANGULY, J.L.BIRMAN: Phys. Rev. **162**, 806 (1967).
2.27. R.M.MARTIN, L.M.FALICOV: *Resonant Raman Scattering*, this volume, p. 79.
2.28. E.BURSTEIN, R.ITO, A.PINCZUK, M.SHAND: J. Acoust. Soc. Am. **49**, 1013 (1971).
2.29. M.CARDONA, F.H.POLLAK: In *The Physics of Opto-Electronic Materials*, ed. by W.A.ALBERS (Plenum Press, New York, 1971), pp. 81–112.
2.30. M.CARDONA: In *Atomic Structure and Properties of Solids*, ed. by E.BURSTEIN (Academic Press, New York, 1972), pp. 514–580).
2.31. M.CARDONA: Surface Sci. **37**, 100 (1973).
2.32. S.BHAGAVANTAM, T.VENKATARAYUDU: *Theory of Groups and Its Application to Physical Problems* (Academic Press, New York, 1969).
2.33. D.C.HAMILTON: Phys. Rev. **188**, 1221 (1969).
2.34. M.I.BELL: *Proc. Conf. Electronic Density of States* (N.B.S., Washington, 1971), pp. 757–766.
2.35. W.L.FAUST, C.H.HENRY: Phys. Rev. Letters **17**, 1265 (1966).
2.36. S.USHIODA, A.PINCZUK, E.BURSTEIN, D.L.MILLS: In *Light Scattering Spectra of Solids*, ed. by G.B.WRIGHT (Springer, New York, 1969), pp. 347–357.
2.37. W.D.JOHNSTON, I.P.KAMINOW: Phys. Rev. **188**, 1209 (1969).
2.38. A.L.MCWHORTER, P.N.ARGYRES: In *Light Scattering Spectra of Solids*, ed. by G.B.WRIGHT (Springer, New York, 1969), pp. 325–333.
2.39. R.A.COWLEY: J. Physique **26**, 659 (1965).
2.40. A.A.MARADUDIN, E.E.FEIN: Phys. Rev. **128**, 2559 (1962).
2.41. H.BENSON, D.L.MILLS: Phys. Rev. B**1**, 1678 (1970).
2.42. H.BENSON, D.L.MILLS: Solid State Commun. **8**, 1387 (1970).
2.43. E.ANASTASSAKIS, D.HWANG, C.H.PERRY: Phys. Rev. B**4**, 2493 (1971).
2.44. T.R.HART, R.L.AGGARWAL, B.LAX: Phys. Rev. B**1**, 638 (1970).
2.45. R.K.RAY, R.L.AGGARWAL, B.LAX: In *Light Scattering in Solids*, ed. by M.BALKANSKI (Flammarion Sciences, Paris, 1971), pp. 288–390.
2.46. F.CERDEIRA, M.CARDONA: Phys. Rev. B**5**, 1440 (1972).
2.47. P.G.KLEMENS: Phys. Rev. **148**, 945 (1966).

2.48. S. BARKER: Phys. Rev. **105**, 917 (1968).
2.49. S. USHIODA, J. D. MCMULLEN: Solid State Commun. **11**, 299 (1972).
2.50. J. M. WORLOCK: In *Structural Phase Transitions and Soft Modes*, ed. by E. J. SAMUELSEN (Universitets Forlaget, Oslo, 1972), pp. 329–370.
2.51. J. F. SCOTT: Revs. Mod. Phys. **46**, 83 (1974).
2.52. A. A. MARADUDIN, I. P. IPATOVA: J. Math. Phys. **9**, 525 (1968).
2.53. A. ZAWADOWSKI, J. RUVALDS: Phys. Rev. Letters **24**, 1111 (1970).
2.54. J. F. SCOTT, T. C. DAMEN, J. RUVALDS, Z. ZAWADOWSKI: Phys. Rev. B**3**, 1295 (1971).
2.55. P. D. LAZAY, P. A. FLEURY: In *Light Scattering in Solids*, ed. by M. BALKANSKI (Flammarion Sciences, Paris, 1974), pp. 406–410.
2.56. D. L. ROUSSEAU, S. P. S. PORTO: Phys. Rev. Letters **20**, 1354 (1968).
2.57. D. T. HON, W. L. FAUST: Appl. Phys. **1**, 241 (1973).
2.58. F. Adar: Ph.D. Dissertation, Dept. of Physics, University of Pennsylvania (1972).
2.59. A. PINCZUK, E. BURSTEIN: In *Light Scattering Spectra of Solids*, ed. by G. B. WRIGHT (Springer, New York, 1969), pp. 429–438.
2.60. P. CORDEN: Ph.D. Dissertation, University of Pennsylvania (1971).
2.61. L. BRILLSON, E. BURSTEIN: Phys. Rev. Letters **27**, 808 (1971).
2.62. M. SHAND, W. RICHTER, E. BURSTEIN: J. Nonmentals **1**, 53 (1972); also M. SHAND, E. BURSTEIN: Surface Sci. **37**, 145 (1973).
2.63. E. ANASTASSAKIS, F. H. POLLAK, G. W. RUBLOFF: *Proc. 11th Intern. Conf. Physics of Semiconductors* (Nauka. Warsaw, 1973), pp. 1187–1194.
2.64. E. ANASTASSAKIS, F. H. POLLAK, G. W. RUBLOFF: Phys. Rev. B**9**, 551 (1974).
2.65. P. B. KLEIN, H. MASUI, J. SONG, R. K. CHANG: Solid State Commun. **14**, 1163 (1974).
2.66. P. Y. YU, Y. R. SHEN, Y. PETROFF, L. M. FALICOV: Phys. Rev. Letters **30**, 283 (1973); and
P. Y. YU, Y. R. SHEN: Phys. Rev. Letters **32**, 373 (1974).
2.67. J. F. SCOTT, T. C. DAMEN, R. C. C. LEITE, W. T. SILFVAST: Solid State Commun. **7**, 953 (1969).
2.68. A. COMPAAN, H. Z. CUMMINS: Phys. Rev. Letters **31**, 41 (1973).
2.69. T. C. DAMEN, J. SHAH: Phys. Rev. Letters **27**, 1506 (1971).
2.70. F. CERDEIRA, W. DREYBRODT, M. CARDONA: *Proc. 11th Intern. Conf. Physics of Semiconductors* (Polish Scientific Publ., Warsaw, 1972), pp. 1142–1147.
2.71. J. L. LEWIS, R. L. WADSACK, R. K. CHANG: In *Light Scattering in Solids*, ed. by M. BALKANSKI (Flammarion Sciences, Paris, 1971), pp. 41–46.
2.72. J. M. RALSTON, R. L. WADSACK, R. K. CHANG: Phys. Rev. Letters **25**, 814 (1970).
2.73. M. I. BELL, R. N. TYTE, M. CARDONA: Solid State Commun. **13**, 1833 (1973).
2.74. T. C. DAMEN, J. F. SCOTT: Solid State Commun. **9**, 383 (1971).
2.75. F. CERDEIRA, W. DREYBRODT, M. CARDONA: Solid State Commun. **10**, 591 (1972).
2.76. W. DREYBRODT, W. RICHTER, M. CARDONA: Solid State Commun. **11**, 1127 (1972).
2.77. P. Y. YU, Y. R. SHEN: Phys. Rev. Letters **29**, 478 (1972).
2.78. W. DREYBRODT, W. RICHTER, F. CERDEIRA, M. CARDONA: Phys. Stat. Sol. (b) **60**, 145 (1974).
2.79. M. A. RENUCCI, J. B. RENUCCI, M. CARDONA: Phys. Stat. Sol. (b) **49**, 625 (1972).
2.80. R. W. RUBLOFF, E. ANASTASSAKIS, F. H. POLLAK: Solid State Commun. **13**, 1755 (1973).
2.81. E. ANASTASSAKIS, C. H. PERRY: In *Light Scattering in Solids*, ed. by M. BALKANSKI (Flammarion Sciences, Paris, 1971), pp. 47–51.
2.82. P. J. COLWELL, M. V. KLEIN: Solid State Commun. **8**, 2095 (1970).
2.83. P. F. WILLIAMS, S. P. S. PORTO: In *Light Scattering in Solids*, ed. by M. BALKANSKI (Flammarion Sciences, Paris, 1971), pp. 70–71.
2.84. R. M. MARTIN, T. C. DAMEN: Phys. Rev. Letters **26**, 86 (1971).

2.85. R.H. CALLENDER, S.S. SUSSMAN, M. SELDERS, R.K. CHANG: Phys. Rev. B**7**, 3788 (1973).
2.86. E. ANASTASSAKIS, E. BURSTEIN: In *Light Scattering in Solids*, ed. by M. BALKANSKI (Flammarion, Paris, 1971), pp. 52–57.
2.87. P.T. WILLIAMS, S.P.S. PORTO: Phys. Rev. B**8**, 1782 (1973).
2.88. C.W. CLENDENING (JR.): Ph.D. Dissertation, Materials Science Center, Cornell University (1974).
2.89. A. PINCZUK, E. BURSTEIN: *Proc. 10th Intern. Conf. Physics of Semiconductors* (U.S. Atomic Energy Commission, Washington, D.C., 1970), pp. 727–735.
2.90. W. RICHTER: J. Phys. Chem. Sol. **33**, 2113 (1972); and *Proc. 11th Intern. Conf. Physics of Semiconductors* (Polish Scientific Publications, Warsaw, 1972), p. 1148.
2.91. K.P. JAIN, G. CHOUDHURY: Phys. Rev. B**8**, 676 (1973).
2.92. R.M. MARTIN: Phys. Rev. B**4**, 3676 (1971).
2.93. Y. YACOBY: Solid State Commun. **13**, 1061 (1973).
2.94. A. PINCZUK, E. BURSTEIN: Surface Sci. **37**, 153 (1973).
2.95. J.M. WORLOCK: In *Light Scattering Spectra of Solids*, ed. by G.B. WRIGHT (Springer, New York, 1969), pp. 411–419.
2.96. E. ANASTASSAKIS, E. BURSTEIN, A. FILLER: In *Light Scattering Spectra of Solids*, ed. by G.B. WRIGHT (Springer, New York, 1969), pp. 421–428.
2.97. M.I. BELL: *Proc. Eleventh Intern. Conf. Physics of Semiconductors* (Polish Scientific Publications, Warsaw, 1972), pp. 845–850.
2.98. S.S. JHA, N. BLOEMBERGEN: Phys. Rev. **171**, 891 (1968).
2.99. D.E. ASPNES, J.E. ROWE: Phys. Rev. B**5**, 4022 (1972). Also D.E. ASPNES: Surface Sci. **37**, 418 (1973).
2.100. J.G. GAY, J.D. DOW, E. BURSTEIN, A. PINCZUK: In *Light Scattering in Solids*, ed. by M. BALKANSKI (Flammarion Sciences, Paris, 1971), pp. 33–38.
2.101. K. PEUKER, F. BECHSTEDT, R. ENDERLEIN: *Proc. 12th Intern. Conf. Physics of Semiconductors* (B. G. Teubner, Stuttgart 1974), pp. 468—472.

3. Resonant Raman Scattering

R. M. Martin and L. M. Falicov*

With 20 Figures

Ordinary Raman scattering studies yield information on low-energy excitations of molecules, liquids, and solids. Such information, contained in the frequency shift, k-vector and polarization of the scattered photon, is in general restricted to states which are typically up to 5×10^{-2} eV from the ground state of the system under study. The frequencies of both the incoming and outgoing radiation are in themselves somewhat irrelevant quantities; only their difference carries the important information.

In this chapter, we study in detail those aspects of the Raman scattering phenomena which are related to resonant behavior. By using resonant characteristics, we involve in the process, in a direct way, well-defined excited states of the material system and therefore we can study their properties. Resonant Raman Scattering (RRS) allows us thus to explore the spectrum of the material in the range of energies of the photon energy itself, typically 1–5 eV for conventional experiments. In this resepect we may say that RRS is to ordinary Raman spectroscopy as optical spectroscopy is to classical optics.

Since the frequencies of the incoming and outgoing photons are of fundamental importance in RRS, and since their difference is a characteristic property of the material, RRS experiments are concerned simultaneously with high and low frequencies strongly coupled in the process. We may say in this respect that RRS is related to modulation spectroscopy, with the important difference, however, that the low frequency modulation is not provided or controlled by the experimenter. The mathematics of the "coupled oscillators" is nevertheless very similar in both cases.

Information closely related to that obtained in RRS can be drawn out of other experiments, in particular absorption, emission and their combined process, known in the literature as fluorescence when applied to molecules and liquids or as luminescence when applied to solids. In particular, luminescence, or more specifically "hot" luminescence, is hard to distinguish from RRS, and the criteria used to differentiate

* Research supported in part by the National Science Foundation through Grant DMR 72-03106-A01.

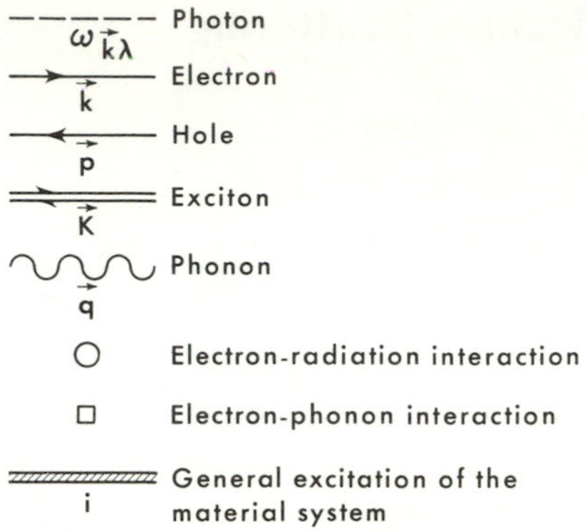

Fig. 3.1. Graphical description of the various excitations and interactions in the radiation field and material system. Time is supposed to evolve from left to right

one from the other are mostly operational, based on theoretical formalisms or experimental convenience; they vary widely from researcher to researcher.

3.1. Fundamental Definitions and Basic Properties

In order to make our language precise and to define properly the terms used in this chapter, we use the rest of this introduction to go back to the fundamental definitions and basic properties of the RRS phenomenon.

The systems we are concerned with consist of the radiation field, described by a radiation Hamiltonian $\mathscr{H}_R$ and a material system (solid, molecule, atom, solution, etc.) described by a Hamiltonian $\mathscr{H}_M$. Radiation and matter interact [3.1] via an interaction Hamiltonian $\mathscr{H}_{MR}$. The radiation Hamiltonian is quantized in the well-known way [3.1], its quanta being photons of wave vector $\boldsymbol{k}$, polarization mode λ, polarization vector $\boldsymbol{\varepsilon}_\lambda$ and angular frequency $\omega_{\boldsymbol{k}\lambda} = ck$. The energy associated with a given eigenstate of $\mathscr{H}_R$, measured from the vacuum level is

$$W(\{n_{\boldsymbol{k}\lambda}\}) = \sum_{\boldsymbol{k},\lambda} n_{\boldsymbol{k}\lambda} \hbar\omega_{\boldsymbol{k}\lambda} . \tag{3.1}$$

In the graphic representation defined in Fig. 3.1, a photon is described by a dashed line.

The material system, as described by $\mathscr{H}_M$, exhibits an energy spectrum $\{E_n\}$. Its ground state energy E_0 is taken to be zero. All other eigenstates n have an excitation energy E_n and, to some degree of approximation, can be described as a superposition of elementary excitations: one-electron

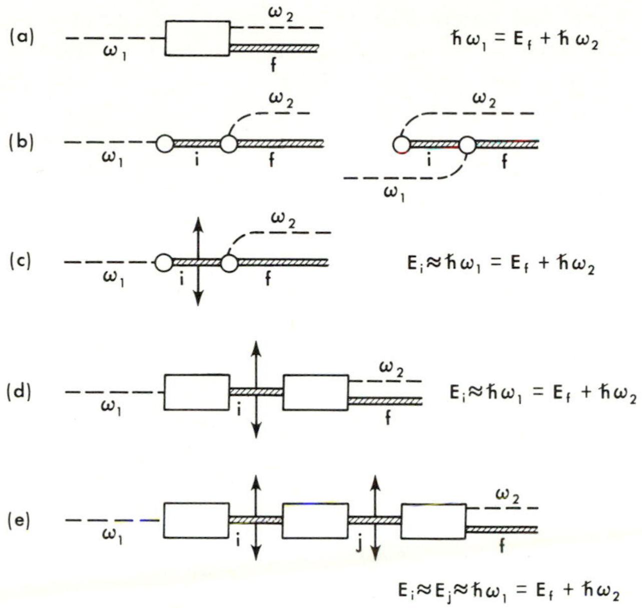

Fig. 3.2a–e. Graphical representations of: (a) a generalized Raman process; (b) the two "basic" Raman matrix elements; (c) the simplest Resonant Raman process; (d) a generalized Resonant Raman process; (e) a generalized double Resonant Raman process

excitations, hole-like excitations, exciton states, phonons, spin-waves, or plasmons in a solid; electronic, vibronic or rotational modes in a molecule or electron excitations in an atom. Graphic representation of some of these excitations is shown in Fig. 3.1.

In Fig. 3.2a we show graphically the Raman process we are concerned with. It involves a change of state of the material system between the initial and final states and an exchange of energy with the radiation field (inelastic scattering). The radiation field may in general either give energy to the material system (Stokes line) or receive energy from it (anti-Stokes line). This last process can only take place if the material system is originally in a state other than the ground state, i.e. in a thermally or otherwise activated state. We restrict ourselves here only to the Stokes line at very low temperatures. The initial state, as shown in Fig. 3.2a, involves the material system in its ground state. It contains, in addition, a photon $(\boldsymbol{k}_1 \lambda_1)$ of frequency ω_1, which becomes scattered in the final state into another photon $(\boldsymbol{k}_2 \lambda_2)$ of frequency ω_2 leaving behind the material system in an excited state of energy E_f. Conservation of energy requires

$$\hbar\omega_1 = \hbar\omega_2 + E_f . \tag{3.2}$$

The differential cross-section $d\sigma$ for the Raman scattering of an initial photon $(\boldsymbol{k}_1 \lambda_1)$ into a photon of frequency between ω_2 and $(\omega_2 + d\omega_2)$ of polarization λ_2 and such that $\boldsymbol{k}_2$ is in a small solid angle $d\Omega$ about the direction $\hat{\theta}$ is given [3.1] by

$$d\sigma(\boldsymbol{k}_1 \lambda_1; \omega_2 \hat{\theta} \lambda_2) = (2\pi/\hbar c) \sum_{\boldsymbol{k}_2}{}' \sum_{f} |K_{2f,10}|^2 \, \delta(\hbar\omega_2 + E_f - \hbar\omega_1), \quad (3.3)$$

where the summation over $\boldsymbol{k}_2$ is restricted over a small volume in k-space, and K is an appropriate "high-order matrix element" of $\mathscr{H}_{\text{MR}}$ connecting the initial and final states (see Fig. 3.2a). The above equation can be simplified by effecting explicitly the summation over $\boldsymbol{k}_2$. This yields

$$d\sigma(\boldsymbol{k}_1 \lambda_1; \omega_2 \hat{\theta} \lambda_2) = [d\Omega/(4\pi^2 \hbar^4 c^4)] \sum_{f}{}' |K_{2f,10}|^2 (\hbar\omega_1 - E_f)^2 , \quad (3.4)$$

where the summation is now restricted to those states for which

$$\hbar\omega_2 < \hbar\omega_1 - E_f < \hbar(\omega_2 + d\omega_2) . \quad (3.5)$$

It is possible now to distinguish three different cases:

i) The final states f which satisfy (3.5) are such that E_f is in the discrete part of the spectrum $\{E_m\}$: such is the case for instance when the material system is a molecule and the state f is a given vibronic mode. If we assume that there is only one such mode, and we write

$$d_1 \, \sigma(\boldsymbol{k}_1 \lambda_1; \hat{\theta} \lambda_2) = \int_{\omega_2} d\sigma$$

we obtain the well-known result [3.1]

$$d_1 \, \sigma(\boldsymbol{k}_1 \lambda_1; \hat{\theta} \lambda_2) = |K_{2f,10}|^2 (\hbar\omega_1 - E_f)^2 \, d\Omega/(4\pi^2 \hbar^4 c^4) . \quad (3.6)$$

The scattered photon is in a very narrow frequency band around $\hbar\omega_2 = \hbar\omega_1 - E_f$. In other words, the outgoing photon spectrum has a delta-function-like line shape.

ii) The final states f which satisfy (3.5) are such that E_f lies in the continuum of the $\{E_m\}$ spectrum; selection rules, however, make K non-zero for only one (or a finite number) of the final states. Such is the case of ordinary first-order Raman scattering in a crystal, in which the final state involves a phonon of wave vector $\boldsymbol{q}$ and branch ν, with energy $E_f = \hbar\omega_\nu(\boldsymbol{q})$. Quasi-momentum conservation requires that

$$\boldsymbol{k}_1 = \boldsymbol{k}_2 + \boldsymbol{q} , \quad (3.7)$$

and since $|k_1| \simeq |k_2|$ and both are much smaller than a typical Brillouin zone wave vector, $\boldsymbol{q} \cong 0$. In this case (3.6) is still valid if E_f is replaced by $E_f = \hbar\omega_\nu(\boldsymbol{k}_1 - \boldsymbol{k}_2) \cong \hbar\omega_\nu(0)$.

iii) In this case, the final states f in (3.4) which satisfy (3.5) are in the continuum part of the spectrum $\{E_m\}$, and an infinite (continuum) subset of them have a non-zero matrix element K connecting them to the ground state. If we now make the dependence of K on E_f explicit and denote by $\varrho(E_f)$ the density of states of the material system, (3.4) reduces to

$$
\begin{aligned}
d\sigma(\boldsymbol{k}_1\lambda_1; \omega_2\hat{\theta}\lambda_2) \\
&= [d\omega_2\, d\Omega/(4\pi^2\hbar c^4)]\,(\omega_2)^2\,|K(E_f = \hbar\omega_1 - \hbar\omega_2)|^2 \\
&\cdot \varrho(E_f = \hbar\omega_1 - \hbar\omega_2)\,.
\end{aligned} \tag{3.8}
$$

Equation (3.8) yields a non-zero cross-section for all scattered frequencies ω_2 such that the product $|K(E_f = \hbar\omega_1 - \hbar\omega_2)|^2\,\varrho(E_f = \hbar\omega_1 - \hbar\omega_2)$ does not vanish. This yields a band spectrum for any direction $\hat{\theta}$. An example of this case is the ordinary second-order Raman scattering in a crystal in which two phonons of branch ν and wave vector $\boldsymbol{q}_1$ and $\boldsymbol{q}_2$ are emitted. Quasi-momentum conservation requires that

$$
\boldsymbol{k}_1 = \boldsymbol{k}_2 + \boldsymbol{q}_1 + \boldsymbol{q}_2 \tag{3.9}
$$

which can be approximated by

$$
\boldsymbol{q}_1 \cong -\boldsymbol{q}_2 = \boldsymbol{q}\,.
$$

If the density of one-phonon states of frequency ω_q and branch ν is denoted by $\varrho_{1\nu}(\hbar\omega_q)$, the total density of final states which enters into (3.8) and takes account of wave-vector conservation is

$$
\varrho(E_f) \equiv \varrho_{1\nu}(\hbar\omega_q = E_f/2)\,.
$$

Formulae (3.6) and (3.8) are the fundamental ones in the theory of Raman scattering. The information we seek is all contained in either $|K|^2$ or in the product $|K|^2\,\varrho$. The study of the matrix element K is therefore paramount and is the subject of our next section.

3.2. The Raman Matrix Elements

Since Raman scattering involves two photons (one absorbed, one emitted), from the point of view of formal quantum electrodynamics [3.1] it is a second-order process in $\mathscr{H}_{\mathrm{MR}}$: the two basic second-order matrix elements which contribute to the cross-section are shown in Fig. 3.2b, and their contribution can be written as

$$K_{2f,10} = \sum_i \left\{ \frac{\langle \omega_2 f | \mathscr{H}_{MR} | 0i \rangle \langle 0i | \mathscr{H}_{MR} | \omega_1 o \rangle}{(\hbar\omega_1 - E_i)} + \frac{\langle \omega_2 f | \mathscr{H}_{MR} | \omega_2 \omega_1 i \rangle \langle \omega_2 \omega_1 i | \mathscr{H}_{MR} | \omega_1 o \rangle}{(-\hbar\omega_2 - E_i)} \right\}, \tag{3.10}$$

where 0, i, and f denote the ground, intermediate, and final states of the material system. The ket $|\omega_s l\rangle$ designates the presence of a photon of frequency ω_s, while $|0l\rangle$ the fact that this photon has been destroyed. It is instructive at this point to compare the above expression with that for a generalized susceptibility, given by

$$\chi(\omega) = \sum_i \left\{ \frac{\langle \omega o | \mathscr{H}_{MR} | 0i \rangle \langle 0i | \mathscr{H}_{MR} | \omega o \rangle}{\hbar\omega - E_i} + \frac{\langle \omega o | \mathscr{H}_{MR} | 2\omega i \rangle \langle 2\omega i | \mathscr{H}_{MR} | \omega o \rangle}{-\hbar\omega - E_i} \right\}. \tag{3.11}$$

The following features are worth noticing:

a) Both expressions consist of terms which contain two matrix elements of $\mathscr{H}_{MR}$ in the numerator and a linear frequency dependent denominator.

b) In (3.10) and (3.11), the first kind of terms can both diverge linearly in the frequency as $\hbar\omega_1$ or $\hbar\omega$ approaches an energy eigenvalue E_i (resonant denominator).

c) The numerators in (3.11) are positive definite quantities; the numerators in (3.10) are arbitrary complex numbers which may interfere with one another so as to cancel partially or totally their individual contributions.

Equation (3.10) gives the lowest order contribution to the matrix element K. It is, in general, not sufficient to use (3.10) to interpret experimental RRS data. This is due to two different causes:

1) The spectrum $\{E_i\}$ and eigenstates $|i\rangle$ of the material system are in general not known, and they should be, therefore, studied by making suitable approximations. These most commonly involve a description in terms of quasi-particles or elementary excitations (electrons, holes, phonons, plasmons, magnons, etc.) and their multiple interactions. In terms of such a description, K may involve various order of perturbation theory in some between quasi-particles interaction. For example, the simple case of one-phonon Raman scattering is at least first-order in the electron-phonon interaction while a two-phonon

process requires at least second-order contributions. The approximate description of the spectrum $\{E_i\}$ also implies that the states $|i\rangle$ which we include in the expression (3.10) are not true eigenstates of $\mathscr{H}_M$ and that consequently such states decay into other states with a given lifetime τ_i. This is equivalent to saying that the energy E_i has a small imaginary component

$$\operatorname{Im}\{E_i\} = \hbar\gamma_i = \hbar/\tau_i \tag{3.12}$$

which describes such finite lifetime effects.

2) The treatment of K in (3.10) in second-order perturbation theory neglects important effects which are relevant in many cases. Two such effects are worth mentioning. First, the multiple interaction of matter and radiation provides the states $|i\rangle$ of the material system with an intrinsic (radiative) lifetime. In other words, there is a contribution to γ_i in (3.12) arising from multiple absorption and emission of photons; even in the hypothetical case in which we could know the spectrum of $\mathscr{H}_M$ exactly, the higher-order effects of $\mathscr{H}_{MR}$ would provide a finite lifetime for each state $|i\rangle$. Second, the multiple interaction of photons with matter produces hybrid modes of excitation which may influence the interpretation of Raman scattering. For instance, an optical phonon and a photon, or an exciton and a photon, interact strongly so as to become a polariton, with a different dispersion relation from either that of a phonon, or an exciton, or a photon. The general interaction of a photon with matter changes its dispersion relation with a consequent change of the parameters which characterize the propagation (wave vector, group velocity, etc.).

It is, however, important to remark that when all these considerations are taken into account, i.e. when $|i\rangle$ is in fact a hybrid mode described only in an approximate way by a superposition of various elementary excitations and E_i contains lifetimes effects in the form of a "rather small" imaginary term, then (3.10) is a very good approximation for describing Raman scattering cross-sections (3.6) and (3.8).

A RRS process is one in which the incident frequency ω_1 is such that

$$\hbar\omega_1 \cong \operatorname{Re}\{E_i\}\,, \tag{3.13}$$

where $|i\rangle$ is a state of the material system accessible from the ground state by electromagnetic radiation absorption. The simplest example of RRS is illustrated in Fig. 3.2c and it corresponds to the first graph in Fig. 3.2b when (3.13) is satisfied. The RRS general process is indicated in Fig. 3.2d in which the boxes contain all the necessary intermediate states which permit a transition from $|0\rangle$ to $|i\rangle$ and from $|i\rangle$ to $|f\rangle$. It is important to notice that, close to resonance, only a few at most of the terms in (3.10) give a large (resonant) contribution. In order to describe the frequency dependence of the matrix elements, it is a good approximation to retain in (3.10) only the terms with small denominators,

$$K_{2f,10} \cong \sum_{\substack{i \\ E_i \cong \hbar\omega_1}} \frac{\langle \omega_2 f | \mathscr{H}_{\mathrm{MR}} | 0 i \rangle \langle 0 i | \mathscr{H}_{\mathrm{MR}} | \omega_1 o \rangle}{\hbar\omega_1 - \mathrm{Re}\{E_i\} - i\hbar\gamma_i}, \tag{3.14}$$

where the sum is restricted to a region of energy near $\hbar\omega_1$. This means that only some, and in general very few, of the many diagrams which contribute to a given Raman scattering process can become resonant. For example, the emission-first-absorption-second diagram in Fig. 3.2b can never be a resonant one.

It is possible, however, to have several states $|i\rangle$, $|j\rangle$, etc. with approximately the same energy $E_i \cong E_j \cong \hbar\omega_1$ which may resonate simultaneously. If an approximate description is used for the states $|i\rangle$ and $|j\rangle$ and a given perturbation connects them in some well-defined way, it is also possible to have successive multiple resonance, as is illustrated in Fig. 3.2e; such processes contain in formal perturbation theory energy denominators in which two factors tend to zero simultaneously. Resonant enhancement in these cases may involve extra features in the spectrum [Ref. 8.1, Sect. 2.2.11].

We now discuss the frequency dependence of K for various cases. As with the final state $|f\rangle$ in Section 3.1, we classify them according to the spectral properties of the intermediate state $|i\rangle$:

I) The resonant state i is such that E_i is in the discrete part of the spectrum $\{E_m\}$;

II) Although E_i falls in the middle of the continuum of $\{E_m\}$, selection rules make only one (or a finite number) of the states $|i\rangle$ have either a non-zero $\langle \omega_2 f | \mathscr{H}_{\mathrm{MR}} | 0 i \rangle$ or a non-zero $\langle 0 i | \mathscr{H}_{\mathrm{MR}} | \omega_1 o \rangle$. For each pair of initial $|\omega_1 o\rangle$ and final $|\omega_2 f\rangle$ states, there is only one resonant intermediate state $|0 i\rangle$;

III) There are an infinite (continuum) number of states in the resonant region (3.13).

In the first case, if only one state exists in the resonant region, and we write

$$\mathrm{Re}\{E_i\} \equiv \hbar\omega_i, \tag{3.15}$$

$$\langle \omega_2 f | \mathscr{H}_{\mathrm{MR}} | 0 i \rangle \equiv M_{fi}, \tag{3.16}$$

$$\langle 0 i | \mathscr{H}_{\mathrm{MR}} | \omega_1 o \rangle \equiv M_{io}, \tag{3.17}$$

substitution in (3.14) yields

$$|K_{2f,10}|^2 \cong \frac{|M_{fi}M_{i0}|^2}{\hbar^2[(\omega_1-\omega_i)^2+\gamma_i^2]}, \tag{3.18}$$

which should be substituted in (3.6) to obtain the cross-section. We can qualitatively distinguish two different resonant regions: very close to resonance $\Delta\omega_1 \equiv |\omega_1-\omega_i| < \gamma_i$, the cross-section is dominated by lifetime effects as in the case of resonance fluorescence [3.1]

$$|K_{2f,10}|^2 \cong |M_f M_0|^2/\hbar^2\gamma_i^2;\ \Delta\omega_i \ll \gamma_i; \tag{3.19}$$

if $\Delta\omega_i > \gamma_i$ but still the contribution of the i state is the dominant one

$$|K_{2f,10}|^2 \cong |M_f M_0|^2/\hbar^2(\omega_1-\omega_i)^2;\ \Delta\omega_i \gg \gamma_i. \tag{3.20}$$

There is, of course, a continuous transition between (3.19) and (3.20), as is indicated by (3.18).

The study of the spectrum, i.e. the frequency ω_1 dependence of the cross-sections (3.6) and (3.8), is closely related to the study of time evolution of the Raman process. As is well known from quantum electrodynamics [3.1], the line broadening parameter $\hbar\gamma_i$ is directly related, through lifetime processes, to the "time delay" between absorption of the incoming photon ω_1 and the emission of the outgoing one ω_2. The process described by (3.18) takes place with a typical exponential time dependence with characteristic time given by $\tau_i = 1/\gamma_i$. Through the usual Fourier transform arguments (or equivalently Heisenberg's uncertainty principle) we may associate the on-resonance region (3.19)—small energy uncertainty—with slow processes, and the off-resonance region (3.20)—larger energy uncertainty—with fast processes. The distinction between slow and fast process only becomes apparent if an outside time scale is imposed on the measurements by experimental conditions. For example, if the incoming photon is produced by a laser with a fast on-off switching time $\tau_L (\gamma_i\tau_L \ll 1)$, which is switched on at time $t=0$ and off at time $t=T_L$, the Raman emission intensity [3.2, 3] as a function of time is at resonance $\omega_1=\omega_i$,

$$I(t) \propto \begin{cases} \gamma_i^2[1-\exp(-t\gamma_i/2)]^2, & t<T_L \\ I(T_L)\exp[-\gamma_i(t-T_L)], & t>T_L \end{cases} \tag{3.21}$$

while far away from resonance $\Delta\omega_i \gg \gamma_i$

$$I(t) \propto \begin{cases} (\Delta\omega_i)^{-2}[1-\exp(-t/\tau_L)]^2, & t<T_L \\ I(T_L)\exp[-2(t-T_L)/\tau_L], & t>T_L. \end{cases} \tag{3.22}$$

In other words, for $\Delta\omega_i < \gamma_i$ all processes are slow, dominated by lifetime effects; for $\Delta\omega_i > \gamma_i$, all processes are fast (instantaneous), dominated by experimental conditions.

A simple extension of the above is for systems which are inhomogenously broadened, i.e., a collection of independent systems (e.g., molecules) which have a distribution of levels ω_i around a mean value ω_{i0}. This randomness in ω_1 is brought about by collisions, random environments, etc. If the probability that a given system has its only resonant state $|i\rangle$ with energy between $\hbar\omega_i$ and $\hbar(\omega_1 + d\omega_i)$ given by $P(\omega_i)\,d\omega_i$, then the weighted average $|K|^2$ is given by

$$|K_{2f,10}|^2 = |M_{fi}M_{i0}|^2 \int \frac{P(\omega_i)\,d\omega_i}{\hbar^2[(\omega_1-\omega_i)^2+\gamma_i^2]}\,. \tag{3.23}$$

It is interesting to note that in this particular case some members of the ensemble may be in the $|\omega_i - \omega_1| \gg \gamma_i$ (fast) regime, while some others may be at resonance or in the $|\omega_i - \omega_1| \ll \gamma_i$ (slow) regime.

If E_i falls in the continuum of $\{E_m\}$ but selection rules make only one such state $|i\rangle$ available as an intermediate state—case II)—(3.18) is still valid. The important distinction in this case, however, is that the pertinent state $|i\rangle$ will in general vary with the k-vector, frequency and polarization of the incoming and outgoing photons as well as with the nature of the final state $|f\rangle$. This dependence may have important consequences in the interpretation of experimental data. For example, if there is a continuum of available final states—case iii) of Section 3.1—and a one-to-one correspondence between final state and intermediate resonant state—case II) in the present context—$f \leftrightarrow i$, then from (3.4) and (3.18) we obtain

$$\begin{aligned} &d\sigma(\boldsymbol{k}_1\lambda_1;\omega_2\hat{\theta}\lambda_2)\\ &= [d\Omega/(4\pi^2\hbar^4c^4)]\sum_f{}' \left[\frac{|M_{fi}M_{i0}(f)|^2(\hbar\omega_1-E_f)^2}{\hbar^2\{[\omega_1-\omega_i(f)]^2+[\gamma_i(f)]^2\}}\right], \end{aligned} \tag{3.24}$$

where the summation is to be carried with the restriction (3.5). The summation in (3.24) is over final states, but because of the one-to-one correspondence $f \leftrightarrow i$, it is also a summation over intermediate resonant states. It is worth remarking here that in this case we are effectively summing $|K|^2$ over intermediate states: since each intermediate state contributes to a *different* final state, the effects are additive and no interference takes place. This result is very similar to the case of inhomogeneous broadening, the difference being that the matrix elements M_{fi} and M_{i0} and the energies $\hbar\omega_i(f)$ may be general functions of the final state. Just as before, there are contributions from both slow and fast scattering with the relative contribution dependent on the specific case.

If several discrete levels exist in the resonant region, interference effects may arise among the various terms in (3.14) which should be *added before squaring K*

$$|K_{2f,1o}|^2 = \left|\sum_i \frac{M_{fi} M_{io}}{\hbar(\omega_1 - \omega_i) - i\hbar\gamma_i}\right|^2 . \tag{3.25}$$

The form of the resonance is determined by the phase and functional dependence of the matrix elements, and the position of the poles. We need not enumerate the many possibilities, but we note that formulas like (3.25)—which are required to determine the RRS cross section over a larger frequency range—can yield a large variety of resonance behaviors.

Finally, for case III), there is a continuum of states i which resonate with each initial $|\omega_1 o\rangle$ and final $|\omega_2 f\rangle$ state pair. In this particular case, the summation in (3.14) can be reduced to an integral

$$K_{2f,1o} \cong \int \frac{M_f(\omega_i)\, M_0(\omega_i)}{\omega_1 - \omega_i - i\gamma(\omega_i)}\, \varrho_i(E = \hbar\omega_i)\, d\omega_i \tag{3.26}$$

which should be evaluated for each particular case. We note in passing that (3.26) also describes the case of discrete intermediate states when ϱ_i consists of one or a sum of delta functions.

It may be of interest to calculate (3.26) for the case of constant $(M_f M_0)$, constant γ, and various forms of ϱ_i. If we take

$$\varrho_i(\hbar\omega) = \begin{cases} 0, & \omega < \Delta; \\ \alpha(\omega - \Delta)^{(d-2)/2}, & \Delta < \omega < \Delta + \omega_c; \\ 0, & \omega > \Delta + \omega_c; \end{cases} \tag{3.27}$$

where d is an index which represents the dimension ($d = 1, 2$, or 3), Δ is the band gap and ω_c is a cut-off energy, we obtain

$$K_{2f,1o} = (M_f M_0 \alpha) \times \begin{cases} (1/a) \ln\left[\dfrac{a + \omega_c^{1/2}}{a - \omega_c^{1/2}}\right] & d = 1 \\ \ln\left[\dfrac{a^2}{a^2 - \omega_c}\right] & d = 2 \\ \left\{-2\omega_c^{1/2} + a \ln\left[\dfrac{a + \omega_c^{1/2}}{a - \omega_c^{1/2}}\right]\right\}, & d = 3 \end{cases} \tag{3.28}$$

where we have written

$$a \equiv (\omega_1 - \Delta - \mathrm{i}\gamma)^{1/2} . \tag{3.29}$$

The resonant effect is quite apparent from (3.28). In the limit of large lifetimes $\gamma \to 0$, as ω_1 approaches the threshold value Δ, this approximation gives for $|K|^2$ an $(\omega_1 - \Delta)^{-1}$ divergence in one dimension, a weak logarithmic divergence in two dimensions, and a convergent result (enhancement but no divergence) in three dimensions. It is important to remark that (3.28) applies only if M_f and M_i can both be considered constant as a function of ω_1. It is also interesting to notice that in the preceding cases, we find

$$K_{2f,10}(\omega_1) \cong (M_f/M_0^*)\,\chi(\omega_1) , \tag{3.30}$$

whenever the matrix elements are taken to be constant; in (3.30) $\chi(\omega_1)$ is the resonant contribution to the susceptibility. This holds for a discrete intermediate state—in which case $\chi(\omega_1)$ is a Lorentzian amplitude—for the case of inhomogeneous broadening—where $\chi(\omega_1)$ also reflects the broadening—and for the continuum case. It should be emphasized that χ in (3.30) is the *total* resonant susceptibility. In the language of quasi-particles it includes not only the direct terms but also the excitation-assisted transitions, e.g., phonon-assisted emission and absorption. If the matrix elements cannot be considered constant, then, of course, (3.30) does not apply. This is evident in (3.25) where the relative phases and magnitudes of the matrix elements change qualitatively the resonance line shape.

In each of the preceding examples the nature of the intermediate states is completely hidden in the matrix elements. The formal assumption of constancy in the matrix elements was essential to the previous analysis; it is, however, justified only in few special cases. To go beyond this assumption in a systematic way, it is convenient to consider a description of $\mathscr{H}_{\mathrm{M}}$ in terms of approximate quasi-particle states and include the quasi-particle-quasi-particle interaction Hamiltonian $\mathscr{H}'_{\mathrm{M}}$ as a perturbation. Consider now the case in which the (approximate) state $|i\rangle$ is coupled to $|o\rangle$ by $\mathscr{H}_{\mathrm{MR}}$

$$M_{i0} \equiv \langle i|\mathscr{H}_{\mathrm{MR}}|o\rangle ,$$

the state $|j\rangle$ is coupled to $|f\rangle$ also by $\mathscr{H}_{\mathrm{MR}}$

$$M_{fj} \equiv \langle f|\mathscr{H}_{\mathrm{MR}}|j\rangle$$

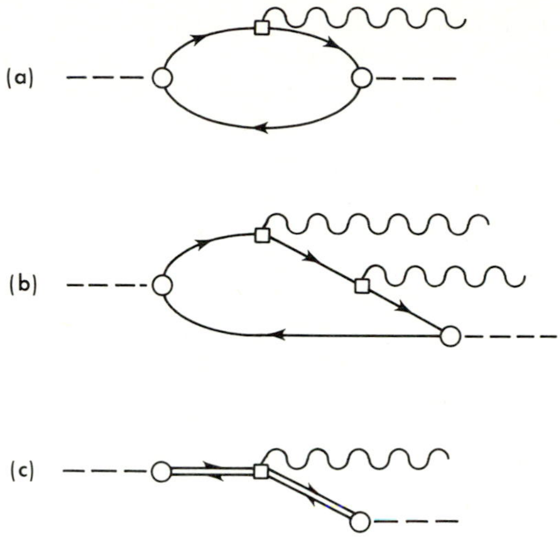

Fig. 3.3a–c. Graphical representation of Raman processes involving phonons: (a) a one-phonon process mediated by an electron-hole pair; (b) a two-phonon process mediated by an electron-hole pair; (c) a one-phonon process mediated by an exciton

and the states $|i\rangle$ and $|j\rangle$ are coupled to one another by $\mathscr{H}'_{\mathrm{M}}$

$$\mathscr{M}_{ji} \equiv \langle j|\mathscr{H}'_{\mathrm{M}}|i\rangle\,.$$

Expansion of eigenfunctions and eigenvalues of $\mathscr{H}_{\mathrm{M}}$ in a perturbation series in $\mathscr{H}'_{\mathrm{M}}$ leads to new expressions for the matrix element K near resonance (3.14). To first order in $\mathscr{H}'_{\mathrm{M}}$ this procedure yields [Ref. 8.1, Eq. (2.160)].

$$K_{2f,10} = \sum_{ij} \left[\frac{M_{fj}\mathscr{M}_{ji}M_{i0}}{\hbar^2(\omega_1-\omega_i-\mathrm{i}\gamma_i)(\omega_1-\omega_j-\mathrm{i}\gamma_j)}\right]; \tag{3.31}$$

to second order in $\mathscr{H}'_{\mathrm{M}}$ it gives

$$K_{2f,10} = \sum_{ijk} \left[\frac{M_{fj}\mathscr{M}_{jk}\mathscr{M}_{ki}M_{i0}}{\hbar^3(\omega_1-\omega_i-\mathrm{i}\gamma_i)(\omega_1-\omega_k-\mathrm{i}\gamma_k)(\omega_1-\omega_j-\mathrm{i}\gamma_j)}\right], \tag{3.32}$$

and so forth to any desired order. In (3.31) and (3.32), as in (3.14), we have kept only the most resonant terms.

Examples of the higher-order diagrams described by these equations are shown in Fig. 3.3. In the three graphs there depicted we have taken the most common case in which the radiation interacts directly only with the electronic part of the material system. Thus the vertices representing $\mathscr{H}_{\mathrm{MR}}$ interactions always involve creation or destruction of electronic states: no other quasi-particles are involved in these $\mathscr{H}_{\mathrm{MR}}$ vertices. Therefore, in (3.31) and (3.32) the state $|i\rangle$ is a purely electronic excited state and the state $|j\rangle$ contains the quasi-particles found in $|f\rangle$ in addition to an excited electronic state. For first-order perturbation in $\mathscr{H}'_{\mathrm{M}}$, as described by (3.31), $\mathscr{M}_{ji}$ must create the quasi-particle (phonon) found in the final state. Therefore, the energy of the $|j\rangle$ state $\hbar\omega_j$ is just the sum of an electronic energy $\hbar\omega_j^{\mathrm{e}}$ and the final state energy $\hbar\omega_f$. The expression (3.31) can thus be rewritten [Ref. 8.1, Eq. (2.160)].

$$K_{2f,10} = \sum_{ij} \left[\frac{M_{fj}\mathscr{M}_{ji}M_{i0}}{\hbar^2(\omega_1 - \omega_i^{\mathrm{e}} - \mathrm{i}\gamma_i)(\omega_2 - \omega_j^{\mathrm{e}} - \mathrm{i}\gamma_j)} \right]. \tag{3.33}$$

Here we have used the equation for conservation of energy (3.2) and we have made explicit that ω_i^{e} and ω_j^{e} are purely electronic excitation energies.

There are two different domains for (3.33): 1) where only one factor in the energy denominator is resonant and the other factor is far from resonance, the latter may be considered constant and taken outside the summation over resonant states; 2) where the simultaneous variation of the two factors in the denominator is essential.

The second case can be called "double resonance"—or "multiple resonance" in the case of higher orders and more factors in the denominator. The first case clearly leads to results of the same form as those given in (3.15)–(3.30). The only difference is that in (3.33) there are two resonant regions, one which involves ω_1, the other involves ω_2. Hence, with the assumption of constant matrix elements for the quasi-particle interactions, we find

$$K_{2f,10} = A_1\,\chi_{\mathrm{e}}(\omega_1) + A_2\,\chi_{\mathrm{e}}(\omega_2)\,. \tag{3.34}$$

Unlike (3.30), χ_{e} here is the resonant contribution to the susceptibility of only *electronic* states

$$\chi_{\mathrm{e}}(\omega) = \sum_i \left[\frac{M_{0i}M_{i0}}{\hbar(\omega - \omega_i^{\mathrm{e}} - \mathrm{i}\gamma_i)} \right], \tag{3.35}$$

where the summation is restricted to only electronic excited states such that $\omega_i^{\mathrm{e}} \cong \omega$. The constants A_1 and A_2 in (3.34) are effective coupling constants. They are in general unrelated to one another. However, for cases of high symmetry (e.g., for photon polarizations $\boldsymbol{\varepsilon}_1$ and $\boldsymbol{\varepsilon}_2$ along equivalent crystal directions) so that $M_{0i} = M_{fj}$, we find $A_1 \cong A_2^*$. Within the approximations made, (3.34) and (3.35) are equally valid for intermediate states either in the continuum or in the discrete part of the spectrum.

The various possible cases of double resonance, e.g., two discrete levels, one discrete and one continuum levels, etc., become very numerous. The discussion here, if carried in a complete fashion, would involve a lengthy enumeration along the lines followed for one resonant set $|i\rangle$. We therefore reduce our explicit study to two cases of special interest; both involve $|i\rangle$ and $|j\rangle$ in the continuum part of the spectrum $\{E_m\}$.

The first case we want to discuss corresponds to taking the product of matrix elements $(M_{fj}\mathscr{M}_{ji}M_{i0})$ to be, within the manifolds $\{|i\rangle\}$ and $\{|j\rangle\}$, independent of the states $|i\rangle$ and $|j\rangle$. This means that we couple any $|i\rangle$ to any $|j\rangle$ with approximately equal strength in addition to coupling $|i\rangle$ to $|o\rangle$ and $|j\rangle$ to $|f\rangle$ with constant strength, independent of $|i\rangle$ and $|j\rangle$. Under these conditions, which we call independent double resonance,

$$K_{2f,10} \tag{3.36}$$

$$\cong (M_f \mathscr{M} M_0) \left(\sum_i \frac{1}{\hbar(\omega_1 - \omega_i^{\mathrm{e}} - \mathrm{i}\gamma_i)} \right) \left(\sum_j \frac{1}{\hbar(\omega_2 - \omega_j^{\mathrm{e}} - \mathrm{i}\gamma_j)} \right)$$

which can be rewritten as

$$K_{2f,10}(\omega_1) \cong (\mathscr{M}/M_f^* M_0^*)\, \chi_{\mathrm{e}}(\omega_1)\, \chi_{\mathrm{e}}(\omega_2) \quad \text{for} \quad \begin{cases} \omega_i^{\mathrm{e}} \cong \omega_1 \\ \omega_j^{\mathrm{e}} \cong \omega_2 \,. \end{cases} \tag{3.37}$$

We can reinterpret (3.36) by saying that for an independent double resonance, the Raman matrix element is proportional to the product of two susceptibility functions. As an example, this may be the appropriate perturbation approximation for RRS from a molecule in which $|i\rangle$ and $|j\rangle$ are states in the dissociation continuum.

A second example of double resonance is what we call coupled double resonance. It corresponds to states $|i\rangle$ and $|j\rangle$ also both in the continuum of $\{E_n\}$, but $\mathscr{M}_{ij}$ is such that selection rules require it to be non-zero only for unique, well-defined pairs $i \leftrightarrow j$. For each i there is only one j such that $\mathscr{M}_{ij} \neq 0$, and vice versa. If we change notation to make this one-to-one relation explicit $i \to ai$, $j \to bi$,

$$K_{2f,10} = \sum_i \left[\frac{M_{f,bi} \mathscr{M}_{bi,ai} M_{ai,0}}{\hbar^2(\omega_1 - \omega_{ai}^{\mathrm{e}} - \mathrm{i}\gamma_{ai})(\omega_2 - \omega_{bi}^{\mathrm{e}} - \mathrm{i}\gamma_{bi})} \right] \tag{3.38}$$

which, with constant matrix elements assumption, reduces to

$$K_{2f,10} \cong (M_f \mathscr{M}_{ba} M_0) \sum_i \left[\frac{1}{\hbar^2 (\omega_1 - \omega_{ai}^{\mathrm{e}} - \mathrm{i}\gamma_{ai})(\omega_2 - \omega_{bi}^{\mathrm{e}} - \mathrm{i}\gamma_{bi})} \right]. \tag{3.39}$$

Consider now the case where $\omega_{ai}^{\mathrm{e}} = \omega_{bi}^{\mathrm{e}}$, that is the quasi-particle scattering is between the same or degenerate electronic states. By means of the identity

$$\frac{1}{(x-A)(x-B)} \equiv \frac{1}{A-B} \left[\frac{1}{x-A} - \frac{1}{x-B} \right] \tag{3.40}$$

and the Definition (3.35), we may now write

$$K_{2f,10} \cong (M_f \mathscr{M}_{ba}/M_0^*) \frac{\chi_{\mathrm{e}}(\omega_1) - \chi_{\mathrm{e}}(\omega_2)}{\hbar(\omega_2 - \omega_1)}. \tag{3.41}$$

Equation (3.41) shows that $K_{2f,10}$ includes the resonant features of the susceptibility at both the incoming frequency ω_1 and the outgoing frequency ω_2. If, in particular, the function $\chi_{\mathrm{e}}(\omega)$ is smoothly varying between ω_2 and ω_1, (3.41) can also be approximated by

$$K_{2f,10} \cong -(M_f \mathscr{M}_{ba}/M_0^* \hbar) \left[\frac{d\chi_{\mathrm{e}}}{d\omega} \right]_{\omega=\omega_i}. \tag{3.42}$$

In other words, the Raman matrix element for a phonon coupled double resonance in the resonant regime and for smoothly varying $\chi(\omega)$ is proportional to the derivative of the susceptibility function. In (3.41) and (3.42), it is often assumed that $(M_f/M_0^*) = 1$, in which case it is possible to measure directly $\mathscr{M}_{ba}$.

Explicit evaluation of (3.39) for various cases and models follows a standard method of calculation: typically the following replacements are made

$$\begin{aligned} \omega_{ai} &\equiv \omega \\ \omega_{bi} &\equiv \Omega(\omega) - (\omega_1 - \omega_2) \end{aligned} \tag{3.43}$$

which yields

$$K_{2f,10} = (M_f \mathscr{M}_{ba} M_0/\hbar) \int \frac{\varrho_a(E_{ai} = \hbar\omega)\, d\omega}{(\omega_1 - \omega - \mathrm{i}\gamma_a)(\omega_1 - \Omega - \mathrm{i}\gamma_b)}. \tag{3.44}$$

In particular, for a linear relationship between Ω and ω

$$\Omega = \Omega_0 + v\omega\,, \tag{3.45}$$

(3.39) becomes

$$K_{2f,10} \cong (M_f \mathscr{M}_{ba} M_0/\hbar) \int \frac{\varrho_a(\hbar\omega)\, d\omega}{(\omega_1 - \omega - \mathrm{i}\gamma)(\omega_1 - \Omega_0 - v\omega - \mathrm{i}\gamma)}\,. \tag{3.46}$$

The example (3.41) is a special case of (3.46) and corresponds to

$$\omega_2 \equiv \omega_1 - \omega_0 = \omega_1 - \Omega_0 \tag{3.47}$$

and $v = 1$. If in this case we take $\varrho_a(\hbar\omega)$ as given by (3.27), the well-known results which apply to one-phonon RRS [3.4–8] follow for $\omega_c \to \infty$ (large cut-off frequency)

$$K_{2f,10}(\omega_1, \omega_2)$$
$$\cong (M_f \mathscr{M}_{ab} M_0 \alpha) \times \begin{cases} (\mathrm{i}\pi/\omega_0)\,[(1/a) - (1/b)], & d = 1; \\ (1/\omega_0) \ln [a^2/b^2], & d = 2; \\ (\mathrm{i}\pi/\omega_0)\,[a - b], & d = 3\,. \end{cases} \tag{3.48}$$

where a is defined by (3.29) and

$$b \equiv (\omega_2 - \Delta - \mathrm{i}\gamma)^{1/2} = (\omega_1 - \omega_0 - \Delta - \mathrm{i}\gamma)^{1/2}\,. \tag{3.49}$$

These results (3.48) can also be obtained from direct application of (3.39), (3.41), and (3.30) to the specific density of states (3.27) and (3.28). We see again that in three dimensions there is enhancement but no divergence, and that there are $(\omega_1^{-1/2})$ and $(\ln \omega_1)$ type divergences for $\omega_1 = \Delta$ and $\omega_1 = \Delta + \omega_0$ in one and two dimensions, respectively. These divergences are only such in the limit $\gamma \to 0$.

In actual examples of RRS it is often a good approximation to consider the Raman matrix elements to be a linear combination of three types of terms: one resonant one of the form similar to (3.34), a second resonant one of the form similar to (3.41), and a third constant, non-resonant term. We thus find to first order in $\mathscr{M}_{ij}$

$$\begin{aligned} K_{2f,10} = {} & (1/2)(A_1 + A_2)\,[\chi_e(\omega_1) + \chi_e(\omega_2)] \\ & + B[\chi_e(\omega_1) - \chi_e(\omega_2)]\,(\omega_1 - \omega_2)^{-1} + C\,. \end{aligned} \tag{3.50}$$

Typical expressions for $\chi_e(\omega)$ are given in (3.28). The coefficient B in (3.50) depends both on the in-the-shell scattering (3.41) and the out-of-the-shell scattering (3.34) via $(A_1 - A_2)$. Note that strictly χ_e is just the resonant contribution to the electronic susceptibility. However, if other contributions to the total susceptibility χ can be considered constant, then (3.50) holds with χ_e replaced by the total $\chi(\omega)$ with a reinterpretation of the constant C.

It is useful to remark that the features of the RRS associated with $\chi(\omega_1)$ are called resonances in the incoming channel, those associated with $\chi(\omega_2)$, resonances in the outgoing channel. In each case considered thus far, the assumptions have led us to resonances which could be described solely in terms of the in and out channels.

There may be other resonances, in general, which are associated only with multiple resonances. This may be seen in (3.46). If in the one-to-one correspondence $i \leftrightarrow j$ of the two successive resonant states the energy E_i increases while E_j decreases and vice versa, a very large enhancement of $K_{2f,10}$ takes place: this extra enhancement does not take place if E_i and E_j change in the same direction. In order to see this effect, it is instructive to take in (3.46) the approximation of constant density of states $\varrho_a(\hbar\omega) = \varrho_a$; the resulting integral evaluated between $(-\infty)$ and $(+\infty)$ is completely straightforward and yields

$$K_{2f,10} \tag{3.51}$$

$$\cong (M_f \mathcal{M}_{ba} M_0 \varrho_a/\hbar) \times \begin{cases} 0 & \text{if} \quad v > 0\,, \\ 2\pi i(1+|v|)^{-1}[(\Omega_1 - \omega_1) + i\gamma]^{-1} & \text{if} \quad v < 0\,; \end{cases}$$

where

$$\Omega_1 \equiv \Omega_0/(1+|v|)\,.$$

Inspection of (3.51) tells us that in this rather unrealistic model there is phase cancellation (total cancellation in this case) when $v > 0$, and that there is an enhancement resulting in a Lorentzian line shape when $v < 0$. The cancellation for $v > 0$ arises from the negative interference of the two poles in (3.46). These two types of resonance behavior are a general feature of a double resonance as was pointed out by Yacoby [3.67].

In order to look into a situation in which we study the application of formula (3.50) to RRS in a crystal, and the appearance of the additional structure predicted in (3.51), let us consider a one-dimensional semiconductor crystal. We assume that an effective-mass electron band and a flat (infinite mass) hole band, as shown in Fig. 3.4, describe the electronic states. The energy of the excitations is only a function of the electron momentum $\hbar k$.

$$E_{ai}(k) \equiv \hbar\omega_{ai} = \hbar\Delta + \hbar^2 k^2/2m\,, \tag{3.52}$$

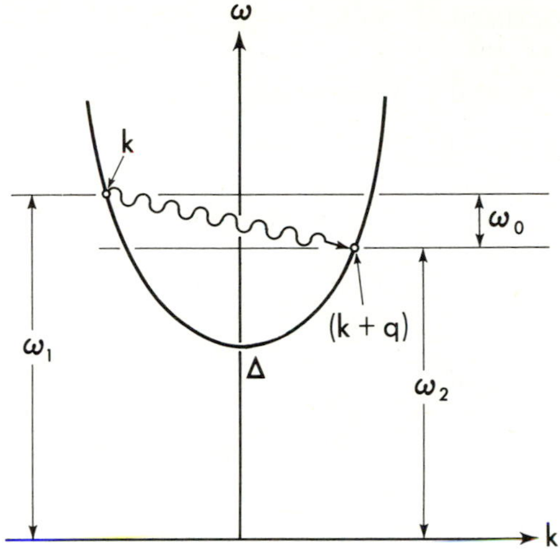

Fig. 3.4. Diagram of first-order RRS corresponding to Fig. 3.3a on an energy-momentum diagram in a crystal. The parabola indicate the dispersion relation of the electronic intermediate state excitation. The scattering shown satisfies the double resonance criteria and is for the maximum wave vector $\boldsymbol{q}$ consistent with simultaneous resonance of the two intermediate states

and the second resonant state $|j\rangle$ is of the same form, but it contains a phonon of well-known frequency $\omega_0(q)$ and wave vector $(-q)$. Wave vector conservation then requires (see Fig. 3.4)

$$E_{bi} \equiv \hbar\omega_{bi} = \hbar(\omega_0 + \Delta) + \hbar^2(k+q)^2/2m\,. \tag{3.53}$$

With the assumption of constant matrix elements $(M_0 \mathcal{M}_{ab} M_f)$, the Definitions (3.52) and (3.53) and with a cut-off in k-space k_c such that $k_c \gg q$, (3.39) reduces now to

$$K_{2f,10}(\omega_1, \omega_2) \cong (M_f \mathcal{M}_{ab} M_0/\hbar^2) \cdot \int\limits_{-\infty}^{\infty} \frac{(L/2\pi)\,dk}{[\omega_1 - \Delta - \mathrm{i}\gamma - \hbar k^2/2m]\,[\omega_2 - \Delta - \mathrm{i}\gamma - \hbar(k+q)^2/2m]} \tag{3.54}$$

where L is a length over which the system is normalized. This integral can be straightforwardly evaluated to yield

$$K_{2f,10}(\omega_1, \omega_2) = (M_f \mathcal{M}_{ab} M_0)(4\pi \mathrm{i} m^2/\hbar^4)(L/2\pi) \cdot \left\{ \frac{1}{\kappa_1[(\kappa_1+q)^2 - \kappa_2^2]} + \frac{1}{\kappa_2[(\kappa_2-q)^2 - \kappa_1^2]} \right\} \tag{3.55}$$

where we have defined

$$\begin{aligned}
&\hbar\kappa_1^2/2m \equiv \omega_1 - \Delta - \mathrm{i}\gamma, \quad \mathrm{Im}\{\kappa_1\} > 0\,,\\
&\hbar\kappa_2^2/2m \equiv \omega_2 - \Delta - \mathrm{i}\gamma = \omega_1 - \omega_0 - \Delta - \mathrm{i}\gamma, \quad \mathrm{Im}\{\kappa_2\} > 0\,.
\end{aligned} \tag{3.56}$$

A simple transformation of (3.55) gives for the Raman matrix element

$$\begin{aligned}
&K_{2f,10}(\omega_1, \omega_2) = (-M_f \mathcal{M}_{ab} M_0)(2\pi \mathrm{i} m^2/\hbar^4)(L/2\pi)\\
&\frac{1}{\kappa_1\kappa_2}\left[\frac{1}{\kappa_1+\kappa_2+q} + \frac{1}{\kappa_1+\kappa_2-q}\right].
\end{aligned} \tag{3.57}$$

This expression exhibits one-dimensional type of resonances $(\omega_1 - \omega_{\mathrm{R}})^{-1/2}$, with a suitable broadening, at resonance frequency $\omega_{\mathrm{R}} = \Delta$, at $\omega_{\mathrm{R}} = \Delta + \omega_0$ and, if

$$\hbar q^2/2m > \omega_0 \tag{3.58}$$

also a Lorentzian resonance for

$$(2m/\hbar)\,[(\omega_{\mathrm{R}} - \Delta)^{1/2} + (\omega_{\mathrm{R}} - \Delta - \omega_0)^{1/2}] = q\,. \tag{3.59}$$

This can be interpreted in the following way: there are, in this one-dimensional case, always two threshold resonances at $\omega_1 = \Delta$ (incoming channel) and $\omega_2 = \Delta$ (outgoing channel). For any q which satisfies (3.58) there is an additional resonance at the value of ω_1 which satisfies (3.59). Conversely, for any $\omega_1 > \Delta + \omega_0$ there is one value of q which satisfies (3.59) and gives an extra resonance. This is, of course, the effect discussed also in (3.50) since, as seen in Fig. 3.4, for large enough (and fixed) q the state on the left decreases in energy when the state on the right increases its value.

A similar double resonance effect occurs in two and three dimensions. The conditions for resonance are identical—(3.58) and (3.59)—but the resonance line shape is changed to inverse-square-root for $d = 2$ and logarithmic for $d = 3$. In general, a double resonance at an ordinary critical point in d dimensions has the form of a simple resonance in $(d-1)$ dimensions.

It is important to notice that the conditions discussed above are satisfied only for wave vectors much greater than photon wave vectors. Thus for typical first-order RRS in which the momenta are determined by wave vector conservation

$$|\boldsymbol{q}| = |\boldsymbol{k}_1 - \boldsymbol{k}_2| \sim \omega_1/c$$

the Raman matrix element is only weakly dependent on the wave vectors $\boldsymbol{k}_1$, $\boldsymbol{k}_2$, and $\boldsymbol{q}$. Such matrix element may be expanded in powers of the wave vector q. For example, (3.57) becomes

$$K_{2f,10} \cong (M_f \mathscr{M}_{ab} M_0)(\pi \mathrm{i} m^2/\hbar^3)(L/2\pi) \cdot \frac{1}{(\omega_1-\omega_2)}\left[\frac{1}{\kappa_1}-\frac{1}{\kappa_2}\right]\left[1+\frac{1}{2}\,\frac{q^2}{(\kappa_1+\kappa_2)^2}+\cdots\right]. \tag{3.60}$$

The first term, independent of q, is just the one-dimensional example of (3.41). This simple example shows that in general the relation (3.50) is expected to describe adequately first-order phonon RRS in crystal, where we neglect the wave vector dependence. The second term in in (3.60)—which may be important in RRS as discussed in Section 3.3—introduces a modified form of the resonance: singularities occur only at the in-channel and out-channel resonances, but the overall line shape is modified by a $(\kappa_1+\kappa_2)^2$ factor. It is only when q can be comparable to $(\kappa_1+\kappa_2)$ that new singularities occur; as is seen below, these new resonance features are manifested clearly in second-order, two-phonon scattering.

Finally we look into case III) in which there is a continuum of final states. Our primary example is two-phonon scattering in crystals. If we consider anharmonic effects in which two phonons are emitted at one vertex (four-quasi-particle vertex), all matrix elements exhibit the same resonance line shape as a function of incident photon frequency as in the one-phonon case. However, when we consider the case of consecutive emission as in Fig. 3.3b, the higher-order matrix element (3.32) is essential; in that case the resonance line shape may be drastically altered. Because of the one-to-one correspondence between intermediate and final states, with the assumption of constant matrix elements (3.32) becomes

$$K_{2f,10} = (M_{fc}\mathscr{M}_{cb}\mathscr{M}_{ba}M_{a0}/\hbar^3) \cdot \sum_i \left[\frac{1}{(\omega_1-\omega_{ai}-\mathrm{i}\gamma_{ai})(\omega_1-\omega_{bi}-\mathrm{i}\gamma_{bi})(\omega_1-\omega_{ci}-\mathrm{i}\gamma_{ci})}\right]. \tag{3.61}$$

Evaluation of (3.61) involves some extra complications, but in general it follows the general procedure described for double resonances. For example, in the case depicted in Fig. 3b, if the first phonon is described by a wave vector $\boldsymbol{q}_1$ and the second one by a wave vector $\boldsymbol{q}_2$, conservation of quasi-momentum requires

$$\begin{aligned} &\boldsymbol{q}_1+\boldsymbol{q}_2 \cong 0 \\ &\boldsymbol{q}_1 \approx -\boldsymbol{q}_2 = \boldsymbol{q}\,. \end{aligned} \tag{3.62}$$

If, in addition, the two phonons belong to the same polarization branch ν

$$\omega_\nu(\boldsymbol{q}) = \omega_\nu(-\boldsymbol{q}) \equiv \omega_q\,, \tag{3.63}$$

conservation of energy is taken into account by

$$\omega_2 - \omega_1 = 2\omega_q\,, \tag{3.64}$$

and we write for intraband scattering

$$\begin{aligned} \omega_{ai} &= \Delta + \hbar\, k^2/2m\,,\\ \omega_{bi} &= \Delta + \hbar(\boldsymbol{k}+\boldsymbol{q})^2/2m + \omega_q\,,\\ \omega_{ci} &= \Delta + \hbar\, k^2/2m + 2\omega_q\,, \end{aligned} \tag{3.65}$$

then (3.61), with constant matrix elements, reduces to

$$\begin{aligned} &K_{2f,10} \\ &= -(M_f \mathscr{M}^2 M_0)(8m^3/\hbar^6) \int \frac{(L/2\pi)^d\, d^d k}{(k^2-\kappa_1^2)\,[(\boldsymbol{k}+\boldsymbol{q})^2-\kappa_2^2]\,(k^2-\kappa_3^2)}\,, \end{aligned} \tag{3.66}$$

where

$$\begin{aligned} \hbar\kappa_1^2/2m &\equiv \omega_1 - \Delta - \mathrm{i}\gamma\\ \hbar\kappa_2^2/2m &\equiv \omega_1 - \Delta - \omega_q - \mathrm{i}\gamma\\ \hbar\kappa_3^2/2m &\equiv \omega_1 - \Delta - 2\omega_q - \mathrm{i}\gamma\\ \operatorname{Im}\kappa_i &> 0\,. \end{aligned} \tag{3.67}$$

This can immediately be reduced to

$$\begin{aligned} K_{2f,10} &= -(M_f \mathscr{M}^2 M_0)(8m^3/\hbar^6)\,[1/(\kappa_1^2-\kappa_3^2)]\\ &\cdot \int \left\{ \frac{(L/2\pi)^d\, d^d k}{(k^2-\kappa_1^2)\,[(\boldsymbol{k}+\boldsymbol{q})^2-\kappa_2^2]} - \frac{(L/2\pi)^d\, d^d k}{(k^2-\kappa_3^2)\,[(\boldsymbol{k}+\boldsymbol{q})^2-\kappa_2^2]} \right\} \end{aligned} \tag{3.68}$$

which are two integrals similar to (3.54)—identical in fact in the case of one dimension $d=1$. We can therefore see that (3.68) shows resonant structure, for fixed q, at values of ω_1 corresponding to Δ, $\Delta+\omega_q$, $\Delta+2\omega_q$

and also at

$$(2m/\hbar)^{1/2}[(\omega_1-\Delta)^{1/2}+(\omega_1-\omega_q-\Delta)^{1/2}]=q\,,$$
$$(2m/\hbar)^{1/2}[(\omega_1-\omega_q-\Delta)^{1/2}+(\omega_1-2\omega_q-\Delta)^{1/2}]=q\,, \tag{3.69}$$
$$\hbar q^2/2m>\omega_q\,.$$

In other words, there are five possible resonant peaks for a fixed wave vector q.

The resonance at Δ and $(\Delta+2\omega_q)$ are the familiar in-channel and out-channel resonances with the same functional form as the susceptibility χ. The resonance at $(\Delta+\omega_q)$ has the same form; it is an intermediate channel resonance and we note that its strength is reduced by the cancellation between the two terms in the integral in (3.68). The last two resonances (3.69) are double resonances not associated with any structure in the susceptibility. If the wave vector of the phonons can be kept fixed in the experiment (e.g., if the dispersion in ω_q is large enough so that a given portion of the spectrum can be assigned to a narrow range of $\boldsymbol{q}$ vectors), then these double resonance enhancements can in principle be measured. Of great interest is the fact that they are more singular than the ordinary threshold resonances and they occur only for phonons with $|\boldsymbol{q}|$ larger than a minimum value given by (3.69).

It should be noted that for fixed ω_1, near resonance, the Raman lineshape as a function of ω_2 is determined by the matrix element $K_{2f,10}$. In the present examples, for constant $\mathcal{M}$, $|K_{2f,10}|^2$ decreases as q^{-4} for large q. Even in three dimensions this decrease in the matrix element overcomes the q^2 increase in the density of final states and consequently produces a peak in the two-phonon Raman intensity. That is, the actual two-phonon Raman spectrum near resonance is distorted from the usual monotonic square-root line shape and exhibits a peak for phonons with $\boldsymbol{q}$ and ω_q such that the double resonance conditions are satisfied. In cases of lower dimensionality the peak should be more pronounced. In addition, if the electron-phonon matrix element $\mathcal{M}$ varies with q as in the Fröhlich interaction [3.9–12] where $\mathcal{M}\sim q^{-1}$ for large q, then $K_{2f,10}$ may be sharply peaked at wave vectors q large compared to photon wave vectors but often small compared to wave vectors at the Brillouin zone boundaries [3.81].

3.3. Hamiltonians, Symmetry, and Selection Rules

In this section we briefly discuss the various Hamiltonians used in studying the RRS phenomenon, and mention their symmetry properties and the selection rules derived from them.

The most important terms in the Hamiltonian—the one in fact which drives the RRS phenomenon—is $\mathscr{H}_{\mathrm{MR}}$, the interaction between matter and radiation. This term can be written [3.1]

$$\mathscr{H}_{\mathrm{MR}} = (|e|/mc) \sum_{\alpha} \boldsymbol{p}_{\alpha} \cdot \boldsymbol{A}(\boldsymbol{r}_{\alpha}) \tag{3.70}$$

where the summation extends over all the electrons α in the material system, $\boldsymbol{p}_{\alpha}$ is the momentum of the αth electron and $\boldsymbol{A}(\boldsymbol{r}_{\alpha})$ is the quantized vector potential of the radiation field measured at the coordinate of the αth electron

$$\begin{aligned} \boldsymbol{A}(\boldsymbol{r}) = {} & [2\pi c^2 \hbar / V_{\mathrm{R}}]^{1/2} \sum_{\lambda} \omega_{\lambda}^{-1/2} \\ & \cdot \boldsymbol{\varepsilon}_{\lambda} \{a_{\lambda} \exp(\mathrm{i}\boldsymbol{k}_{\lambda}\boldsymbol{r}) + a_{\lambda}^{\dagger} \exp(-\mathrm{i}\boldsymbol{k}_{\lambda}\boldsymbol{r})\}\,. \end{aligned} \tag{3.71}$$

In (3.71) V_{R} is the volume over which the radiation field is normalized, λ indicates a photon mode of wavevector $\boldsymbol{k}_{\lambda}$ and polarization $\boldsymbol{\varepsilon}_{\lambda}$, and a_{λ} and $a_{\lambda}^{\dagger}$ are operators which destroy and create, respectively, one photon of mode λ. The only non-vanishing matrix elements of a_{λ} and $a_{\lambda}^{\dagger}$ are

$$\langle n_{\lambda} | a_{\lambda}^{\dagger} | n_{\lambda} - 1 \rangle = \langle n_{\lambda} - 1 | a_{\lambda} | n_{\lambda} \rangle = (n_{\lambda})^{1/2}\,. \tag{3.72}$$

In (3.70), we have neglected terms proportional to $|\boldsymbol{A}(\boldsymbol{r}_{\alpha})|^2$ which are not resonant. In taking matrix elements of $\mathscr{H}_{\mathrm{MR}}$, the radiation field provides a factor which is either zero or $(n_{\lambda})^{1/2}$; the material part of the system contributes another factor which is either

$$\boldsymbol{P}_{12,\lambda} = \left\langle 1 \left| \sum_{\alpha} p_{\alpha} \exp(\mathrm{i}\boldsymbol{k}_{\lambda}\boldsymbol{r}_{\alpha}) \right| 2 \right\rangle \tag{3.73}$$

for photon absorption, or

$$\boldsymbol{P}_{12,\lambda}^{\dagger} = \left\langle 1 \left| \sum_{\alpha} p_{\alpha} \exp(-\mathrm{i}\boldsymbol{k}_{\lambda}\boldsymbol{r}_{\alpha}) \right| 2 \right\rangle \tag{3.74}$$

for photon emission.

Since the wavelength of light is large compared with atomic dimensions, the matrix elements can be evaluated by making a multipole expansion of the exponential factors in (3.73) and (3.74). For example, the fundamental absorption matrix element in (3.10) is given by

$$\begin{aligned} \langle 0i | \mathscr{H}_{\mathrm{MR}} | \omega_1 o \rangle = {} & (|e|/mc)\,[2\pi c^2 \hbar / V_{\mathrm{R}}]^{1/2} \omega_1^{1/2} \\ & \cdot \sum_{\alpha} \{\boldsymbol{\varepsilon}_1 \cdot \langle i | \boldsymbol{p}_{\alpha} | o \rangle + \mathrm{i}\boldsymbol{\varepsilon}_1 \cdot \langle i | \boldsymbol{p}_{\alpha}\ \ \boldsymbol{r}_{\alpha} | o \rangle \cdot \boldsymbol{k}_1 + \cdots\}\,, \end{aligned} \tag{3.75}$$

The first term in the expansion in (3.75) is the electric dipole, the second the electric quadrupole and magnetic dipole, etc. Usually [3.1, 9] only the electric dipole term is retained since it is typically $\sim\alpha^{-1} = 137$ times larger than the higher-order terms. Under these conditions $\langle 0i|\mathscr{H}_{\mathrm{MR}}|\omega_1 o\rangle$ is just a matrix element of the $\boldsymbol{\varepsilon}_\lambda$ component of the total momentum operator; it depends on the photon polarization $\boldsymbol{\varepsilon}_\lambda$ but not on the direction of propagation or wavelength of the photon.

Within the dipole approximation $\mathscr{H}_{\mathrm{MR}}$ does not couple states of the same parity. If there is a center of inversion this limits the states which can serve as intermediate states in RRS. In addition, the bilinear dependence of $K_{2f,10}$ on $\mathscr{H}_{\mathrm{MR}}$ requires the final state to have the same parity as the ground state, i.e. $|f\rangle$ cannot be a state of odd parity. In general, the bilinear dependence of $K_{2f,10}$ on polarization vectors can be used to define a Raman tensor $\underset{\sim}{\boldsymbol{R}}$ which determines the matrix element for any polarization directions of the incoming and outgoing photons

$$K_{2f,10} = \boldsymbol{\varepsilon}_1 \cdot \underset{\sim}{\boldsymbol{R}} \cdot \boldsymbol{\varepsilon}_2 \,. \tag{3.76}$$

The tensor $\underset{\sim}{\boldsymbol{R}}$ is a function of the photon frequencies ω_1 and ω_2 and the properties of the final state $|f\rangle$. The possible non-zero components of $\underset{\sim}{\boldsymbol{R}}$ corresponding to different symmetries of the state $|f\rangle$ and each crystal class in the dipole approximation have been extensively discussed and tabulated by LOUDON [3.9]. All this information is implicitly contained in the matrix elements M and $\mathscr{M}$ of the last section.

The electric quadrupole term in (3.75) gives rise to different selection rules. It does couple states of the same parity. If in the general expression (3.10) one electric quadrupole and one electric dipole matrix elements are used for $\mathscr{H}_{\mathrm{MR}}$, then the Raman final state $|f\rangle$ may have opposite parity to $|o\rangle$, i.e., it may be odd. For this particular approximation we may define two third-rank Raman tensors $\underset{\sim}{\boldsymbol{R}}^1(f,\omega_1,\omega_2)$ and $\underset{\sim}{\boldsymbol{R}}^2(f,\omega_1,\omega_2)$ such that

$$K_{2f,10} = \sum_{\alpha,\beta,\gamma=1}^{3} \varepsilon_{1\alpha}\varepsilon_{2\beta}(R^1_{\alpha\beta\gamma}k_{1\gamma} + R^2_{\alpha\beta\gamma}k_{2\gamma}) \,. \tag{3.77}$$

These definitions can be straightforwardly extended to higher multipoles.

The rules we have just described are, in the case of RRS, "soft" selection rules. One reason for their breakdown lies in the very nature of the resonance enhancement mechanism: the enhancement close to resonance of a "forbidden" electric quadrupole or magnetic dipole transition may be in general larger than the reduction factor $\sim(137)^2$ characteristic of the ratio of dipole-to-quadrupole intensities, and therefore normally "forbidden" lines may appear with an unusually strong intensity. In other words, RRS intensities of a "forbidden" line may be (and frequently are) stronger than ordinary "allowed" lines off-resonance at the same incident frequency. This breakdown of the selection rules also applies for zero-intensity polarization modes of allowed Raman lines in forbidden polarization configurations.

A second reason for the breakdown of the normal selection rules is that often in RRS the intermediate states $|i\rangle$ have a large spatial extent a, e.g., Wannier excitons. In these cases the expansion parameter (ka) may be large: this leads to spatial dispersion in both the susceptibility and the Raman matrix elements. Examples are discussed in Section 3.4.

The discussion of the Hamiltonian related to the material system $\mathscr{H}_{\mathrm{M}}$ depends, of course, on the system under consideration and cannot be sensibly discussed in a general fashion without further specification. In most common cases $\mathscr{H}_{\mathrm{M}}$ consists of several terms which include:

a) An electronic part which represents the quasi-particles (electrons and holes in a solid) or single-particle excitations of the system.

b) An electron-electron interaction term which is responsible for the formation of excitons and other more involved many-particle and collective modes.

c) A nuclear motion term which describes phonons in a solid and rotational, translational and vibrational modes in a molecule or a liquid.

d) A spin (magnon) Hamiltonian which describes the magnetic properties of a solid or liquid in terms of localized atomic or molecular spins.

e) A collection of interaction terms: electron-phonon, electron-magnon, phonon-phonon, phonon-magnon, etc.

Which ones of these terms are considered "basic" and which ones are considered "perturbations" depends crucially on the nature of the problem under study. The most commonly seen cases are those involving phonons in an insulating crystal. There the "unperturbed" part of $\mathscr{H}_{\mathrm{M}}$ consists of the electronic contribution (electrons and holes), the phonon contribution and in some instances the electron-hole interaction (excitons). The "perturbation" part of $\mathscr{H}_{\mathrm{M}}$ is the electron-phonon interaction which in general can be written as

$$\begin{aligned}\mathscr{H}_{\mathrm{M}}^{1}(e-p) &= V_{\mathrm{c}}^{-1/2}\\ &\cdot \sum_{\nu\alpha} \{\theta(\nu, \boldsymbol{r}_{\alpha}) \exp(\mathrm{i}\boldsymbol{q}_{\nu}\boldsymbol{r}_{\alpha})\, b_{\nu} + \theta^{*}(\nu, \boldsymbol{r}_{\alpha}) \exp(-\mathrm{i}\boldsymbol{q}_{\nu}\boldsymbol{r}_{\alpha})\, b_{\nu}^{\dagger}\}\,,\end{aligned} \tag{3.78}$$

where $\boldsymbol{r}_{\alpha}$ is the coordinate of the αth electron, V_{c} is the volume of the crystal, ν is the phonon mode of wavevector $\boldsymbol{q}_{\nu}$, and the operators b_{ν} and $b_{\nu}^{\dagger}$ destroy and create, respectively, a phonon of mode ν. The quantity

$\theta(v, \boldsymbol{r})$ is a function of the electron coordinate and depends on the wavevector and polarization mode of the phonon, its properties are essentially determined by the spatial range of the interaction. For a given optic phonon branch in an insulators, θ can be [3.13, 14] the usual deformation potential (DP), θ_{DP}, or the so-called Fröhlich (F) interaction [3.9–12], θ_{F}. In the DP approximation, $\theta = \theta_{\mathrm{DP}}$, the potential is short-range; in that case, all wave vector dependence in the calculation of the electron-phonon matrix elements can be ignored. Then the selection rules are determined by the symmetry of the $\boldsymbol{q}=0$ phonon and the $\mathscr{H}_{\mathrm{MR}}$ matrix elements—and their multipole expansion—as discussed above.

The F interaction on the other hand is long range and may be written in $\boldsymbol{q}$ space as [Ref. 8.1, Sect. 2.2.8].

$$\theta_{\mathrm{F}}(v, \boldsymbol{r}) = (\gamma_v/q)\, \mathrm{e}^{\mathrm{i}\boldsymbol{q}\cdot\boldsymbol{r}}\,, \tag{3.79}$$

where γ_v is determined by the macroscopic electric field which accompanies the phonon. The coefficient γ_v is zero unless the phonon mode v has a longitudinal component. The importance of (3.79) is that, in addition to the usual interband contribution, it also provides an additional intraband matrix element. When inserted in the Raman matrix element the first term in the expansion in powers of q, proportional to (q^{-1}), always cancels identically due to equal and opposite sign contributions of the electron and hole created by the photon. This merely results from the fact that a photon can only create neutral (uncharged) excitons in the solid. The linear term describes the interband electrooptic contribution. The interesting term, the quadrupole in the multipolar expansion of the exponential in (3.79), is of the form

$$\theta_{\mathrm{F}} \rightarrow -(\gamma_v/q)\,(\boldsymbol{q}\cdot\boldsymbol{r})^2\,,$$

which is linear in q. This term does not cancel when both the electron and hole contributions are taken into account, and it may provide a strong q dependent contribution to the Raman matrix element [3.10–12]. In this case, if both $\mathscr{H}_{\mathrm{MR}}$ matrix elements are electric dipole allowed the Raman matrix element may be written in the form

$$K_{2f,10} = \sum_{\alpha\beta\gamma=1}^{3} P^1_{\alpha\beta\gamma}\varepsilon_\alpha\varepsilon_\beta q_\gamma\,. \tag{3.80}$$

Since in first-order phonon RRS

$$\boldsymbol{q} = \boldsymbol{k}_1 - \boldsymbol{k}_2\,,$$

this predicts a scattering cross section which depends crucially on the scattering angle of the photons.

As a specific example of this effect, consider the one-dimensional case given by (3.60). Now $\mathcal{M}_{ab} \sim (\gamma/q)$ and the matrix element takes the form

$$K_{2f,10} = \frac{1}{2}(\gamma a)(q a)\left(\frac{M_f}{M_0^*}\right)\frac{\chi_e(\omega_1) - \chi_e(\omega_2)}{\hbar(\omega_1 - \omega_2)}, \tag{3.81}$$

where a is a characteristic length

$$\begin{aligned} a &= (\kappa_1 + \kappa_2)^{-1} \\ &= (\hbar/2m)^{1/2}[(\omega_1 - \Delta - \mathrm{i}\gamma)^{1/2} + (\omega_2 - \Delta - \mathrm{i}\gamma)^{1/2}]^{-1}. \end{aligned} \tag{3.82}$$

The matrix element (3.81) exhibits the same in and out resonances as does (3.41), the only difference being the prefactors in (3.81). The frequency dependence of the characteristic length a given in (3.82) shows an obvious resonance character. The maximum value of $|a|$ is obtained for $\omega_2 < \Delta < \omega_1$. For an electron band mass of 0.1 m_e (light mass) and a phonon energy 0.025 eV,

$$|a|_{\max} \cong (\hbar/2m\omega_0)^{1/2} \cong 35\ \text{Å}.$$

With this large characteristic length

$$qa = 2\pi a/\lambda \sim 0.2$$

and we therefore expect large wave-vector-dependent interactions. The long range nature of the Fröhlich interaction is evidenced in (3.81), where we see that the Fröhlich coupling energy (γa) also scales with the characteristic length a. For this reason the Fröhlich interaction is expected to contribute greatly to RRS near critical points in polar materials. Similar arguments apply to RRS near large exciton states, where the matrix element has the form (3.81), but with a having a frequency dependence different from (3.87). In that case, near an exciton resonance a becomes the exciton radius [3.10–12].

3.4. Discussion of Specific Cases

The previous sections have been devoted to laying the theoretical foundations for understanding the interaction of light with matter. From the general theory of the interaction of matter and radiation systems, we extracted the cross sections for inelastic light scattering and examined the cross sections in cases where the photons are at or near resonance with excited electronic states of the matter. The natural classification scheme for such resonance Raman scattering (RRS) was shown to be based upon the division into discrete and continuum states of both the final and the resonant intermediate states of the matter.

In this section we discuss specific experimental and theoretical examples of RRS. The examples are classified according to the scheme based on discrete vs. continuum states rather than any divisions based on type of material. Thus, we group together under each category experimental and theoretical work on scattering from perfect crystals, impurity states in crystals, and molecules in the gaseous phase. The emphasis here is upon scattering from solids. However, understanding the basic elements of RRS is facilitated by drawing together the common features of RRS from each possible source. It is, of course, impossible to discuss all relevant experiments and theories in detail; hence, we attempt to choose only those examples which best illustrate the salient features of each category.

We follow the diagrammatic approach of the previous sections to illustrate and discuss the resonance interactions. Definitions of the diagrammatic symbols are given in Fig. 3.1 and the basic diagrams for Raman scattering are shown in Fig. 3.2. In Section 3.2 theoretical analysis has been presented for several interesting cases. In this section we relate experiments to those theoretical results in many cases. In other instances, it is essential to sketch extensions of the theoretical formulations in order to interpret specific examples.

3.4.1. The Approach to Resonance

In Section 3.2 general expressions have been given for the Raman tensor in terms of the matrix elements of the material-radiation interaction Hamiltonian. The expressions were compared with the similar ones for the linear dielectric susceptibility, cf. (3.10) and (3.11), where it is clear that the two expressions involve the same intermediate states of the material system, but with different matrix element. The comparison and contrast of $K(\omega)$ and $\chi(\omega)$ is one of the primary techniques for interpreting RRS in the present subsection.

Consider the transparent region of photon frequencies below all electronic excitation frequencies in an insulator. As is pointed out following (3.11), $\chi(\omega)$ involves positive definite squares of matrix elements divided by energy denominators which, in this case, are all positive. Hence, $\chi(\omega)$ is a monotonically increasing function of photon frequency ω. The Raman tensor, on the other hand, involves products of different

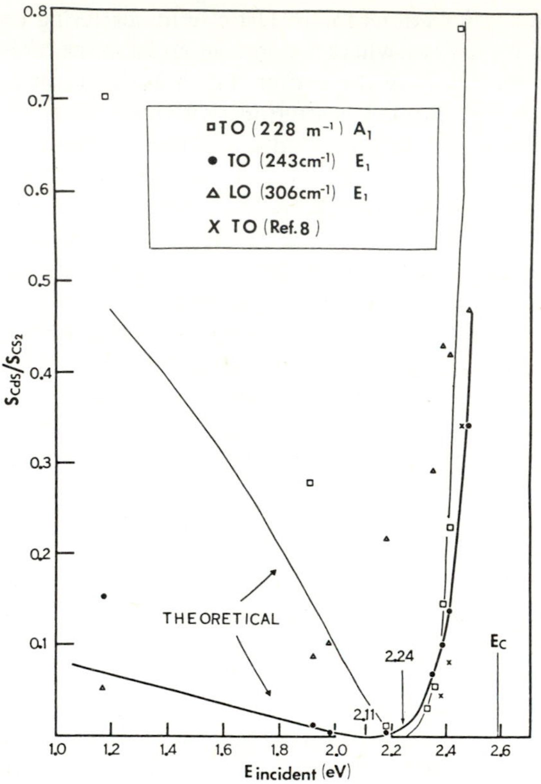

Fig. 3.5. The Raman scattering efficiencies of CdS (80 K) normalized to CS_2 as a function of incident photon energy. E_c is the electron-energy gap of CdS. The solid curves are computed from a simple dispersion formula discussed in the text. The data of Ref. [8] of [3.15] (normalized at 2.47 eV) are included. (From RALSTON et al. [3.15])

matrix elements which are in general complex and may have any phase. Consequently, even in the transparent region of an insulator, the Raman tensor may have interferences so that the Raman cross section may have a non-monotonic dependence upon ω. In the frequency region far from any resonances, such structure in the Raman cross section is essentially independent of the detailed nature of the intermediate states. Therefore, before proceeding to examples of the detailed resonances in $K(\omega)$, we discuss the overall resonance behavior in the the transparent region.

There are many examples of either monotonic or non-monotonic variations in Raman cross sections as a function of light frequency ω. Perhaps the best example in which a variety of different resonance behaviors has been observed is scattering from the various phonons in CdS. The first observations of a null point in the cross section was that of RALSTON et al. [3.15]. Results from their work are shown in Fig. 3.5, from which we see that the two transverse optic (TO) phonons have similar, but not identical, cancellation in the cross section at a photon energy about 0.8 of the band gap. In contrast, the longitudinal (LO) phonon has a monotonically increasing intensity. They have interpreted their cross sections as due to the interference of a constant term with a frequency dependent term representing the contribution of the lowest bands. For the dispersion they chose K of the form of (3.50) with $A_1 = A_2 = 0$ and $[\chi(\omega_1) - \chi(\omega_2)]$ given by the free electron expression (3.48) for 3 dimensions. The adjusted theoretical curves are shown in Fig. 3.5. The predicted curves are not sufficiently sensitive to the form of the resonance enhancement to distinguish between different types of band or exciton intermediate states. However, the analysis establishes the qualitative fact that for the A_1 and E_1 TO phonons the matrix elements in (3.10) change sign between the lowest band and some average of the higher bands. Furthermore, at long wavelengths the higher bands dominate the scattering intermediate states. On the other hand, no cancellation occurs for the LO phonon. Since it differs from the TO only by the presence of a macroscopic electric field, the additional Fröhlich interaction [3.9–12] between the electrons and LO phonons caused by the electric field must qualitatively change the matrix elements in (3.10) so that there is apparently no cancellation effect.

Similar observations have been made by DAMEN and SCOTT [3.16] for the lower frequency E TO mode in CdS in which there is a cancellation. Also, CALENDER et al. [3.17] have reported extensive measurements in CdS and ZnO. By comparing LO and TO in ZnO, they [3.17] have been able to establish that the deformation potential and interband Fröhlich contributions to LO scattering in ZnO tend to cancel. The competition between the two scattering mechanisms causes sign changes in the matrix elements in (3.10) for LO Raman scattering and a cancellation point in the LO cross section for photons below the band gap. Resonance effects, including cancellations, for photons near or at intermediate state frequencies depend upon the detailed nature of the intermediate states and will be considered in the following subsections.

Before proceeding to the examples of resonance effects it is also appropriate to comment on some aspects of resonance scattering which are not emphasized here but which should be kept in mind in the following discussions. The first aspect is the effect of dielectric dispersion, i.e., polariton effects. For strong interactions it may be essential to carry the perturbation in $\mathscr{H}_{MR}$ to high-orders. This has been done for several specific cases [3.18–21]. We shall not consider higher-order effects except insofar as to incorporate the exact dispersion, i.e., index of refraction, for photons in the crystal [Ref. 8.2, Chap. 7].

Because the fundamental resonances occur at frequencies where the light is absorbed, possible consequences of the finite penetration depth and the nature of the material within the penetration volume must be considered. The distinction may be important in crystals because the restriction of scattering to a surface region breaks the momentum conservation. This has been shown by WILLIAMS and PORTO [3.47], and MARTIN [3.22] to qualitatively change the scattering in some cases. Also, the fact that the surface has different symmetry from the bulk may change the polarization selection rules. This has been observed in many semiconductors [3.7, 23] in which Schottky barriers give rise to large surface electric field extending into the sample distances comparable to typical penetration depths of the light. The possibility of such surface effects must be recognized in RRS, but they are not emphasized here since they are merely one example of impurity or inhomogeneity-induced effects in solids [Ref. 8.1, Fig. 2.40].

3.4.2. Single Discrete Intermediate and Final States

The most straightforward theoretical example of RRS is one in which there is a single discrete final state and a single intermediate state which dominates the scattering matrix element over some range of incident photon frequencies. As was shown in Section 3.2, the Raman intensity as a function of incident photon frequency in this range has a Lorentzian resonance, (3.18). The resonance in the complex Raman matrix element (and Raman tensor) $K(\omega)$ is the same as that in the dielectric susceptibility $\chi(\omega)$, i.e. it is centered at the absorption frequency and has a width determined by the lifetime of the intermediate state. It is interesting to note [3.24] that the Lorentzian formula for the scattered intensity is indistinguishable from a two-step process of absorption of the initial photon followed by decay of the intermediate state via any possible paths, one of which is the emission of the scattered light.

We shall refer to all scattering, both at resonance and off resonance in the wings of the Lorentzian as RRS. The only distinction is that at resonance the intensity is determined by the lifetime, whereas off resonance it is determined by the frequency separation from resonance. In the domain, this in turn leads to the result (see Section 3.2) that far from resonance the scattering is essentially instantaneous, whereas at resonance the scattered light is emitted with an exponential time delay determined by the lifetime.

Distinctions have been made in which scattering at resonance is termed "resonance fluorescence" (RF) [3.1, 2] or "hot luminescence"

(HL) [3.25, 26]. We shall reserve such terms for cases in which it is *essential* (not merely possible) to describe the scattering as a multiple step process with a loss of phase memory between the steps. The formal description of the distinction between RRS and RF or HL has been presented by SHEN [3.26]. He introduced random fields which can scatter the intermediate state. The light emitted from states so randomized is termed HL. Each experimental case must be examined to determine if such random fields are present or absent. We note that the distinction is unimportant in the present example in which there is only one intermediate state. The phase of the state is essential only in cases where there are degenerate or nearly degenerate intermediate states. Then the different contributions to the scattering amplitudes must be added with the proper phase relations.

The fact that one intermediate state dominates the scattering may occur for two reasons: I) there is only one state of the system near resonance, or II) only one state couples to the light because of selection rules. The former condition may be found for isolated impurity states in solids or (approximately) for discrete molecular excited states. The later condition may isolate one intermediate state from a continuum in a perfect crystal, for example, wave-vector conservation limits the interaction of a photon of wave vector $\boldsymbol{k}$ to excited states of the same wave vector. Solids at low temperatures have well-defined discrete exciton bands. If the photon has a well-defined wave vector, then scattering near resonance with an exciton band often can be described in terms of a single intermediate state. Similarly, in a crystal the final state may be isolated from a continuum by the overall wave-vector conservation in the scattering process. Note that, in fact, the intermediate state may be degenerate with other states and in the case of an exciton is essentially degenerate with other states having slightly different total wave vector $\boldsymbol{k}$. In order to compare results with the simple Lorentzian formula, we must ensure that we consider only those aspects of the light scattering which can be described in terms of a single intermediate state.

First-Order RRS in Cu_2O [Ref. 8.1, Sect. 2.3.6]

The conditions outlined above for a single intermediate state are realized in Cu_2O, as shown in the experimental work of COMPAAN and CUMMINS [3.27] and the theoretical analysis of BIRMAN [3.28]. COMPAAN and CUMMINS have measured first-order Raman intensities using a dye laser excitation source which can be tuned through the absorption line for the lowest "yellow" exciton line in Cu_2O. The 1*s* "yellow" exciton at the zone center has even parity and hence is electric dipole forbidden.

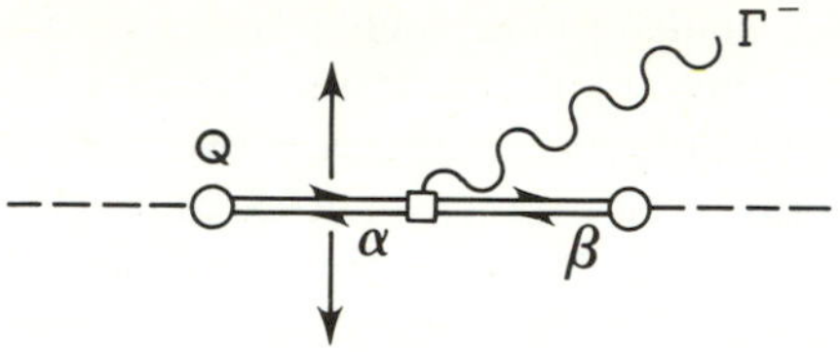

Fig. 3.6. Diagram of quadrupole-dipole RRS in Cu_2O. The $1s$ yellow exciton is denoted α and other non-resonant excitations β. The electron-photon interaction for the α exciton is quadrupole (Q) and the resonance with the incoming photon is denoted by the arrows. The phonon is any odd-parity phonon

The interaction with radiation $\mathscr{H}_{MR}$ for this state is a weak quadrupole coupling. The absorption is sufficiently weak that polariton effects are negligible and the photon wave vector is well-defined, yet the coupling is strong enough compared to the exciton lifetime so that the resonance enhancement is clearly observable. Also, because the absorption is weak, the scattering occurs in the bulk of the crystal and polarization selection rules can be unambiguously applied to test theoretical predictions.

The fundamental diagram for the scattering is given in Fig. 3.3c which represents creation of the exciton at the wave vector of the incident photon $\boldsymbol{k}_1$ via the quadrupole interaction, followed by a phonon assisted radiative decay. The resonant state is the $1s$ "yellow" exciton which we shall refer to as the α exciton. For ω_1 sufficiently close to the α exciton absorption line, this diagram dominates the scattering despite the small quadrupole interaction. The observability of such processes depends upon the lifetime of the intermediate state, which is quite long in Cu_2O. The quadrupole absorption of the α exciton in Cu_2O is well-known [3.29] and the phonon assisted emission of light from the α exciton has been studied [3.30], hence the combination of these two steps in RRS has many possible checks for internal consistency and for comparison of detailed resonance curves.

The quadrupole interaction has been discussed in Section 3.3 [see (3.75) and (3.77)]. Because one of the interactions between light and matter is quadrupolar- and hence of even parity—*only odd parity* phonons are involved. Specific polarization selection rules have been derived by BIRMAN [3.28] which predict that all odd parity phonons should be observable. The quadrupolar interaction depends upon the wave vector of the incident photon $\boldsymbol{k}_1$: For each wave vector $\boldsymbol{k}_1$ a Raman tensor coupling the polarization eigenvector of the incident and scattered light can be given. The dependence of this tensor upon the direction of the incident photon $\boldsymbol{k}_1$ was given explicitly by BIRMAN [3.28].

The actual electron-phonon couplings responsible for the phonon-assisted emission can be made explicit by expanding in powers of the electron-phonon interaction. The diagram for the case where the incident photon is at resonance is shown in Fig. 3.6. There the quadrupolar interaction is denoted by Q and account has been taken that the odd parity Γ^- phonon must scatter the α exciton to some non-resonant electron state denoted β. Since the β state is non-resonant, it merely provides an effective coupling mechanism for the phonon assisted recombination and in no way affects the resonance enhancement. The relative cross sections for the different Γ^- phonons are estimated by considering the possible β states which are different for the different phonons. BIRMAN [3.28] concluded that one expects the largest coupling for Γ_{12}^- phonons in agreement with measurements of phonon assisted emission [3.30].

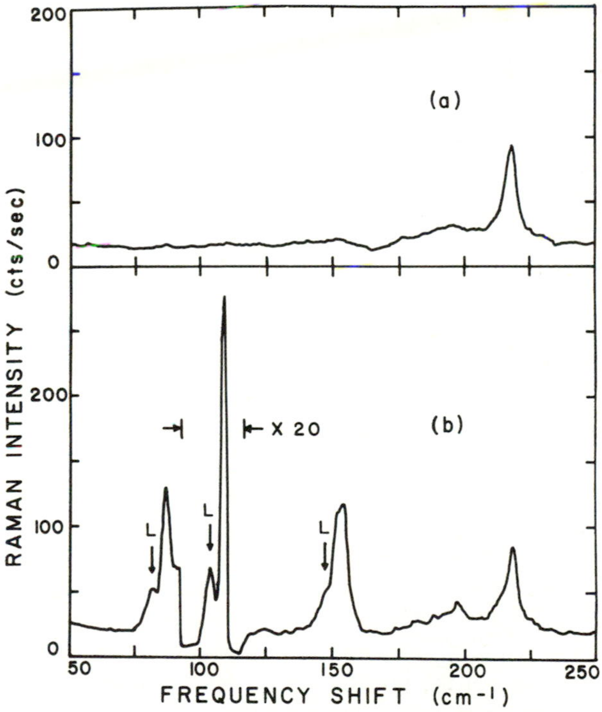

Fig. 3.7a and b. Raman spectra of Cu_2O at 4 K with 12 mW incident laser power. Incident polarization, [001]; scattered polarization, [001] + [110]. Instrumental resolution, 2 cm^{-1}. (a) Laser frequency 10 cm^{-1} above the 1*s* yellow exciton frequency. (b) Laser in resonance with the 1*s* yellow exciton. The features labeled L are phonon-assisted luminescence as discussed in the text. (From COMPAAN and CUMMINS [3.27])

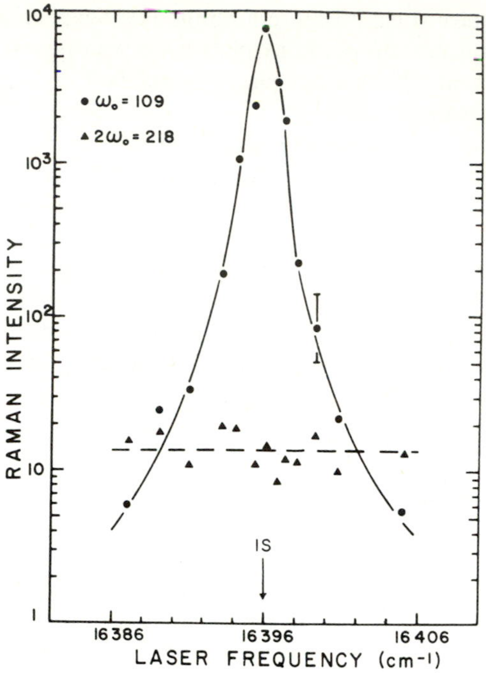

Fig. 3.8. Resonance enhancement of Cu_2O of the first-order 109-cm^{-1} Γ_{12}^{-}-Raman line and its 218-cm^{-1} second-order feature. The smooth curves through the data points are included for visibility. The frequency of the 1*s* exciton, $16396 \pm 0.5\ cm^{-1}$, was measured in direct luminescence. (From COMPAAN and CUMMINS [3.27])

The experimental results of COMPAAN and CUMMINS [3.27] confirm each aspect of the theory for one particular propagation direction $\boldsymbol{k}_1$. In Fig. 3.7 we show their data for scattering both far from the resonance and at resonance. In the former case only one two-phonon peak (of even parity) is discernible. At resonance, however, *all* odd parity phonons appear in the first-order Raman spectrum. In Fig. 3.8 the resonance enhancement of the strongest RRS feature, that of the Γ_{12}^{-} phonon, is reproduced. The data indicate an approximately Lorentzian line of about $1.5\ cm^{-1}$ half width, which is somewhat larger than the half width of $0.75\ cm^{-1}$ measured for the absorption line [3.29]. A detailed comparison of the line shapes for RRS and the optical susceptibility has not been presented. Data are also presented in Fig. 3.8 which show the complete absence of enhancement of the even-parity second-order feature involving two Γ_{12}^{-} phonons.

The explicit polarization selection rules for one photon direction $\boldsymbol{k}_1$ are in good agreement with the theory. However, no measurements have been made to test explicitly the dependence of RRS upon the direction of the incident photon relative to the crystal axes. Since the absorption is weak, Cu_2O appears to be an ideal case to test predicted dependence.

It is possible at this point to make an explicit distinction between RRS and hot luminescence and to show that the scattering described above is RRS. The distinction rests upon the fact that in a cubic crystal the exciton states are degenerate. If the exciton were sufficiently long lived that it was "depolarized", e.g. by elastic impurity scattering to the degenerate exciton states with different polarizations, then the polarization of the emitted light would have no relation to that at the incident light. No essential depolarization is observed in the experiments, hence in this particular case there is no evidence for a decoupling of the states, i.e. loss of phase memory in the intermediate state.

There is light emitted, however, which cannot be described as simple RRS. These are the sharp peaks denoted in Fig. 3.7 by L for phonon-assisted luminescence. This luminescence results from recombination of α excitons which have absorbed energy to thermalize $\simeq kT$ above the band minimum where the absorption occurs. In Cu_2O the widths of the α exciton and the final state phonons are narrow compared to kT even at 4 K. Thus, for RRS with excitation within the line width of the α exciton, the luminescence emissions are always separate from the RRS sidebands. The total luminescence process is fundamentally different from the particular cases of RRS considered here in that it must involve absorption of low energy acoustic phonons. These phonons, which are not explicitly detected in the experiment, provide the loss of phase which make it essential to regard this scattering as a two- (or more-) step process.

RRS in I_2 Molecules

The excitation spectrum of a diatomic molecule consists of both discrete states and bands [3.31]. The discrete states are the vibrational-rotational levels of the bound molecule and the bands are the continuum states of the dissociated molecule associated with each electronic level, respectively. In Raman scattering, the initial and final states of the molecule are different vibrational-rotational levels of the ground electronic state. Restricting our attention to only the discrete levels of the ground electronic state, the Raman spectrum consists of a series of sharp lines representing all possible allowed transitions. Because of the unequal spacing of both the vibrational and rotational levels, each line is associated with the transition between particular intermediate and final states. Let us focus attention upon one such line and denote the initial state for that line by *a* and the final state by *f*. This line in the Raman spectrum is near resonance if the initial photon energy is close to the energy for an allowed transition from state *a* to an intermediate state of the molecule labeled *i*. The state *i* in general involves an electronic excitation and may be either a discrete or a continuum vibrational-rotational state of the excited electronic state. In this section we consider only the case where *i* is a discrete level. The appropriate diagram is shown in Fig. 3.2c where *i* is the resonant state.

The "discrete" excited state is, of course, broadened by collisions, Doppler shifts, hyperfine splittings, etc. Hence we expect that the theoretical discussion in terms of a single inhomogeneously broadened intermediate state [see (3.23)] should describe the RRS. Our interest is primarily the division of the scattered light into the "slow" and "fast" components and the nature of the RRS in this frequency range. As was remarked in Section 3.2, we expect both types of time dependent RRS given, respectively, in (3.21) and (3.22) to be present simultaneously.

As an example, we consider the I_2 molecule, which has discrete absorption lines conveniently located so that they may be scanned by selecting the possible modes of an Ar^+ ion laser within the doppler broadened gain profile [3.32]. Raman scattering as a function of the laser frequency has been studied in this way by several groups [3.3, 33–36]. Traditionally, in this case, the spectrum is called resonance fluorescence (RF) [3.3, 33–36] because the spectrum appears to be determined by the lifetime of the states; in practice this means that the spectrum is depolarized and changes markedly as a function of the partial pressures of various gases [3.33]. This operational approach, however, is not sufficient to uniquely determine the role of the lifetime in RRS. First, the depolarization is not a unique aspect of RF. Simple kinematics of a rotating molecule shows that the emitted light is depolarized in cases in which a single rotational transition involving a state of $l \neq 0$ is enhanced by resonance effects. This is independent of whether or not the emitted light is lifetime limited. Also, increase in gas pressure causes both homogeneous and inhomogeneous Lorentzian broadening so that the dependence of the RRS upon lifetime is not directly measured.

The ultimate test of the role of lifetime in RRS is time-resolved Raman spectroscopy to distinguish the lifetime-limited exponential time dependence discussed in Section 3.2. WILLIAMS et al. [3.3] have performed an elegant first experiment to detect the delay. They used a 100 nsec laser pulse excitation and examined the Raman shifted light emission as a function of time during and after the pulse. Examples of the time dependence are shown in Fig. 3.9. On resonance with the absorption line (curve labelled 0 GHz) the light emitted has a slow rise and decay constant of $\sim 1\,\mu$sec consistent with previous measurements

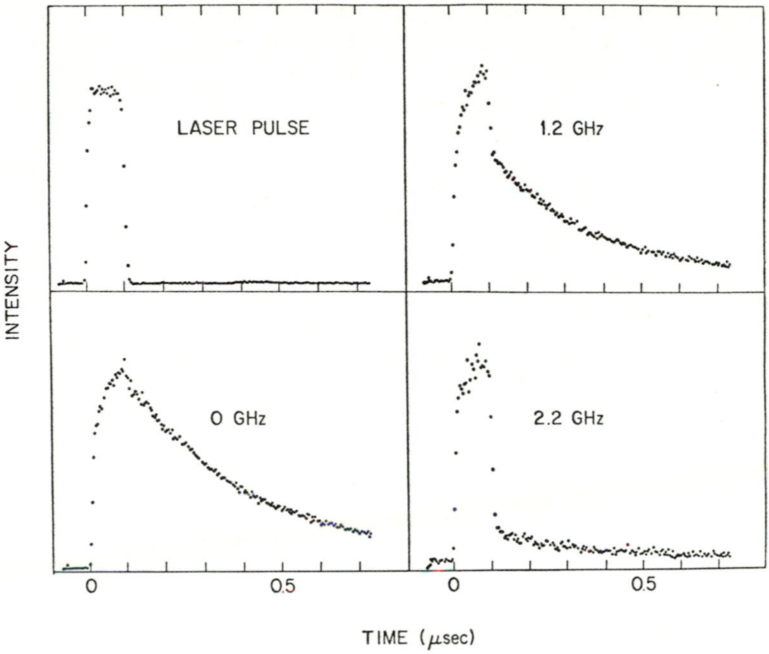

Fig. 3.9. Temporal response of the radiation from the Q branches of the $P(13)$ and $R(15)$ $\Delta v = 1$ transitions in iodine. The various incident laser frequency shifts are relative to the fluorescence maximum. The incident pulse is shown at the upper left and it has a time width of about 100 nsec. On resonance (curve labeled 0 GHz) the emitted light has an observable time delay whereas off resonance (curves labeled 1.2 GHz and 2.2 GHz) a portion of the light is emitted with a delay shorter than 10 nsec. (From WILLIAMS et al. [3.3])

of the lifetime in RF. With the excitation shifted off resonance (cf., the curve shown for 1.2 GHz), there is both a fast (< 10 nsec) component and the same slow component reduced in amplitude. They conclude that because of the inhomogeneous distribution of oscillators in the "discrete" line, the scattered light contains both a portion which is independent of the lifetime and a portion which is lifetime limited, i.e., RF. This is further supported by examination of the time resolved scattering at higher gas pressures, in which case the slow component grows in intensity. They interpret the growth in intensity of the slow component to be a manifestation of increasing inhomogeneous broadening with increasing gas pressure, although the actual lifetime, as measured by the time delay of the emitted light, is relatively insensitive to pressure.

In the frequency range considered both the fast and slow components of the scattering are found to be fully depolarized. As mentioned above, this is merely a manifestation of the fact that specific rotational states dominate the intermediate states for both the fast and slow processes. Only far from resonance where the intermediate states must be summed over a large number of rotational states is the RRS polarized so that it obeys molecular selection rules.

The theoretical description in Section 3.2 is sufficient to extract the primary points. For inhomogeneous broadening the normalized distribution function $P(\omega_1)$ in (3.23) is peaked at some frequency ω_i^0, has a characteristic width $\Delta\omega$, and decreases rapidly for $|\omega_i - \omega_i^0| \gg \Delta\omega$. We assume the inhomogeneous broadening is much greater than the lifetime broadening, $\gamma \ll \Delta\omega$. Then the lifetime limited (i.e. slow) contribution to the matrix element squared (3.23) becomes

$$|K_{2f,10}|_s^2 \simeq \hbar^{-2} |M_{fi} M_{i0}|^2 \frac{\pi}{2} \frac{P(\omega_1)}{\gamma(\omega_1)} . \tag{3.83}$$

The integrated intensity for the slow component of the scattering is proportional to (3.83) and hence measured directly the ratio of the distribution function to the lifetime at the exciting frequency ω_1. With the simultaneous direct measurement of $\gamma(\omega_1)$, the inhomogeneous distribution is determined directly from such time resolved experiments. The "fast" component is a more complicated function of the distribution $P(\omega_1)$. It simplifies far from resonance, $|\omega_1 - \omega_i^0| \gg \Delta\omega$, to the simple formula for a single oscillator

$$|K_{2f,10}|_f^2 \cong \hbar^{-2} |M_{fi} M_{i0}|^2 \frac{1}{(\omega_1 - \omega_i^0)^2} . \tag{3.84}$$

In the case of inhomogeneous Lorentzian broadening it is easy to see that the two contributions (3.83) and (3.84) are comparable in magnitude. Therefore, unless $P(\omega_i)$ and $\gamma(\omega_i)$ can be determined independently, a time resolved experiment such as that performed by WILLIAMS et al. [3.3] is essential to completely analyze the RRS and to extract such information as the inhomogeneous distribution appropriate to the given excited intermediate state. With such information one can, in principle, observe the molecular kinetics of each discrete excited molecular state individually which is coupled to the ground electronic state by the material-radiation interaction $\mathscr{H}_{MR}$.

First-Order RRS in CdS

The example of RRS which has received the most attention is scattering from CdS [3.10–12, 37–45]. This is because of the fortunate occurence of a number of lines from the Ar^+ ion laser close to the lowest band gap of CdS at ~2.6 eV. The results are representative of scattering from other semi-conductors [3.17, 37, 38, 46–49] and we shall discuss only CdS. Several complications make scattering from CdS intrinsically much more difficult than the previous example of Cu_2O. The primary reason is that the resonant optical transition is electric dipole allowed. The strong exciton-photon coupling leads to polariton effects and a finite penetration depth of light at resonance. Also, the presence near the band edge of impurity states having "giant" [3.50] oscillator strengths complicates analysis of the resonance behavior. On the other hand, the fact that the optical transition is allowed leads to intense first-order RRS which extends over a much larger energy range than is the case in Cu_2O.

The primary interest in the first-order scattering in CdS is the longitudinal optic (LO) Raman sideband. The LO cross section increases rapidly near resonance and is observed for photon polarizations not allowed by the usual Raman selection rules [3.10–12]. The scattering is apparently an intrinsic feature of RRS in polar semiconductors [3.11, 39–41] although similar impurity induced LO scattering may also be important [3.43–45]. The basis for the interpretation of the 1 LO–RRS is 1) qualitative arguments based on the correspondence of RRS and luminescence [3.39–41] and 2) quantitative calculations [3.10–12, 43] of the RRS cross section in terms of the Fröhlich electron-phonon interaction.

It is well-known that the dominant electron-phonon interaction near the band edge in polar semiconductors is the intraband Fröhlich interaction [3.9, 12] discussed before, (3.79). By this mechanism the electron or hole is scattered within the same band by the macroscopic electric field of an LO phonon. It is this interaction that leads to polaron effects, lattice screening of the exciton binding interaction, etc. The strong 1-LO phonon-assisted luminescence [3.51] has been interpreted in terms of the Fröhlich interaction. Also, the dynamics of electrons created well above the band gap has been shown to be determined primarily by electron-LO scattering by this mechanism [3.52–53]. We have emphasized, in particular, the close relation of RRS and phonon-assisted luminescence (cf. the correspondence in Cu_2O discussed under *First-order* RRS *in* Cu_2O, in Subsection 3.4.2). Therefore, we anticipate that the anomalous LO RRS can be understood in terms of the Fröhlich electron-phonon interaction.

The lowest-order diagram for the scattering is illustrated in Fig. 3.3c, where we choose the incident photon to be near resonance with the exciton resonance. It is a good approximation to consider only the dipole electron-photon interaction and to assume that the exciton envelope function is an eigenfunction of parity. Then each exciton intermediate state has s symmetry and intraband Fröhlich's matrix elements vanish in the long wavelength limit [dipole approximation, as discussed following (3.79)]. At small finite wave vector q the interaction is linear in q (quadrupole) and has a peak at $q \simeq a_0^{-1}$, where a_0 is the exciton radius [3.11]. The characteristic expansion parameter for the strength of the quadrupole interaction is $(qa_0)^2$. As is discussed in Section 3.3, this wave-vector-dependent Fröhlich interaction may be comparable to other electron-phonon coupling mechanisms even for the small wave vector $\boldsymbol{q} = (\boldsymbol{k}_1 - \boldsymbol{k}_2)$ involved in first-order RRS.

The scattering is analogous to Cu_2O except that here the quadrupole wave-vector-dependent interaction is in the electron-phonon instead of the electron-photon interaction. The polarization selection rules for the present case are very simple: the intraband Fröhlich interaction has Γ_1 symmetry since it connects s states. Therefore, the RRS tensor has the same symmetry as the dielectric tensor. In particular, for CdS it is diagonal and has resonances at the same frequencies as do the respective principal components of the dielectric tensor. The new RRS selection rules apply only for LO phonons and only for photons near resonance with excitons or band states [3.10–12].

Two experiments have supported the existence of strong first-order LO Raman scattering which is wave-vector-dependent. The most direct observation was that of Gross et al. [3.39–41], who used a monochromator to excite Raman scattering essentially at the exciton resonance. Because of large polariton effects the wave vector of the photons inside the sample—and consequently the wave vector q of the scattered phonon—can be varied as a function of incident frequency ω_1. The relation of q and ω_1 is readily derived from the known exciton parameters. Within the small frequency range considered, all Raman features are expected to have a resonance enhancement which is approximately Lorentzian, but the 1 LO scattering has an additional variation caused by the change in q. For example, the ratio of intensities $I_{1\,\mathrm{LO}}/I_{2\,\mathrm{LO}}$ should eliminate all other resonance factors and scale simply with wave vector, i.e. $I_{1\,\mathrm{LO}}/I_{2\,\mathrm{LO}} \propto q^2$.

This relation has been tested experimentally with the results shown in Fig. 3.10. The strong electron-phonon and electron-photon interactions in this case make it difficult to separate RRS and luminescence. The measured ratio in Fig. 3.10 spans the range from RRS to imcompletely thermalized luminescence [3.39]. Although the intensities vary greatly, the ratio $I_{1\,\mathrm{LO}}/I_{2\,\mathrm{LO}}$ varies simply as q^2 [Ref. 8.2, Chap. 7].

The anomalous 1 LO-RRS has also been analysed by Martin and Damen [3.11]. They reported measurements of the resonance enhancement of the 1 LO scattering in a configuration forbidden for usual Raman scattering, but allowed for the modified RRS selection rules.

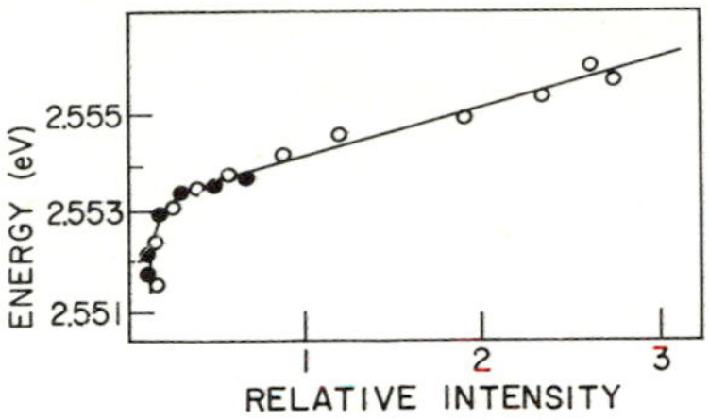

Fig. 3.10. Dependence of the relative intensity 1 LO and 2 LO inelastic polariton scattering $S = I_{1\,LO}/I_{2\,LO}$ on the polariton energy. CdS, $T-4.2$ K. Filled and open circles are experimental points obtained by two different methods (see text). Solid line is the theoretical curve from the polariton dispersion relation. (From GROSS et al. [3.41])

The frequencies were sufficiently far from the absorbing region so that the scattering could be done in the bulk with right-angle scattering geometry and only minor corrections to the intensities caused by absorption [3.11]. The anomalous increase in intensity of the "forbidden" 1 LO is illustrated in Fig. 3.11. In the lower traces we see that far from resonance the forbidden scattering is very small compared to the allowed 1 LO scattering. Nearer resonance, however, the intensity of the forbidden scattering increases greatly while the allowed scattering actually decreases, as shown in the upper traces.

The theoretical third-order perturbation expression (3.33) for the RRS cross section has been evaluated for this case by MARTIN [3.12], who assumed isotropic hydrogenic exciton intermediate states and found that an accurate calculation of the absolute cross section at finite wave vector q requires a summation over all intermediate states including continuum states of the exciton. Only in the case where one photon is very near resonance can one of the sums over intermediate states be restricted to one term. Even then the second sum over intermediate states for the other photon must be carried over all states. The results are shown in Fig. 3.12 together with the experimental points [3.11]. The strongly resonant nature of this "forbidden" scattering is clear from the rapid decrease in the cross section away from resonance compared to an allowed cross section [also shown in Fig. 3.12].

A stringent test of the interpretation is the absolute magnitude of the scattered intensity. Since the electron-phonon coupling parameter is accurately known for the Fröhlich interaction and all exciton parameters are known approximately [3.12], the absolute cross section can be predicted. The theoretical cross section for right angle scattering with the laser line closest to resonance leads to a scattering rate per unit length of $2.8 \times 10^{-6}\,(\mathrm{cm\,sr})^{-1}$ [3.12], which is approximately twice as

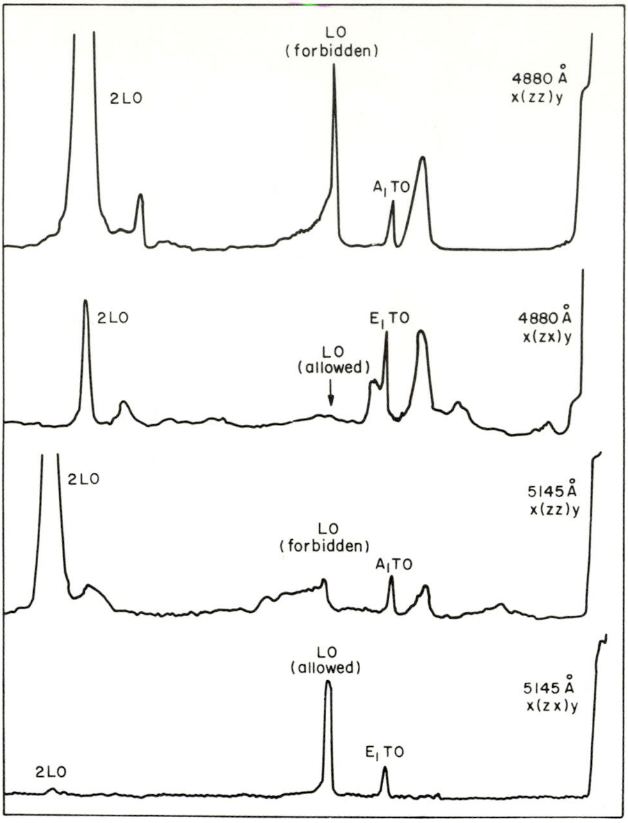

Fig. 3.11. Uncorrected experimental traces of the Raman scattered light near resonance in CdS for various incident frequencies and polarizations. Note that near resonance the allowed $x(zx)y$ 1 LO line is weak whereas the forbidden $x(zz)y$ line is very strong. The resonance occurs at the $1s$ B exciton at 4827 Å. Note also the strong 2 LO line. (From MARTIN and DAMEN [3.11])

large as the experimental value of $1.3 \pm 0.1 \times 10^{-6}$. A similar comparison holds for the other laser frequencies. Considering the uncertainties in the exciton parameters [3.12] the agreement is very good.

It should be emphasized that despite the agreement between theory and experiment for selection rules and intensities, there have been no direct experimental tests of the dependence of the RRS cross section upon the directions of propagation of the photons. Since $\boldsymbol{q} = \boldsymbol{k}_1 - \boldsymbol{k}_2$ the "forbidden" 1 LO scattering is predicted to vanish in the forward scattering direction and be maximum for backward scattering.

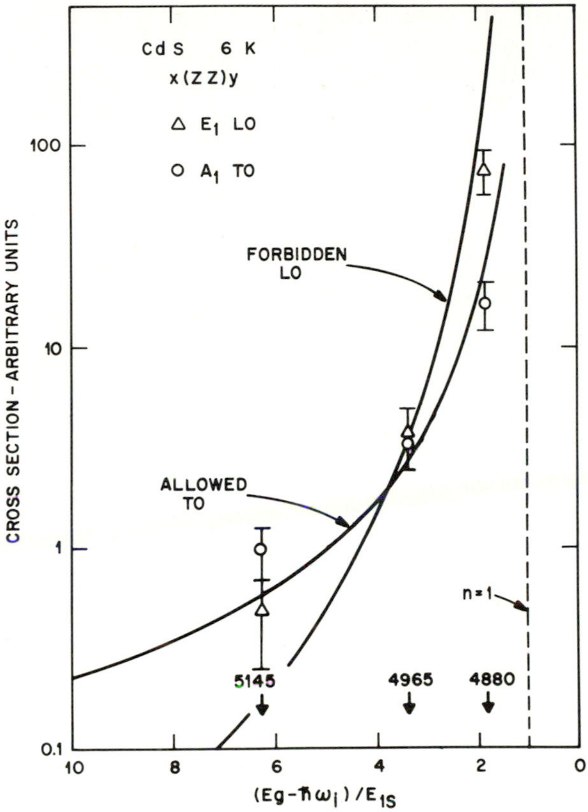

Fig. 3.12. Comparison with experiment (from [3.11]) for $x(zz)y$ geometry in CdS at three incident photon energies. In this geometry all LO scattering is forbidden. Experimental absorption corrections have been made. Relative LO and TO cross sections are found experimentally and the relative theoretical cross sections have been scaled to fit experiments. Only the B valance band is included in the calculation. The influence of the C band can account for the discrepancy in A_1 TO, but this has not been done here. (From MARTIN [3.12])

The dominance of impurity states in the optical properties of semiconductors near the band edge suggests an alternative mechanism for large "forbidden" scattering. The lack of translation invariance for impurity states breaks the momentum conservation so that the phonon wave vectors are not determined by the photons. Involvement of phonons with $q \gg |\boldsymbol{k}_1 - \boldsymbol{k}_2|$ greatly enhances the q-dependent 1 LO scattering via the intraband Fröhlich interaction with little effect upon other RRS features. The dependence upon the photon polarizations of this impurity assisted scattering is the same as for the intrinsic scattering discussed above, although, of course, there is no relation of the scattering intensity to the photon wave vectors.

COLWELL and KLEIN [3.43] have examined impurity induced scattering in terms of the well-known bound exciton states [3.50, 54]. The four primary features of scattering by this mechanism are: 1) the lack of dependence upon the photon wave vectors; 2) the involvement of LO phonons of wave vectors of order the inverse of the impurity state radius (This is typically much greater than photon wave vectors but much less than the BZ boundaries.); 3) sharp resonances at the energies of the bound excitons which form discrete levels below the intrinsic exciton resonance; and 4) dependence upon impurity concentrations. COLWELL and KLEIN [3.43] have observed 1 LO scattering from CdS consistent with this mechanism. In particular, the measured intensity was much greater (relative to the allowed 2 LO) than that reported by MARTIN and DAMEN [3.11] suggesting feature 4 and they established there that was no assymmetry between forward and backward scattering, i.e. feature 1.

Resonances (feature 3) in the impurity induced RRS at the bound exciton energies have been observed DAMEN and SHAH [3.45], using a pulsed dye laser tuned through the prominent peaks. The observed resonances appear to be approximately Lorentzian and are centered at the absorption peaks for the impurity states. The intensities correlate as expected with impurity concentrations. A detailed examination of these resonances would presumably be similar to that of the inhomogeneously broadened molecular resonances.

DAMEN and SHAH [3.45] also observed that the impurity resonances were superposed upon a frequency dependent cross section which was independent of sample preparation. This contribution is presumably intrinsic and must be explained by an intrinsic mechanism such as the wave-vector-dependent scattering discussed above.

From these experiments we conclude that first-order RRS in polar semi-conductors is dominated by forbidden 1 LO scattering, which may be used to study either extrinsic or intrinsic intermediate states. In each case the scattering is approximately described by the single state Lorentzian resonances over some energy range. Outside these ranges in general both types of scattering occur and accurate descriptions require sums over a range of intermediate states. Because the intrinsic exciton binding energies and binding to the impurities are both small, the only experiments in which a simple resonance behavior has been observed are those in which the incident photon frequency ω could be adjusted continuously to achieve the desired resonance. In addition to the frequency dependence of the cross section, the nature of the intermediate states is manifested in the dependence upon the wave vector of the phonons.

3.4.3. Continuum of Intermediate States: One per Final State

In this subsection we consider cases in which there is a distribution or a continuum of intermediate states, but with the restriction that in any given Raman matrix element only one intermediate state need be considered. The theoretical reason for singling out this restricted class is clear: if there is only one dominant intermediate state there can be no interference in the Raman matrix element. The general formulation for the RRS cross section is given in (3.24). The important aspect of (3.24) is that the total cross section is a weighted sum over positive definite contributions from a distribution of scatterers.

For this category of RRS the important problem is the determination of which intermediate state dominates the scattering at a given excitation frequency. Because the intermediate state may vary as a function of frequency, the resonance spectrum may bear no resemblance to the Lorentzian form for a single discrete state. Unlike the example of a discrete state or an inhomogeneously broadened state considered in Subsection 3.4.2, there is no universal functional form and the nature of the resonance enhancement for specific cases must be considered individually.

In the previous subsection we utilized crystalline selection rules to single out one discrete intermediate state for first-order RRS in Cu_2O. In this section we consider the same example for higher-order scattering. A schematic diagram of the scattering is shown in Fig. 3.13, which is essentially the same as Fig. 3.3b with the addition of specific labels for the present case. As was discussed in Section 3.2, conservation of momentum in a crystal requires that, for fixed photon and final state phonon wave vectors, the momentum of each intermediate state is fixed. Furthermore, for exciton bands in crystals the set of intermediate states of a given wave vector form a set of discrete levels. For photon and phonon energies such that one discrete intermediate state is near resonance, then in the formal expression for the Raman tensor only one intermediate

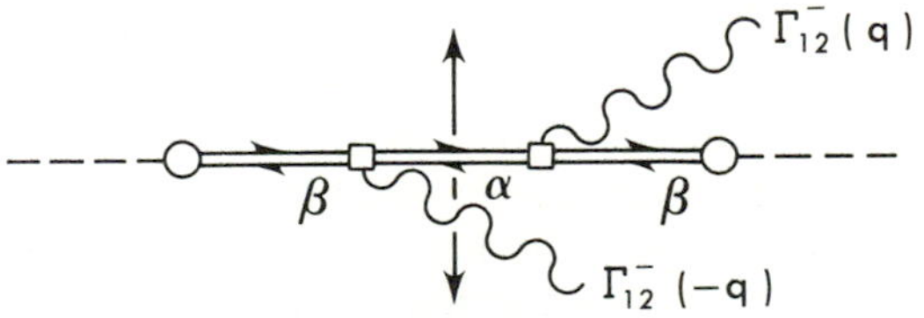

Fig. 3.13. Second-order RRS in Cu_2O. The notation is the same as in Fig. 3.6. The resonant state is the α exciton. Other intermediate states β are non-resonant. The phonons are odd-parity

state need be considered for each final state. Of course, the individual phonon momenta are not fixed in the experiments and the sum over final states must be carried out as shown in (3.24).

Let us now restrict ourselves to specific assumptions valid for the case of second-order scattering in Cu_2O reported by YU et al. [3.55]. Because the 1s "yellow" exciton is dipole forbidden, the dominant second-order scattering corresponds to both phonon-assisted absorption and phonon-assisted emission of the photons, i.e., the middle intermediate state in Fig. 3.13 is the resonant (α) exciton and the other two intermediate states are non-resonant (β) excitations. The resonant intermediate state α forms a band with lowest frequency being the absorption frequency discussed under "*First-order* RRS *in* Cu_2O", in Subsection 3.4.2 and having exactly one state for each value of the wave vector.

As is essentially always the case in second-order scattering, the photon wave vectors are negligible so that the phonons have wave vectors $\pm \boldsymbol{q}$. For the case considered by YU et al., dispersion is negligible ($\omega_0(\boldsymbol{q}) \simeq \omega_0$) so that only the total integrated intensity of the sharp second-order Raman feature could be measured. Using an isotropic parabolic dispersion relation $\omega_\alpha(\boldsymbol{q}) - \omega_\alpha(0) \propto q^2$ for the exciton band to convert the sum over final states $\boldsymbol{q}$ in (3.24) to an integral over frequency of the intermediate state ω_α, we find

$$d\sigma(\omega_1) \propto \int\limits_{\omega_\alpha(0)}^{\infty} d\omega_\alpha \frac{[\omega_\alpha - \omega_\alpha(0)]^{1/2}}{(\omega_1 - \omega_0 - \omega_\alpha)^2 + \{\gamma[\omega_\alpha - \omega_\alpha(0)]\}^2}, \tag{3.85}$$

where the inverse lifetime γ has been allowed to be an arbitrary function of the frequency of the intermediate state measured relative to the bottom of the exciton band $\omega_\alpha(0)$. The integral is readily performed in the limit of long lifetimes, $\gamma[\omega_\alpha - \omega_\alpha(0)] \ll \omega_\alpha - \omega_\alpha(0)$, with the result

$$d\sigma(\omega_1) \propto \begin{cases} \sim 0, & \omega_1 - \omega_0 < \omega_\alpha(0) \\ \dfrac{[\omega_1 - \omega_0 - \omega_\alpha(0)]^{1/2}}{\gamma[\omega_1 - \omega_0 - \omega_\alpha(0)]}, & \omega_1 - \omega_0 > \omega_\alpha(0). \end{cases} \tag{3.86}$$

In this case, unlike that of inhomogeneous broadening, the density of states is smoothly varying as a function of frequency. Then the lifetime-limited contribution to the scattering greatly dominates over the "fast" contributions arising from virtual intermediate states. Thus, following YU et al. [3.55], it is a good approximation to neglect all virtual contributions for photons both above and below the absorption edge $\omega_1 = \omega_\alpha(0) + \omega_0$. The predicted scattering corresponds to resonance fluorescence and is lifetime limited in intensity and in the temporal dependence of the emitted light. We expect (3.85) to adequately describe all scattering except very near the onset of the RRS at $\omega_1 - \omega_0 \simeq \omega_\alpha(0)$, where the integral (3.85) needs to be treated more exactly.

The experimental results of YU et al. [3.55] are reproduced in Fig. 3.14. The Raman feature observed is the second-order overtone of the Γ_{12}^- phonon band. The fact that this is the strongest feature in the RRS spectrum supports the theoretical interpretation since Γ_{12}^- is known to dominate both phonon-assisted absorption and emission to the 1*s* "yellow" (α) exciton. In agreement with the theory, the intensity increases rapidly at the threshold where $\omega_1 = \omega_\alpha(0) + \omega_0$. The decrease in intensity at higher energies is attributed to a rapid decrease in the lifetime, especially for $\omega_1 > \omega_\alpha(0) + 3\omega_0$ where the exciton can decay via emission of two Γ_{12}^- phonons. In Fig. 3.14a is shown the remarkable agreement between the experimental intensities and the theoretical fit based on a parametrized expression for the lifetimes. The changes in the intensity displayed in Fig. 3.14b at higher temperature has not been analyzed but appears to be consistent with a variation in lifetime as a function of temperature [3.55].

The expressions for the scattering cross-section, (3.24) and (3.86), can be interpreted in an intuitively understandable form: inserting the matrix elements and density of states factors into (3.86), we find the simple result that

$$d\sigma(\omega_1) \propto \alpha_p(\omega_1) \frac{\gamma_R}{\gamma[\omega_1 - \omega_0 - \omega_\alpha(0)]} . \tag{3.87}$$

Here $\alpha_p(\omega_1)$ is the absorption constant ($\propto [\omega_1 - \omega_0 - \omega_\alpha(0)]^{1/2}$) for phonon-assisted absorption to the α exciton; γ is the total lifetime broadening of the exciton which depends upon the exciton energy above the band minimum $\omega_1 - \omega_0 - \omega_\alpha(0)$; and γ_R is the probability of phonon-assisted radiative recombination of a given exciton state, which is very nearly independent of the energy of the exciton. Equation (3.87) is just the absorption probability per unit length multiplied by the branching ratio that describes the fraction of exciton that recombine radiatively without suffering other inelastic scattering. Such branching ratio arguments describe RRS in cases in which all the scattering is lifetime-limited and there are no interferences between different intermediate states.

In the experiment described above, the wave vectors of the phonons involved could not be measured experimentally and were inferred from the theoretical calculations. Cases in which the phonons have larger dispersion would allow the wave vectors to be determined experimentally from the dispersion relation $\omega_0(\boldsymbol{q})$. This has been observed very recently

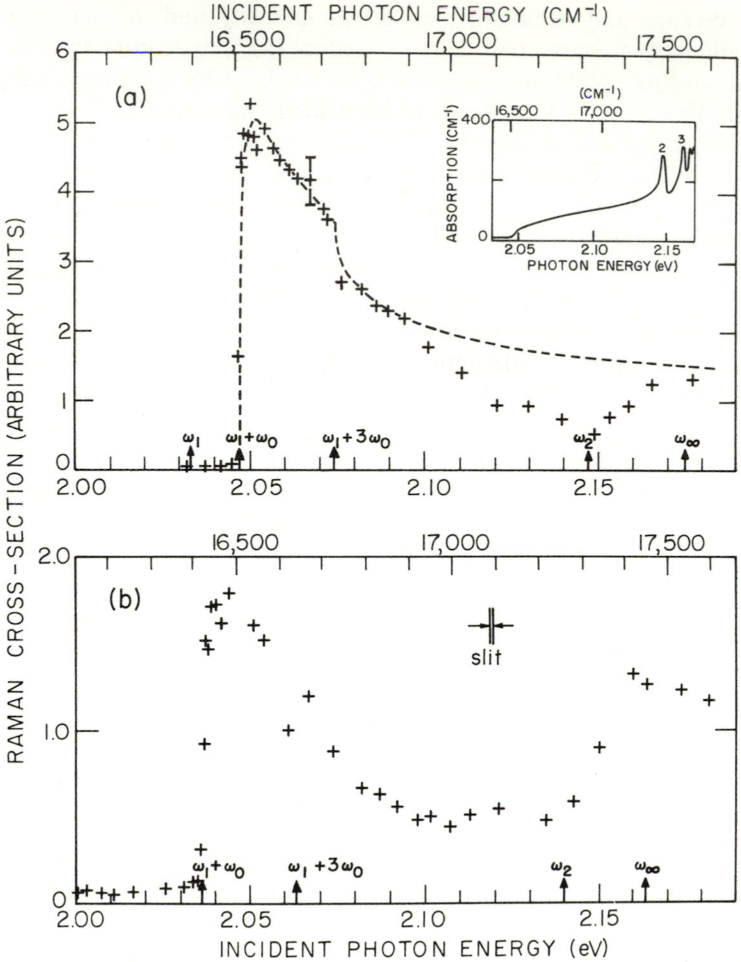

Fig. 3.14a and b. Raman cross section of the 220-cm^{-1} line of Cu_2O at two different temperatures: (a) $\sim$16 K and (b) $\sim$80 K. [These temperatures represent those of the lattice obtained in (a) from the line shape of the phonon-assisted free-exciton recombination spectra and in (b) from the position of the 1s yellow exciton in the luminescence.] Inset in (a), absorption spectra of Cu_2O at 4.2 K taken from P. W. BAUMEISTER [Phys. Rev. **121**, 359 (1961)]. Dashed curve in (a), plot of the theoretical expression described in the text. (From YU et al. [3.55])

in Cu_2O by YU and SHEN [3.56]. They observed RRS peaks for which the energy transfer $\omega_1-\omega_2$ varied as a function of frequency ω_1. In particular the peak position for $\Gamma_{12}^-+\Gamma_{15}^-$(TO) shifted to lower energy and that for $\Gamma_{12}^-+\Gamma_{15}^-$(LO) to higher energy as ω_1 was increased above

$\omega_1 = \omega_\alpha(0) + \omega_0(\Gamma_{12}^-)$. The shift was found to be linear in ω_1 as required by the parabolic dispersion relations of both the exciton and the phonons. The shifts measured were compatible with the exciton dispersion deduced from other measurements of YU and SHEN described below.

YU and SHEN [3.56] also have interpreted peaks in their RRS measurements as due to three and four phonons—two Γ_{12}^- phonons plus one and two acoustic phonons, respectively. For the second-order RRS discussed above scattering of the exciton by acoustic phonons was merely a decay or loss mechanism. The present case is a result of the fact that those decay products can, in turn, emit light. Theoretical analysis is straightforward and can be cast in the form of a branching ratio. For a third-order process involving one acoustic phonon, the light emitted at frequency ω_2 is determined by

$$\begin{aligned} & d\sigma(\omega_1, \omega_2) \\ & \propto \alpha_p(\omega_1) \left\{ \sum_{q_a} \frac{\gamma_a(\omega_1 - \omega_0 - \omega_\alpha(0), q_a)}{\gamma(\omega_1 - \omega_0 - \omega_\alpha(0))} \delta(\omega_1 - \omega_2 - 2\omega_0 - \omega_\alpha(q_a)) \right\} \\ & \cdot \frac{\gamma_R}{\gamma(\omega_1 - \omega_0 - \omega_\alpha(q_a) - \omega_\alpha(0))} . \end{aligned} \tag{3.88}$$

The first branching ratio is the probability of emission of an acoustic phonon of energy $\omega_a(q_a) = \omega_1 - \omega_2 - 2\omega_0$, where ω_0 is the energy of the (dispersionless) Γ_{12}^- phonon. The total branching ratio in brackets has been written explicitly as a sum over the ratios for emission of acoustic phonons of specific wave vectors. The final ratio is the probability of light emission from the exciton state after being scattered by the acoustic phonon.

In general, the curly bracket in (3.88) involves a complicated average. The dominant feature, however, is that γ_a varies smoothly as a function of q_a up to $q_a = q_a(\max)$ given by

$$\begin{aligned} q_a(\max) = & [2M(\omega_1 - \omega_0 - \omega_\alpha(0))]^{1/2} \\ & + [2M(\omega_1 - \omega_0 - \omega_a(q_a) - \omega_\alpha(0))]^{1/2} . \end{aligned} \tag{3.89}$$

For larger q_a, γ_a abruptly drops to zero since the exciton cannot decay by emission of phonons with $q_a > q_a(\max)$. Here M is the total mass of the exciton. Scattering via an acoustic phonon of the maximum wave vector can be illustrated by Fig. 3.4, if we identify the parabola as the exciton band and replacing $\boldsymbol{q} \to \boldsymbol{q}_a$, $\omega_0 \to \omega_a(\boldsymbol{q})$, $\omega_1 \to \omega_1 - \omega_0$, $\omega_2 \to \omega_2 + \omega_0$. If we also assume an isotropic velocity of sound, $\omega_a(\boldsymbol{q}_a) = v_s q_a$, then the density of states of the acoustic phonons causes a peak in the scattering at q_a(max); i.e. the RRS cross-section is strongly peaked at

$$\omega_1 - \omega_2 = 2\omega_0 + v_s \times q_a(\max)\,. \tag{3.90}$$

Combining (3.89) and (3.90) we see that the peak scattering intensity for this 3-phonon scattering varies as a function of incident frequency ω_1 in a way determined by the sound velocity v_s and the exciton mass.

From the variations measured in Cu_2O and an approximate isotropic analysis of the acoustic phonons, YU and SHEN extracted a value of $M = 3.0 \pm 0.2\, m_e$ for the exciton mass. This value is in disagreement with other measurements indicating further analysis is needed, but nevertheless the experiment clearly demonstrated the fact that higher-order RRS can probe in detail electronic states of crystals at wave vectors much greater than the wave vectors of light. In particular, YU and SHEN [3.56] estimated that both phonons and excitons involved in the scattering can be systematically probed up to about 1/4 of a Brillouin zone boundary.

3.4.4. Continuum of Intermediate States with a Discrete Final State

In this subsection we turn to a regime of Raman scattering which is qualitatively different in the theoretical interpretation from the preceding examples. Here we consider cases in which it is essential to sum over many intermediate states in the Raman scattering amplitude, (3.10). Indeed, the major features of RRS in this case result from the constructive and destructive interferences in the amplitude. We first consider the simplest cases of scattering from a solid with photons resonant with the absorption continuum. As in the previous subsections, molecular RRS also provides a particularly clear example of this category of RRS.

Continuum Bands in Solids

The analysis of RRS is greatly facilitated by comparison with the dielectric susceptibility. The central equation is (3.50) which was derived for first-order scattering. Through the one-dimensional example, we showed that this relation to the dielectric tensor is expected to hold for first-order allowed RRS in crystals, significant corrections being needed only if there are non-degenerate bands which happen to be close in energy, e.g. spin-orbit split bands—in which case slightly more complicated formulas are needed [3.8, 57–60]. It is also useful to note that formula (3.50) holds also if excitons are taken into account within the effective mass approximation. The only necessary assumption is that of constant electron-phonon matrix elements. Thus formulas in terms of χ, in principle, have all broadening and exciton effects included.

There have been many reports of RRS from semiconductors in which the experimental RRS and dielectric functions are compared [3.7, 8, 57–62]. In each of these references the general expected critical point features have been observed. However, upon careful examination in many cases [3.58–62] there are definite discrepancies between the critical point positions given by RRS and modulation spectroscopy. For our examples we choose experiments which have been pursued to a sufficient extent to test the simple theory.

The most directly interpretable experiment is that of BELL et al. [3.58], who have observed RRS in the vicinity of the lowest direct gap in GaP. In this case the critical point is well understood—it is an M_0 point—and the spin-orbit split wave functions are known for GaP and other semiconductors. This appears to be an ideal case to compare experimental RRS intensities with the dielectric susceptibility relations. Furthermore, exciton effects are small and theoretical expressions for the susceptibility in terms of free-electron expressions are expected to be sufficient.

The only complication in this case is that the spin-orbit splitting of the valence bands causes two critical points separated by only 0.08 eV. Since the phonon energy is $\omega_0 \sim 0.05$ eV, the structure from the critical points at ω_1 and $\omega_2 = \omega_1 - \omega_0$ will overlap and interfere. This has been accounted for by CARDONA [3.8], in the approximation in which the phonon frequency, electron lifetimes, and scattering to other bands are neglected. The theoretical formula is a special case of (3.91) given below, where the constant term is set equal to zero and the constants A and B are related by a simple argument [3.8].

The results of BELL et al., for the TO RRS intensity are shown in Fig. 3.15. The experimental data indicate structure near each critical point and there is a hint of the predicted decrease between the critical points. The theoretical results are shown as the solid curve. The general agreement is very good and the proper inclusion of lifetime and the phonon frequency would round the sharp theoretical singularities to improve the agreement. We conclude that for this straightforward case the experimental RRS is well-described by the simple theory and that with proper inclusion of lifetime, phonon frequency, exciton effects, etc., the theoretical analysis of RRS can determine the electron-phonon interactions [3.8, 57].

Let us turn to the set of experimental measurements of RRS [3.59–62] near the E_1 and $E_1 + \Delta_1$ critical points in semiconductors [3.63]. This critical point is very anisotropic, having one heavy mass, and is con-

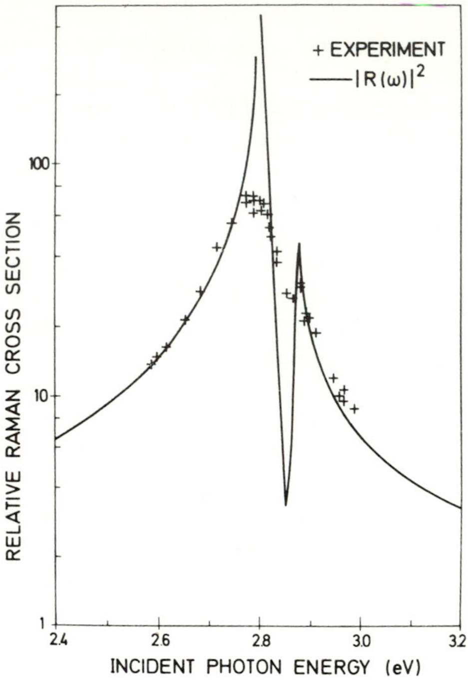

Fig. 3.15. Raman cross section as a function of incident photon energy for allowed, first-order TO-phonon scattering in GaP at room temperature. The experimental results (crosses) have been adjusted to agree with the theory (solid curve—see text) at 2.64 eV. (From BELL et al. [3.58])

structed of states at arbitrary points along a symmetry line in the Brillouin zone. In order to describe the susceptibility it is apparently essential to include exciton effects [3.63–65]. The band and exciton states for E_1 and $E_1 + \Delta_1$ critical points are not as well understood as for the E_0 and $E_0 + \Delta_0$ states and theoretical predictions are more tenuous. There are, however, comparisons of RRS intensities and experimental structure in the susceptibility.

The most studied case is that of RRS from InSb which has been reported by two groups [3.60, 61]. The two sets of results are shown in Figs. 3.16 and 3.17. They are in essential agreement but different features are emphasized in the two figures. DREYBRODT et al. [3.60] have measured the RRS intensity for the TO phonon over an energy range including both E_1 and $E_1 + \Delta_1$ with the results shown in Fig. 3.16. We see that the experimental data indicates the two well-separated critical points with a smaller intermediate peak. They have compared their results

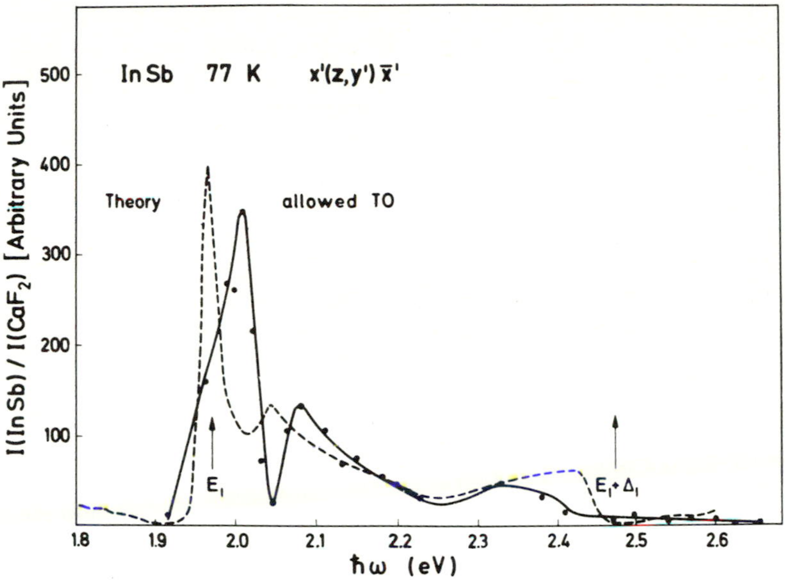

Fig. 3.16. Measured resonance in the Raman cross section of InSb for allowed TO scattering at 77 K. The dashed line was calculated in the manner described in the text. (From DREYBRODT et al. [3.60])

with the predicted RRS intensities (dashed curve) based on the experimental susceptibility χ. Thus all lifetime and exciton effects are implicitly included. However, they do make the approximation of neglecting the phonon frequency. Their expression for the Raman matrix element has the form [Ref. 8.1, Sect. 2.2.6b].

$$K \propto A(\chi^+ - \chi^-) + B\,\frac{d}{d\omega}(\chi^+ + \chi^-) + C\,. \tag{3.91}$$

Here χ^+ and χ^- are, respectively, the spin-orbit split band contributions to χ. In this case the spin-orbit splitting is large and the two contributions can be isolated. Consider the lower E_0 peak for which $\chi^- \approx 0$. Then the form (3.91) reduces to

$$K \propto A\chi + B\,\frac{d}{d\omega}\,\chi + C\,. \tag{3.92}$$

This is a general linear combination of the three types of terms given in (3.50) with the neglect of the phonon frequency. The sharp peak in the theoretical curve in Fig. 3.16 results from the $d\chi/d\omega$ term in (3.92) and the smoother structure results from the other terms. The constant C has been adjusted so that the real part of the matrix element has a zero just above the E_1 peak. This gives the minimum in the theoretical RRS intensity, in general agreement with the measured intensity.

The most prominent feature of the comparison between theory and experiment is that the sharp peak in the RRS occurs at an energy *higher* than that predicted from the critical points measured optically [3.65, 66]. The shift is also observed for the resonance enhancement of other phonon sidebands. This is best illustrated in the work of Yu and Shen [3.61] in Fig. 3.17. There are shown RRS intensity variations for several Raman features. All show similar sharp peaks reminiscent of the peak in $|d\chi/d\omega|^2$ but shifted to higher energy by ~ 0.05 eV. Similar shifts have been observed in other materials near the E_1 critical point [3.59, 62]. In addition, Yu and Shen [3.61] have shown that the RRS peak does not shift in exactly the same way with temperature as does the optical critical point.

Shifts in the peaks in RRS spectra to higher energies than those measured in modulation spectroscopy have also been observed in other similar semiconductors [3.59, 62]. For example, for RRS near E_1 and $E_1 + \Delta_1$ in Ge [3.59], only a single peak between E_1 and $E_1 + \Delta_1$ is observed. This is explained by assuming the intraband scattering term proportional to A in (3.91) is negligible small. The peak position, however, is shifted by about 45 meV above that predicted from (3.91). All examples of the shifts reported to date involve the E_1 or $E_1 + \Delta_1$ gaps. It is not clear whether these are general properties of higher band gaps or are the results of specific features of the E_1 gaps.

There is as yet not satisfactory explanation for the shifts. It appears that the inclusion of the correct phonon energy (0.02 eV) in the theoretical comparison could not bring theory and experiment into agreement. As we have pointed out, excitons are already properly included in the experimental susceptibilities, so that the shift cannot be simply ascribed to "exciton effects". Dreybrodt et al. [3.60] have shown that cancellation between different terms in the Raman matrix element can introduce structure into the RRS intensity, but it appears that the shift in the main peak cannot be produced in this way. (A quantitative explanation is still missing in 1982!).

It has been proposed [3.61] that the shift is caused by a double resonance effect dependent upon the finite wavevector of the phonon. However, as was first pointed out by Yacoby [3.67] and discussed in more detail in Section 3.2 and in [3.22, 68, 69], the double resonance in this case does not lead to additional enhancement. This is shown explicitly in the one-dimensional example, (3.52)–(3.59), where we find that additional peaks can occur only for wave vectors exceeding a

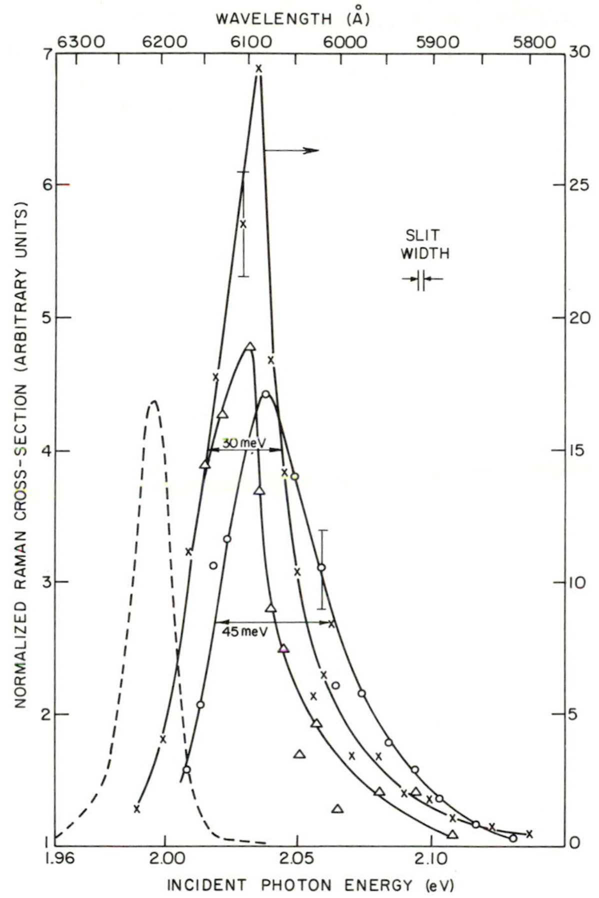

Fig. 3.17. Experimental Raman scattering cross section of InSb as a function of incident photon frequency. Crosses, surface-field-induced LO scattering ($T = 12$ K); triangles, allowed LO ($T = 14$ K); and circles, allowed TO ($T = 10$ K). Polarization and surface geometries are described in [3.61]. Dashed curve, $|d\chi/d\omega|^2$ obtained from the experimental wavelength-modulated reflectivity spectrum of InSb ($T = 5$ K) in [3.66]. (From Yu and Shen [3.61])

minimum value. The arguments are readily extended to the present case and one finds that typical wavevectors of the phonons are much smaller than this minimum value.

Continuum Bands in Molecules

Among the possible intermediate states of a diatomic molecule discussed under "RRS *in* I_2 *molecules*" in Subsection 3.4.2 are the continuum bands which are the vibrational-rotational states of the unbound molecule. For photons degenerate with such a band, these are expected to be the dominant intermediate states [3.70–72]. The scattering process is very similar to that involving continuum bands in a solid except that in this case 1) selection rules are much less restrictive since there is no translation invariance, and 2) the strong electron-vibrational coupling necessitates a non-perturbative approach for quantitative calculations.

The expression for the matrix element is the basic one, (3.10), and the diagrammatic representation is given in Fig. 3.2c. The matrix elements of $\mathscr{H}_{MR}$ needed in (3.10) are overlaps of the nuclear wave functions weighted by the dipole radiation matrix element between ground and excited electron states. Within the Franck-Condon approximation, these are calculable from the equation of motion of the atoms with potential energy functions $V_g(r)$ and $V_{ex}(r)$ for ground and excited electronic states, respectively. Here r is the internuclear separation in diatomic molecules. It is thus tedious but straightforward to calculate the contribution of any intermediate state to the Raman matrix element.

The primary point for our discussion is the sum over intermediate states in the continuum. The essential features are evident in the approximation for treating the electron-vibrational coupling as a perturbation $\mathscr{H}'_m$. Then the zeroth-order approximate eigenstates are the decoupled atoms which have a simple kinetic energy and square root density of states just as for a three-dimensional effective mass band in a solid. Making the constant matrix element approximation, we arrive at the "independent double resonance" expressions, (3.36) and (3.37). The expressions are closely related to those for a solid, the only difference being that the lack of translation invariance leads to the expression (3.37) as a product of "in" and "out" susceptibilities rather than the difference expression, (3.41) found for the "coupled double resonance" in a solid.

As a specific example we consider the case of the I_2 molecule which has been extensively studied [3.33, 70, 73, 74]. In this case, unlike that discussed in Subsection 3.4.2 of resonance at discrete states, the enhancement occurs simultaneously for many possible Raman transitions. The spectrum is very complex consisting of many vibrational orders, each split by rotational levels and unequal spacings of vibrational levels. Examples of the fine structure in specific vibrational orders are given in Fig. 3.18, which is taken from WILLIAMS and ROUSSEAU [3.70]. Each peak in the Raman spectrum, nevertheless, corresponds to specific initial and final states and may be considered individually.

Also shown in Fig. 3.18 are the theoretical calculations of WILLIAMS and ROUSSEAU [3.70], who computed the matrix elements properly

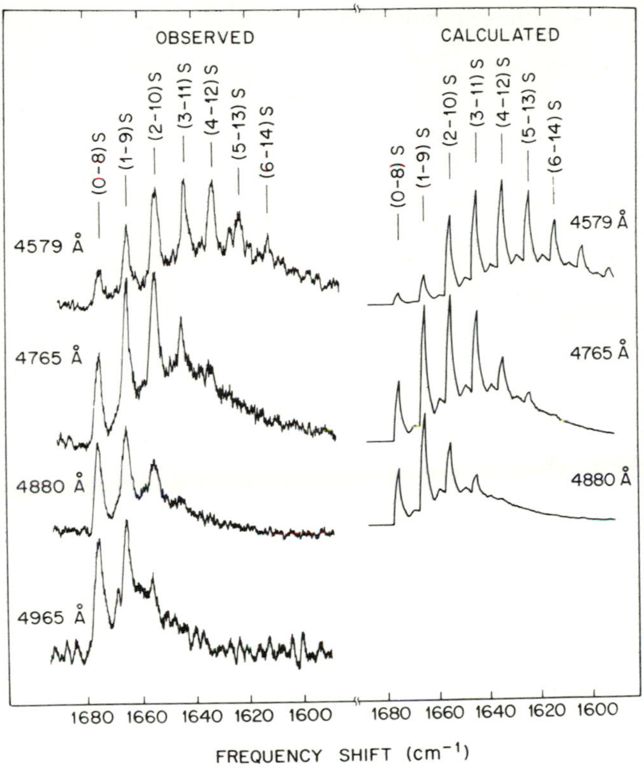

Fig. 3.18. Experimentally observed and theoretically calculated iodine resonant Raman spectra for the $\Delta v'' = 8$ transition. S numbers refer to the initial- and final-vibrational-state assignments of the S branches. (From WILLIAMS and ROUSSEAU [3.70])

within the Franck-Cordon approximation and performed the integral over the continuum numerically. Their numerical results for the relative amplitudes of the different Raman features as a function of frequency are sensitive functions of the choice of the potentials $V_g(r)$ and $V_{ex}(r)$, which have been adjusted to fit the experimental data. The RRS for the continuum bands thus provides information about the internuclear interactions $V_{ex}(r)$ for specific electronic states, in addition to that derived from resonance with discrete states. The continuum states are especially sensitive to the form of $V_{ex}(r)$ at both small and large r. It should be noted that this procedure can also determine $V_{ex}(r)$ for unbound electronic states for which there is only the continuum and no discrete states.

3.4.5. Continuum Intermediate and Final States

In this subsection we turn to the final category of RRS, cases in which it is essential to sum over a continuum of final states and for each final state to evaluate the matrix element by summing over a continuum of intermediate states. All theoretical calculations are closely related to those needed for the previous subsection, but the sum over final states change qualitatively the comparison with experiment. Here we must consider both the intensity as a function of incident photon frequency ω_1 and the line shape of the Raman spectrum as function of the scattered photon frequency ω_2 for fixed ω_1. The simplest examples are ones in which the resonance factors are essentially constant for all final states, in which case all considerations derived for a single final state apply with only the modifications that the Raman line shape is determined by the characteristic density of final states. We shall not consider any such cases and shall restrict consideration to cases in which the resonance factors vary greatly for the different possible final states.

Multiple-order RRS in crystals is the primary theoretical and experimental example for this category. Momentum conservation requires that the total momentum of the final state be fixed by the experiment. A continuum of possible final states exists for a given total momentum, corresponding to the momenta of the individual excitations. The intermediate states similarly are labeled by their momenta. The analytic energy-momentum relations for the excitations of a crystal lead to resonance effects which may vary drastically for different final states.

The most direct example of resonance as a function of wave-vector is the double resonance that can occur in the continuum. This is discussed in Section 3.2. It was shown there that for second-order scattering with photons in the absorption continuum, the RRS intensity is greatly enhanced for phonons satisfying the double resonance criterion. In the one-dimensional crystalline case treated explicity in Section 3.2, the double resonance is manifested as a Lorentzian peak in the cross section as a function of wave vector q of each phonon. In this case the double resonance may completely dominate the second-order scattering—the momentum of the phonons can be restricted to a narrow range which is a function of the frequency of the incident photon.

The results from three-dimensional calculations [3.22, 68] are also mentioned in Section 3.2. The double resonance there leads only to a weak logarithmic peak. Whether or not a small range of phonons is particularly enhanced in three dimensions depends upon lifetime of the electrons and the nature of the interaction. The q-dependent Fröhlich interaction, which is important in RRS, enhances small-wave-vector scattering ($q \ll$ BZ boundary). This can lead to strong RRS involving

LO phonons only near the zone center. Therefore, we expect that in polar semiconductors, the Fröhlich interaction can lead to sharply peaked, strong RRS dominated by the double (multiple) resonance phenomena. For deformation potential scattering the line shapes will be much less drastically altered, although peaks not associated with critical points can occur [3.22, 68]. (See [Ref. 8.1, Sect. 2.2.10].)

Let us proceed to the experimental results where the qualitative features of the scattering are exhibited clearly. First, consider a case in which the photons are near but not at resonance with the excitonic and band states in a semiconductor. The resonance enhancement of the second-order LO phonon RRS has been considered by many workers [3.17, 37–46, 57, 75–77]. In this case strongly peaked spectra are observed. The matrix element effects which restrict the phonons to near the zone center appear to be so strong that the density of states factors cannot be observed. An example near resonance is shown in Fig. 3.11, where we see that the 2 LO peak is almost as narrow as the first-order peak. For photons above the band gap it becomes even narrower [3.42]. Similarly, LO + TO combinations have been observed. The most thoroughly studied case is GaP in which the total intensities have been measured by Weinstein and Cardona [3.57] for photons near the lowest direct gap. The peaks for 2 LO and LO + TO not associated with any critical point are shown in Fig. 3.19. Zeyher [3.68] has calculated the integrated intensities from the three-dimensional calculations referred to above. The agreement is very good for both LO + TO and 2 LO. Zeyher [3.68] has predicted the detailed line shapes, but the lack of dispersion in the phonon branches has apparently prevented experimental determination of the line shapes. No additional peaks for 2 TO phonons near the zone center have been reported for this case. Note, however, that near the two-dimensional critical point E_1 (where double-resonance factors are more important as discussed in Section 3.2) additional peaks for 2TO phonons near the zone center have been observed in Ge [3.78, 81].

Similar modifications of the two-phonon spectra occur at indirect resonances; in these cases they involve phonons of wave vectors which bring one energy denominator in the theoretical expression (3.32) into resonance. Since only one denominator is resonant, the sums over intermediate states have the forms given in (3.26)–(3.29). The result is that the two-phonon spectrum is modified in the same manner as for double resonances, i.e. a pole in one-dimension, inverse square root in two, and a logarithmic divergence in three dimensions. Examples of indirect resonances have been observed [3.77, 79]. There are changes in the shape of the second-order spectra but no additional peaks have been observed.

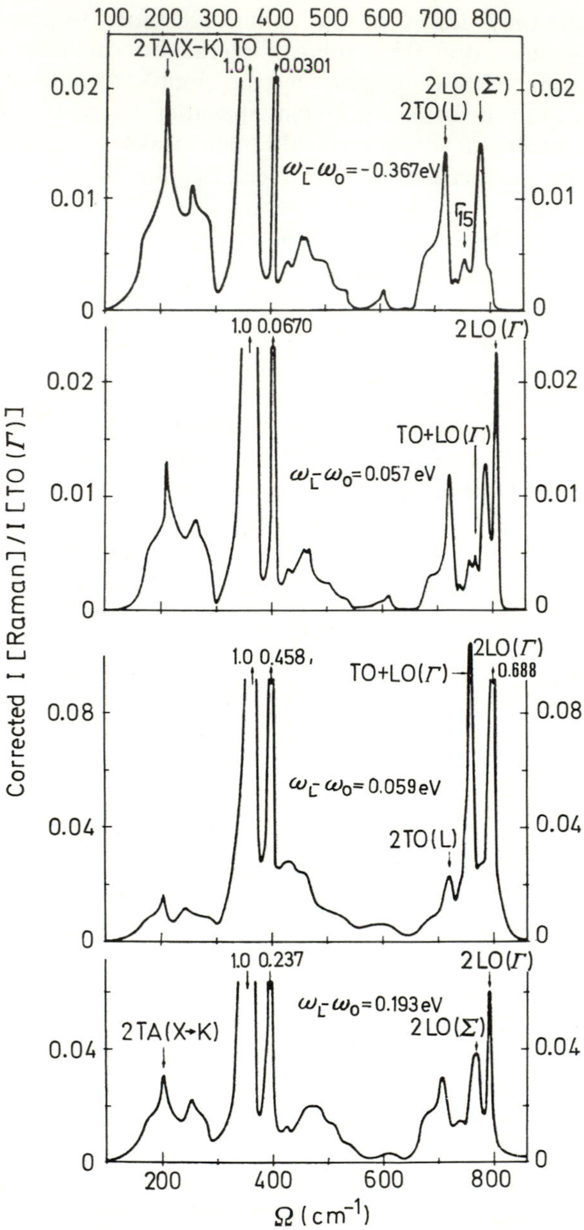

Fig. 3.19. Raman spectra of GaP normalized to the first-order TO intensity. The notation $\omega_L - \omega_0$ denotes the separation of the laser energy from the lowest direct gap. The features labelled TO + LO (Γ) and 2 LO (Γ) are much more strongly resonant than other features. (From WEINSTEIN and CARDONA [3.57])

It is interesting to note that the theoretical formulation of scattering at an indirect resonance is the same as that for scattering near a forbidden transition. The analysis applied to the 2-phonon scattering in Cu_2O (Subsection 3.4.3 and [3.55]) applies equally well to an indirect exciton; conversely, the preceeding analysis applies to resonance with the continuum states for the lowest (α) band in Cu_2O.

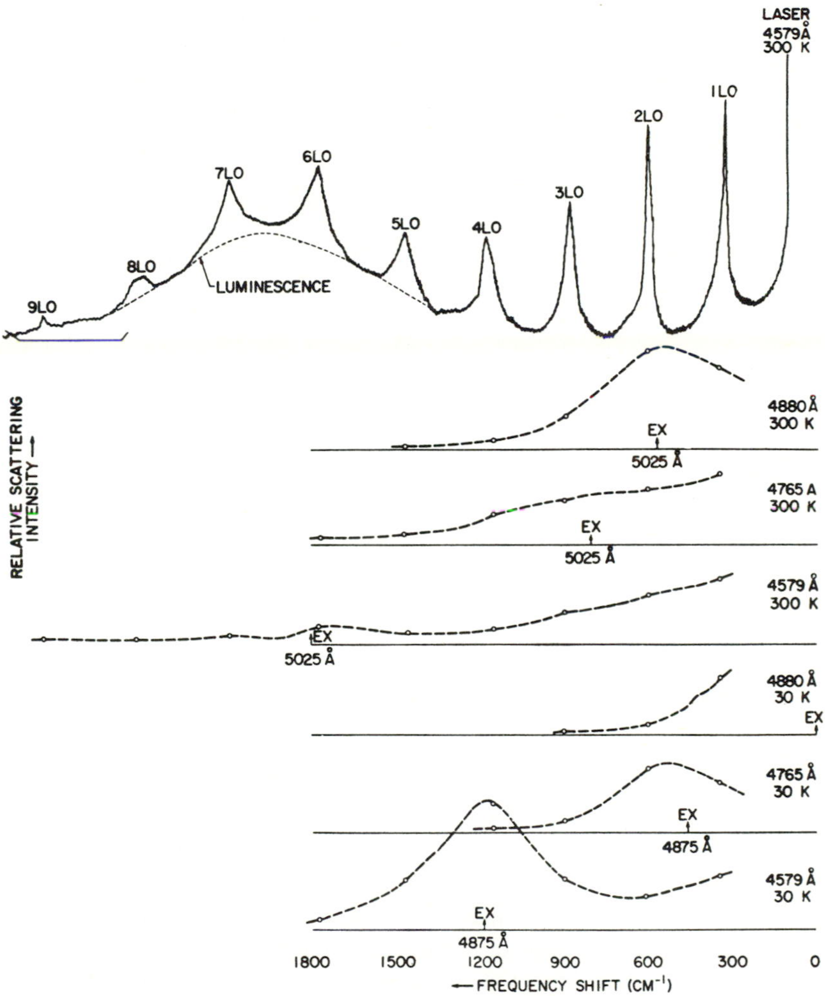

Fig. 3.20. Top trace: Uncorrected spectrum of CdS at ~300 K with excitation at 4579 Å. Below: Corrected intensities of multiphonon scattering features in CdS as a function of temperature and laser frequency. The arrows indicate the approximate energy of the $n=1$ free and I_2, I_5 bound excitons. (From LEITE et al. [3.75])

For photons above the band gap in the absorption continuum the higher-order scattering is even more striking. As in previous cases, the best example is CdS [3.42, 75, 76]. LEITE et al. [3.75], and KLEIN and PORTO [3.76] have each presented the RRS spectra for CdS using, several excitation lines from the Ar^+ ion laser which lie in the electronic band continuum. We reproduce in Fig. 3.20 the results of LEITE et al. [3.75]. There we show sharp n-LO peaks for $n=1$ through $n=9$. No other Raman features are detectable in Fig. 3.20 although TO + LO combinations also can be observed [3.42] in some cases. Similar results are found in many other polar semiconductors [3.37, 38, 46], so that it is clear that the intense, sharp multiple LO lines are an intrinsic aspect of RRS in these materials [Ref. 8.1, Sect. 2.3.5].

The most striking aspect of the scattering above the band gap is the intensity of the higher-order scattering, which does not decrease as a simple perturbation analysis would predict. This has been interpreted by MARTIN and VARMA [3.80] as a cascade process very similar to that of the higher-order processes in Cu_2O discussed in Subsection 3.4.3. Analysis in this case is in general very complicated because of the interference between different intermediate states in the continuum, which must be summed properly. The suggestion of a cascade description has been supported by the analysis of ZEYHER [3.69] and MARTIN [3.22], who concluded that a branching ratio description holds approximately for order $n>3$, although not for $n=2$. In this case the RRS is explicitly dependent upon the lifetime and there is a characteristic time delay as in resonance fluorescence, but the calculation of the Raman emission probabilities themselves must take into account the interference between the different intermediate states in the continuum [3.22, 69, 80].

3.5. Conclusion

The basic theoretical formulas describing RRS have been derived and their consequences explored for representative cases. The important formula is the second-order perturbation expression (3.10) for the Raman matrix element in terms of the material-radiation interaction. Expanding that expression in terms of approximate eigenstates of the material system leads to the often-used expression (3.33) and similar higher-order expressions in terms of the interactions between the approximate eigenstates. In each case the fundamental theoretical distinctions are between discrete and continuum intermediate and final states, the combinations of which form the categories discussed in Section 3.2.

Two concepts were found to be of central importance in the interpretation of RRS: 1) the comparison and contrast of the Raman matrix element with the dielectric susceptibility and 2) double (multiple) resonances which describe the evolution of the relevant states of the material system in terms of sucessive scattering between approximate eigenstates. The basic expression (3.10) shows that poles in the Raman matrix element occur at the same frequencies as do those in the dielectric susceptibility, but with different material-radiation matrix elements. This is of great utility in the interpretations of cases where the variations in matrix elements is small compared to that of the density of states. Double or multiple resonance describes cases in which the formal material-radiation matrix element is itself a very complicated function of the photon frequencies. In such situations the dependence of RRS upon the intermediate states is quite different from that of the susceptibility.

Application of the basic formulas to specific cases for each category was discussed in Section 3.4 with the aid of results derived in Sections 3.2 and 3.3. In most cases experimental results and comparisons with theoretical calculations could be used to illustrate the dominant features of the theoretical formulas. These are the various ramifications of the Lorentzian resonance of a discrete intermediate state discussed in Subsections 3.4.2 and 3.4.3 and the possible interferences among continuum intermediate states considered in Subsections 3.4.4 and 3.4.5. In the final category, continuum intermediate and final states, the experimental results were not sufficient to quantify the resonance changes in the line shapes. In that case, a simple one-dimensional crystalline model for higher-order phonon scattering, described in Section 3.2, suffices to illustrate additional structure in Raman lineshapes which varies as a function of phonon energy and depends upon the nature of the resonant intermediate states.

References

3.1. W. Heitler: *The Quantum Theory of Radiation*, 3rd ed. (Oxford at the Clarendon Press, Oxford 1957).

3.2. M. M. Sushchinskii: *Raman Spectra of Molecules and Crystals* (Israel Program for Scientific Translations, New York, 1972), pp. 35, 44ff.

3.3. P. F. Williams, D. L. Rousseau, S. H. Dworetsky: Phys. Rev. Letters **32**, 196 (1974).

3.4. R. Loudon: Proc. Roy. Soc. (London) A **275**, 218 (1963).

3.5. A. K. Ganguly, J. L. Birman: Phys. Rev. **162**, 806 (1967).

3.6. J. L. Birman: In *Light Scattering in Solids*, ed. by M. Balkanski (Flammarion, Paris, 1971), p. 15.

3.7. A. Pinczuk, E. Burstein: In *Proc.* 10th *Intern. Conf. Physics of Semiconductors.* Cambridge, Mass. 1970 (U.S. AEC, Tech. Infor. Div., Oak Ridge, Tenn., 1970), p. 727.

3.8. M. CARDONA: Surface Sci. **37**, 100 (1973).
3.9. R. LOUDON: Advan. Phys. **13**, 423 (1964).
3.10. R. M. MARTIN: In *Light Scattering in Solids*, ed. by M. BALKANSKI (Flammarion, Paris, 1971), p. 25.
3.11. R. M. MARTIN, T. C. DAMEN: Phys. Rev. Letters **26**, 86 (1971).
3.12. R. M. MARTIN: Phys. Rev. B**4**, 3676 (1971).
3.13. B. BENDOW, J. L. BIRMAN: Phys. Rev. B**1**, 1678 (1970).
3.14. G. L. BIR, G. E. PIKUS: Fiz. Tverd. Tela **2**, 2287 (1960) [Sov. Phys. Solid State **2**, 2039 (1961)].
3.15. J. M. RALSTON, R. L. WADSACK, R. K. CHANG: Phys. Rev. Letters **25**, 814 (1970).
3.16. T. C. DAMEN, J. F. SCOTT: Solid State Commun. **9**, 383 (1971).
3.17. R. H. CALLENDER, S. S. SUSSMAN, M. SELDERS, R. K. CHANG: Phys. Rev. B**7**, 3788 (1973).
3.18. R. BERKOWICZ, D. H. R. PRICE: Solid State Commun. **14**, 195 (1974).
3.19. D. L. MILLS, E. BURSTEIN: Phys. Rev. **188**, 1465 (1969).
3.20. B. BENDOW: Phys. Rev. B**2**, 5051 (1970).
3.21. R. ZEYHER, C. TING, J. L. BIRMAN: Phys. Rev. B**10**, 1725 (1974).
3.22. R. M. MARTIN: Phys. Rev. B**10**, 2620 (1974).
3.23. R. C. C. LEITE, J. F. SCOTT: Phys. Rev. Letters **22**, 130 (1969).
3.24. M. V. KLEIN: Phys. Rev. B**8**, 919 (1973).
3.25. K. REBANE: *Impurity Spectra of Solids; Elementary Theory of Vibrational Structure* (Plenum Press, New York, 1970).
3.26. Y. R. SHEN: Phys. Rev. B**9**, 622 (1974).
3.27. A. COMPAAN, H. Z. CUMMINS: Phys. Rev. Letters **31**, 41 (1973).
3.28. J. L. BIRMAN: Solid State Commun. **13**, 1189 (1973).
3.29. E. F. GROSS, A. G. ZHILICH, B. P. ZAKHARCHENYA, A. V. VARFALOMEEV: Fiz. Tverd. Tela **3**, 1445 (1961) [Sov. Phys. Solid State **3**, 1048 (1961)].
3.30. A. COMPAAN, H. Z. CUMMINS: Phys. Rev. B**6**, 4753 (1972).
3.31. G. HERZBERG: *Molecular Spectra and Molecular Structure*, Vol. 1 (van Nostrand-Reinhold, New York, 1950).
3.32. J. L. STEINFELD, J. D. CAMPBELL, N. A. WEISS: J. Mol. Spectrosc. **29**, 204 (1969).
3.33. W. HOLZER, W. F. MURPHY, H. J. BERNSTEIN: J. Chem. Phys. **52**, 399 (1970).
3.34. M. KROLL, D. SWANSON: Chem. Phys. Letters **9**, 115 (1971).
3.35. D. G. FOUCHE, R. K. CHANG: Phys. Rev. Letters **29**, 536 (1972).
3.36. R. L. ST. PETERS, S. D. SILVERSTEIN, M. LAPP, C. M. PENNEY: Phys. Rev. Letters **30**, 191 (1973).
3.37. J. F. SCOTT, R. C. C. LEITE, T. C. DAMEN: Phys. Rev. **188**, 1285 (1969).
3.38. R. C. C. LEITE, T. C. DAMEN, J. F. SCOTT: In *Light Scattering in Solids*, ed. by G. B. WRIGHT (Springer-Verlag New York 1969), pp. 359.
3.39. E. F. GROSS, S. A. PERMOGOROV, V. V. TRAVNIKOV, A. V. SEL'KIN: Fiz. Tverd. Tela **13**, 699 (1971) [English Transl.: Sov. Phys.-Solid State **13**, 578 (1971)].
3.40. S. A. PERMOGOROV, V. V. TRAVNIKOV: Fiz. Tverd. Tela **13**, 709 (1971) [English Transl.: Sov. Phys.-Solid State **13**, 586 (1971)].
3.41. E. F. GROSS, S. A. PERMOGOROV, V. V. TRAVNIKOV, A. V. SEL'KIN: In *Light Scattering in Solids*, ed. by M. BALKANSKI (Flammarion, Paris, 1971), pp. 238–243.
3.42. T. C. DAMEN, R. C. C. LEITE, J. SHAH: In Proc. 10th International Conference on the Physics of Semiconductors, Cambridge, Mass., ed. by S. P. KELLER, J. C. HENSEL, F. STERN (USAEC Div. of Technical Information Extension, Oak Ridge, Tenn., 1970), p. 735.
3.43. P. J. COLWELL, M. V. KLEIN: Solid State Commun. **8**, 2095 (1969).
3.44. P. F. WILLIAMS, S. P. S. PORTO: In *Light Scattering in Solids*, ed. by M. BALKALNSKI (Flammarion, Paris, 1971), pp. 70, 71.

3.45. T.C. Damen, J. Shah: Phys. Rev. Letters **27**, 1506 (1971).
3.46. J.F. Scott, T.C. Damen, W.T. Silfvast, R.C.C. Leite, L.E. Cheesman: Opt. Commun. **1**, 397 (1970).
3.47. P.F. Williams, S.P.S. Porto: Phys. Rev. B**8**, 1782 (1973).
3.48. T. Fukumoto, H. Yoshida, S. Nakashima, A. Mitsuishi: J. Phys. Soc. Japan **32**, 1674 (1972).
3.49. E.F. Gross, A.G. Plyukhin, L.G. Suslina, E.B. Shadrin: Zh. Eksper. I. Teor. Fiz. Pisma **15**, 312 (1972) [English Transl.: Sov. Phys.—JETP Letters **15**, 220 (1972)].
3.50. E.I. Rashba, G. Gurgenishvili: Fiz. Tverd. Tela **4**, 1029 (1962) [English Transl. Sov. Phys.—Solid State **4**, 759 (1962)].
3.51. W.C. Tait, R.L. Weiher: Phys. Rev. **166**, 769 (1968).
3.52. J. Conradi, R.R. Haering: Phys. Rev. **185**, 1088 (1969).
3.53. G.P. Vella-Coleiro: Phys. Rev. Letters **23**, 697 (1969).
3.54. D.G. Thomas, J.J. Hopfield: Phys. Rev. **128**, 2135 (1962).
3.55. P.Y. Yu, Y.R. Shen, Y. Petroff, L.M. Falicov: Phys. Rev. Letters **30**, 283 (1973).
3.56. P.Y. Yu, Y.R. Shen: Phys. Rev. Letters **32**, 939 (1974).
3.57. B.A. Weinstein, M. Cardona: Phys. Rev. B**8**, 2795 (1973).
3.58. M.I. Bell, R.N. Tyte, M. Cardona: Solid State Commun. **13**, 1833 (1973).
3.59. F. Cerdeira, W. Dreybrodt, M. Cardona: Solid State Commun. **10**, 591 (1972).
3.60. W. Dreybrodt, W. Richter, M. Cardona: Solid State Commun. **11**, 1127 (1972).
3.61. P.Y. Yu, Y.R. Shen: Phys. Rev. Letters **29**, 468 (1972).
3.62. G.W. Rubloff, E. Anastassakis, F.H. Pollak: Solid State Commun. **13**, 1755 (1973).
3.63. J.C. Phillips: In *Solid State Physics*, ed. by F. Seitz and D. Turnbull, Vol. 18, p. 56. (Academic Press, New York 1966).
3.64. E.O. Kane: Phys. Rev. **180**, 852 (1969).
3.65. K.L. Shaklee, J.E. Rowe, M. Cardona: Phys. Rev. **174**, 828 (1968).
3.66. R.R.L. Zucca, Y.R. Shen: Phys. Rev. B**1**, 2668 (1970).
3.67. Y. Yacoby: Solid State Commun. **13**, 1061 (1973).
3.68. R. Zeyher: Phys. Rev. B**9**, 4439 (1974).
3.69. R. Zeyher: Solid State Commun. **16**, 49 (1975).
3.70. P.F. Williams, D.L. Rousseau: Phys. Rev. Letters **30**, 951 (1973).
3.71. M. Berjot, M. Jacon, L. Bernard: Optics Commun. **4**, 246 (1971).
3.72. J. Behringer: Z. Physik **229**, 209 (1969).
3.73. W. Kiefer, H.J. Bernstein: J. Mol. Spectrosc. **43**, 366 (1972).
3.74. M. Berjot, M. Jacon, L. Bernard: Opt. Commun. **4**, 117 (1971).
3.75. R.C.C. Leite, J.F. Scott, T.C. Damen: Phys. Rev. Letters **22**, 780 (1969).
3.76. M.V. Klein, S.P.S. Porto: Phys. Rev. Letters **22**, 782 (1969).
3.77. P.B. Klein, H. Masui, J. Song, R.K. Chang: Solid State Commun. **14**, 1163 (1974).
3.78. M.A. Renucci, J.B. Renucci, R. Zeyher, M. Cardona: Phys. Rev. B**10**, 4309 (1974).
3.79. G.L. Bottger, C.V. Damsgard: Solid State Commun. **9**, 1277 (1971).
3.80. R.M. Martin, C.M. Varma: Phys. Rev. Letters **26**, 1241 (1971).
3.81. D. Olego, M. Cardona: Solid State Commun. **39**, 1071 (1981).

4. Electronic Raman Scattering

M. V. KLEIN

With 15 Figures

Semiconductors contain a wide variety of electronic excitations, and many of them can be studied by Raman scattering techniques. They form the subject of this article. Excluded is the class of deep-level electronic transitions associated with transition-metal or rare earth atoms in semiconductors or oxides. Included are almost all the excitations of shallow levels in semiconductors that have been observed experimentally or proposed theoretically, with the important exception of excitations in a magnetic field. Thus spin-flip excitations in a magnetic field will not be discussed, except for a few comments at the end in Section 4.6. See [Ref. 8.8, Chap. 8].

After this introduction, we begin in Section 4.1 by discussing light scattering by carriers in simple semiconductors via a coupling of fluctuations in carrier density to the optical radiation fields. Long range coulomb interactions are included via the random phase approximation. This leads to a discussion of scattering by plasmons and (in polar crystals) by plasmon-LO phonon coupled modes [Ref. 8.8, Chap. 8].

Section 4.2 is devoted to scattering by (weakly) bound electrons and holes. The effective mass approximation provides a good basis for the discussion of several experimentally observed strongly allowed Raman scattering processes. The theory of acceptor transitions given here is new.

In Section 4.3 Raman scattering from a variety of coupled electron-phonon excitations is discussed. Some are localized in space; some are not, but they share features common to coupled mode problems, namely frequency shifts, line shape changes, and interference effects. The theoretical discussion of this topic contains some new material.

In Section 4.4 we return to the discussion of light scattering by single particle excitations that was begun in Section 4.1. The emphasis is on the nature of the fluctuations, such as spin density fluctuations, or energy density fluctuations, to which the scattering process can couple, on the strength of the scattering processes, and on the shape of the spectra.

In Section 4.5 scattering from multicomponent carriers is discussed. The emphasis is on multivalley *n*-type semiconductors and on the possibility of scattering of light via intervalley density fluctuations.

Acoustic plasmon scattering will be mentioned. Unlike the other sections, this one is mainly theoretical.

Section 4.6 closes the article with a few comments about possible areas of emphasis in the future and about spin-flip Raman scattering.

4.1. Light Scattering from Free Carriers in Semiconductors

We consider first those light scattering phenomena that are straightforward generalizations of scattering properties of a free electron gas. The recent book by PLATZMAN and WOLFF [4.1] is a good general reference for this subject.

4.1.1. Theory of Light Scattering by a Free Electron Gas [4.1, 2]

The Hamiltonian for the interaction of an electron charge $-e$ at position $\boldsymbol{r}$ with a radiation field having vector potential $\boldsymbol{A}(\boldsymbol{r}, t)$ is

$$\mathscr{H}_{\text{el-rad}} = e^2 |A|^2/(2mc^2) + \tfrac{1}{2} e(\boldsymbol{p} \cdot \boldsymbol{A})/mc + \tfrac{1}{2} e(\boldsymbol{A} \cdot \boldsymbol{p})/mc\,. \tag{4.1}$$

Let $A_1 \boldsymbol{e}_1 \exp[\mathrm{i}(\boldsymbol{k}_1 \cdot \boldsymbol{r} - \omega_1 t)]$ + Hermitian conjugate (H.C.) and $A_2 \boldsymbol{e}_2 \cdot \exp[\mathrm{i}(\boldsymbol{k}_2 \cdot \boldsymbol{r} - \omega_2 t)]$ + H.C. denote the vector potential of incident and scattered photons, respectively. The $|\boldsymbol{A}|^2$ term in (4.1) caused a Lorentz force on the electron due to the vector product of the velocity acquired in the $\boldsymbol{E}$ field with the $\boldsymbol{B}$ field. Its cross terms dominate the scattering process [4.2]. The scattering Hamiltonian is then

$$\mathscr{H}' = r_0 (\boldsymbol{e}_1 \cdot \boldsymbol{e}_2)\, A_1 A_2^\dagger \exp[\mathrm{i}(\boldsymbol{q} \cdot \boldsymbol{r} - \omega t)] + \text{H.C.}\,, \tag{4.2}$$

where $\boldsymbol{q} = \boldsymbol{k}_1 - \boldsymbol{k}_2$ and $\omega = \omega_1 - \omega_2$, and where $r_0 = e^2/(mc^2) = 2.83 \times 10^{-13}$ cm is the classical radius of the electrons. The force on the electron implied by (4.2) is along $\boldsymbol{q}$. If $|i\rangle$ and $|f\rangle$ denote initial and final plane wave states for the electron, the power scattering cross-section that results from the application of time-dependent perturbation theory (the "golden rule") using the perturbation (4.2) is

$$d^2\sigma/d\omega\, d\Omega = (\omega_2/\omega_1)^2\, r_0^2 (\boldsymbol{e}_1 \cdot \boldsymbol{e}_2)^2 \sum_f |\langle f | \exp(\mathrm{i}\boldsymbol{q} \cdot \boldsymbol{r}) | i \rangle|^2 \cdot \delta(\omega - \omega_f + \omega_i)\,. \tag{4.3}$$

The *photon* scattering cross-section is (ω_1/ω_2) times (4.3).

For a collection of electrons $\exp(i\boldsymbol{q}\cdot\boldsymbol{r})$ is replaced by an operator which is essentially the Fourier component of the number density:

$$\varrho_q \equiv \sum_j \exp(i\boldsymbol{q}\cdot\boldsymbol{r}_j) = \sum_K C^\dagger_{K+q} C_K , \tag{4.4}$$

where $C^\dagger_{K+q}$ and C_K are creation and annihilation operators, and the volume has been equal to one.

Equation (4.2) becomes

$$\mathscr{H}' = r_0(\boldsymbol{e}_1 \cdot \boldsymbol{e}_2)\, A_1 A_2^\dagger \exp(-i\omega t)\, \varrho_q + \text{H.C.} \tag{4.5}$$

When used with the "golden rule", Eq. (4.5) gives the following result for the "scattering efficiency" or cross section (equal for unit volume) from n electrons per unit volume, see (2.16),

$$d^2R/d\omega\, d\Omega = n d^2\sigma/d\omega\, d\Omega = (\omega_2/\omega_1)^2\, r_0^2 (\boldsymbol{e}_1 \cdot \boldsymbol{e}_2)^2\, S(q,\omega) , \tag{4.6a}$$

where

$$S(q,\omega) = Av_i \left[\sum_f |\langle f|\varrho_q|i\rangle|^2\, \delta(\omega_{fi} - \omega) \right] \tag{4.6b}$$

is the "dynamical structure factor", and where $Av_i[\ ]$ represents a statistical average over state $|i\rangle$. The states $|i\rangle$ and $|f\rangle$ are now exact many-body states [4.3]. $S(q,\omega)$ is an equilibrium property of the electron system in the absence of the perturbation $\mathscr{H}'$. In such a case the mean value $\langle \varrho_q \rangle$ of the number density fluctuation is zero. The total mean square number density fluctuation is given by

$$S(q) \equiv Av_i \langle i | \varrho_q|^2 i \rangle , \tag{4.7a}$$

which can be written

$$\begin{aligned} S(q) &= Av_i \sum_f \langle i|\varrho_q^*|f\rangle \langle f|\varrho_q|i\rangle \\ &= \int Av_i \sum_f |\langle f|\varrho_q|i\rangle|^2\, \delta(\omega_{fi} - \omega)\, d\omega \\ &= \int S(q,\omega)\, d\omega . \end{aligned} \tag{4.7b}$$

Equations (4.6a) and (4.6b) and (4.7a) and (4.7b) show that the light scattering spectrum is proportional to the dynamical structure factor $S(q,\omega)$ which in turn can be interpreted as the power spectrum of the density fluctuation at angular frequency ω.

It is not possible to calculate $d^2R/d\omega\, d\Omega$ from (4.6) using the unknown exact many-electron states $|i\rangle$ and $|f\rangle$. A very useful approximation, the RPA (random phase approximation), may be easily obtained if we first use the "fluctuation-dissipation theorem" [4.4]. This says that $S(q, \omega)$, the power spectrum of the density fluctuations, equals $(-\pi^{-1})$ multiplied by the imaginary part of a density-density response function, which we shall denote by $\bar{F}(q, \omega)$, multiplied by a temperature-dependent factor $(1+n_\omega)$, to be given below. $\bar{F}(q, \omega)$ describes the driven response of the system to an external time-varying potential $\varphi_{\text{ext}} \equiv \exp(-i\omega t)$. the perturbing Hamiltonian is assumed to be of the form (4.5), namely $\mathscr{H}' = -e\varphi_{\text{ext}}\varrho_q + \text{H.C.}$ Due to its presence there will be a non-zero induced charge density $\varrho_{\text{ind}} = \langle \varrho_q \rangle$, which in the lowest order of time-dependent perturbation theory will be proportional to φ_{ext}. The coefficient defines $\bar{F}(q, \omega)$

$$\varrho_{\text{ind}} = -e\varphi_{\text{ext}}\bar{F}(q, \omega) .$$

Many-electron effects due to the Coulomb interaction are to be included in $\bar{F}$. The RPA results if we neglect all such interactions, but a self-consistent one between ϱ_q and φ_{ind}, the electric potential associated with the induced charge density $-e\varrho_{\text{ind}}$. φ_{ind} may be calculated from ϱ_{ind} by Poisson's equation. This interaction adds a term $\mathscr{H}'_{\text{ind}} = -e\varphi_{\text{ind}}\varrho_q + \text{H.C.}$ to the perturbing Hamiltonian. The linear response to $\mathscr{H}' + \mathscr{H}'_{\text{ind}}$ may then be determined by perturbation theory using free-electron states, giving [4.1, 4, 5]

$$\varrho_{\text{ind}} = -e(\phi_{\text{ext}} + \phi_{\text{ind}})F(q, \omega) . \tag{4.8a}$$

Here $F(q, \omega)$ is the free-electron response function and is given by

$$\hbar F(q, \omega) = \sum_{k} [n(\boldsymbol{k}) - n(\boldsymbol{k}+\boldsymbol{q})]\,[\omega + \mathrm{i}0^+ + \omega(\boldsymbol{k}) - \omega(\boldsymbol{k}+\boldsymbol{q})]^{-1} , \tag{4.8b}$$

where $n(k)$ is the thermal occupation number (per unit volume).

The dielectric response function of the free electron gas is

$$\varepsilon(q, \omega) = 1 - 4\pi e^2 F(q, \omega)/q^2 . \tag{4.9}$$

The light scattering efficiency is then given by [4.5], see (2.17),

$$d^2R/d\omega\,d\Omega$$
$$= (\omega_2/\omega_1)^2 (\boldsymbol{e}_1 \cdot \boldsymbol{e}_2)^2 r_0^2 (1+n_\omega)\hbar\pi^{-1}(-\operatorname{Im}\{F\})\,|\varepsilon|^{-2} \tag{4.10a}$$
$$= (\omega_2/\omega_1)^2 (\boldsymbol{e}_1 \cdot \boldsymbol{e}_2)^2 r_0^2 (1+n_\omega)\,[\hbar q^2/(4\pi e^2)]\operatorname{Im}\{-1/\varepsilon\} , \tag{4.10b}$$

where $n_\omega = [\exp(\hbar\omega/k_B T) - 1]^{-1}$.

If the RPA correction is neglected, ε in (4.10a) is replaced by unity, and we have $d^2R/d\omega d\Omega \propto \pi^{-1} \operatorname{Im}\{F\}$, the free particle excitation spectrum. This is screened by a factor of $|\varepsilon|^{-2}$ when Coulomb interactions are considered within the RPA. The nature of the screening depends on the values of q and ω. For small q and at low frequencies where $\omega \gtrsim q v_e$ (v_e is either the Fermi velocity v_f or a thermal electron velocity $v_{th} = (2k_B T/m)^{\frac{1}{2}}$), ε is approximately

$$\varepsilon \cong 1 + (q_s^2/q^2) . \tag{4.11}$$

The screening wave vector q_s is given by the Fermi-Thomas result ($k_B T \ll E_f$)

$$q_{FT} = (6\pi n e^2/E_f)^{\frac{1}{2}} = \sqrt{3}\,\omega_p/v_f \tag{4.12a}$$

or the Debye-Hückel result

$$q_D = (4\pi n e^2/k_B T)^{\frac{1}{2}} = \sqrt{2}\,\omega_p/v_{th} = \sqrt{3}\,\omega_p/v_{rms} \tag{4.12b}$$

in the degenerate-quantum and classical limits, respectively.

In the degenerate, small q, $T=0$ case, the free particle excitation spectrum $(-\operatorname{Im}\{F\})$ is proportional to ω from $\omega = 0$ to $\omega = q v_f$. It then drops to zero at $\omega = q v_f + \hbar q^2/2m$ [4.4]. A finite value of T will cause additional rounding near $\omega = q v_f$, as will damping due to a finite electron lifetime. This latter effect was discussed by Mermin [4.6]. At nonzero temperatures $(-\operatorname{Im}\{F\})$ is nonzero for negative ω. This allows nonzero anti-Stokes light scattering.

In the classical high temperature limit, there is a Maxwellian excitation spectrum

$$-\operatorname{Im}\{F\} \propto \exp[-\omega^2/(q v_{th})^2] .$$

In both degenerate and classical cases the screening by the factor $|\varepsilon|^{-2}$ is very large in the limit $q \ll q_s$ for then $\varepsilon \approx q_s^2/q^2$, which gives, according to (4.10a)

$$d^2R/d\omega\, d\Omega \propto (-\operatorname{Im}\{F\})\,(q/q_s)^4 . \tag{4.13}$$

It is known that $S(q,\omega)$ obeys several exact sum-rules [4.3, 4], one of which leads to the following exact low temperature result for the scattering cross-section per electron

$$\int_0^\infty \omega(d^2\sigma/d\omega\, d\Omega)\, d\omega = (\omega_2/\omega_1)^2\, (\boldsymbol{e}_1 \cdot \boldsymbol{e}_2)^2\, r_0^2\, \hbar q^2/(2m) . \tag{4.14}$$

When (4.13) is integrated, it gives a much smaller result for small q than is predicted by (4.14). The discrepancy is removed by using (4.10b), which equals (4.10a) when $\text{Im}\{F\} \neq 0$, but which contributes an extra term to the scattering efficiency from regions where ε itself (or more properly, its real part) is zero. Such behavior is found at the plasma frequency ω_{pq}, whose dispersion relation is approximately [4.3]

$$\omega_{pq} = \omega_p(1 + 3q^2 v_f^2 \omega_p^{-2}/10), \tag{4.15a}$$

where

$$\omega_p = (4\pi n e^2/m)^{\frac{1}{2}}, \tag{4.15b}$$

is the plasma frequency for $q=0$. For ω near ω_p one can show that [4.3]

$$\varepsilon(q \to 0, \omega) = 1 - \omega_p^2/(\omega^2 + i\omega\Gamma). \tag{4.16}$$

Here we have added a phenomenological electron lifetime Γ^{-1}. If this quantity obeys $\omega_p \gg \Gamma$, (4.10b) then yields a Lorentzian line at the plasma frequency with an integrated cross-section per electron that to lowest order in q is $(\omega_2/\omega_1)^2 (\boldsymbol{e}_1 \cdot \boldsymbol{e}_2)^2 r_0^2 \hbar^2 q^2/(2m\omega_p)$, in agreement with (4.14).

If q is sufficiently large, ω_{pq} given by (4.15) will be degenerate with the free particle excitation spectrum. The plasmon ceases to exist as an elementary excitation. This effect is sometimes called "Landau damping".

4.1.2. Light Scattering from Bloch Electrons

Electrons (or holes) moving through a perfect crystal within the conduction (valence) band will undergo intraband scattering of light via the $|A|^2$ terms of (4.1) and (4.2). The $\boldsymbol{p} \cdot \boldsymbol{A}$ term in (4.1) used with second-order perturbation theory produces virtual intraband and virtual interband transitions. The former are negligible for small electron velocities [4.7]. The effect of the virtual interband transition is to produce real intraband and interband scattering of a Bloch electron (hole) accompanied by the absorption of one photon and the emission of another. We treat the scattering produced by the real intraband transition. The scattering amplitude is proportional to the sum of two terms having denominators of the form $\omega_G - \omega_1$ and $\omega_G + \omega_2$, where $h\omega_G$ is the appropriate energy gap for the interband transitions. The result has been discussed by WOLFF [4.7] for a simple case and more generally by JHA [4.8] and by BLUM [4.9]. For small q and for ω_1 sufficiently less than ω_G

the denominators can be replaced by ω_G^{-1}, and the resulting expressions then combine with the contribution of the $|A|^2$ term, through the $k \cdot p$ effective mass expression, to give a scattering Hamiltonian [Ref. 8.1, Eq. (2.144)]

$$\mathscr{H}'_{\varrho} = r_0 \mu A_1 A_2^{\dagger} \exp(-\mathrm{i}\omega t)\, \varrho_q + \mathrm{H.C.}, \tag{4.17}$$

for carriers in a single band, where $\mu \equiv \boldsymbol{e}_1 \cdot \overleftrightarrow{\boldsymbol{\mu}} \cdot \boldsymbol{e}_2$, with $\overleftrightarrow{\boldsymbol{\mu}}$ given by m times the reciprocal effective mass tensor. When the photon energy $\hbar\omega_1$ is close to the band gap, (4.17) must be multiplied by a resonance enhancement factor R_{12}^2, which for light mass carriers is:

$$R_{12} \cong E_G^2 (E_G^2 - \hbar^2 \omega_1^2)^{-1} \,. \tag{4.18}$$

With these changes the RPA results (4.10a) and (4.10b) still hold for the scattering efficiency [4.9] (see [Ref. 8.8, Sect. 2.3.2b])

$$\begin{aligned} &d^2 R/d\omega\, d\Omega \\ &= (\omega_2/\omega_1)^2 R_{12}^2 \mu^2 r_0^2 (1+n_\omega)\, [\hbar q^2/(4\pi e^2)]\, \varepsilon_s^2 \mathrm{Im}\{-1/\varepsilon\}\,, \end{aligned} \tag{4.19}$$

where ε is given by an appropriate modification of (4.9) to include screening of free carriers by virtual interband transitions in the crystal. Equation (4.19) is often incomplete, since it does not include phonon effects.

The polarization selection rules are hidden in the definition of μ. Note that the rule $(\boldsymbol{e}_1 \cdot \boldsymbol{e}_2) \neq 0$ implied by (4.10) is replaced by $\mu \neq 0$ implied by (4.17). For simple mass ellipsoids, this gives "parallel" selection rules with respect to the principal axes of the ellipsoids.

4.1.3. Experimental Results for GaAs [Ref. 8.8, Chap. 2]

Mooradian performed a series of experiments on scattering of 1.06 μm light from a YAG : Nd^{3+} laser by free carriers in GaAs [4.10–13]. For an electron concentration $n < 10^{16}\,\mathrm{cm}^{-3}$ at room temperature the plasmon does not exist. The free particle excitations produced a symmetric Gaussian spectrum centered at zero frequency shift, as expected. Such scattering will be discussed in more detail in Section 4.4.

For $n > 10^{16}\,\mathrm{cm}^{-3}$ a plasma line emerges from the free-particle spectrum. Figure 4.1 provides a good example. The notation is conventional and gives incident and scattered propagation directions $\hat{k}_1$ and $\hat{k}_2$ and polarization directions $\boldsymbol{e}_1$ and $\boldsymbol{e}_2$ as follows: $\hat{k}_1(\boldsymbol{e}_1, \boldsymbol{e}_2)\hat{k}_2$. The plasmon is seen at $130\,\mathrm{cm}^{-1}$ in the (zz) spectrum. The free particle spectrum is also visible, as are the sharp transverse optical phonon (TO)

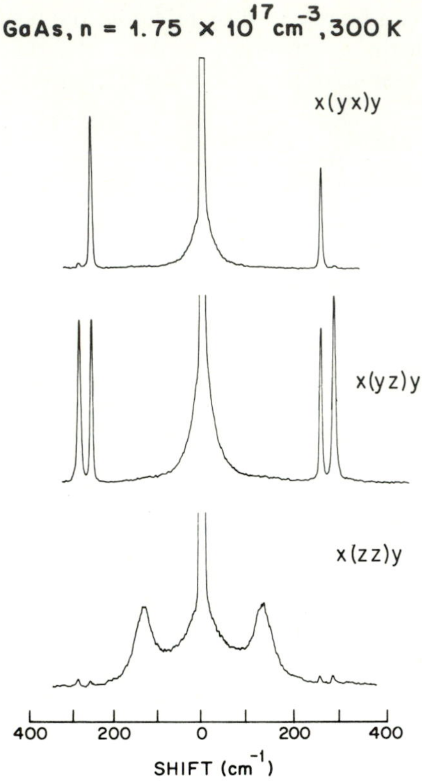

Fig. 4.1. Polarized Raman spectra from n-type GaAs at room temperature [4.13]

at $\omega_t = 272\,\mathrm{cm}^{-1}$ and the sharp longitudinal optical phonon (LO) at $\omega_l = 296\,\mathrm{cm}^{-1}$. The position of the plasmon is proportional to $n^{\frac{1}{2}}$ until it reaches the vicinity of ω_t. Then for a crystal lacking inversion symmetry, such as GaAs, the plasmon interacts with the LO phonon producing coupled modes L^+ and L^- at frequencies ω^+ and ω^- [4.14–16]. These modes can be detected in the lower two traces of Fig. 4.2, which shows the scattered light spectrum from a more heavily doped crystal at 2 K. The TO phonon is unaffected, but the LO phonon and the plasmon have ceased to exist.

Figure 4.2 reveals other new phenomena. The single particle spectrum shows the expected cut-off near $q v_f$, but it appears in the (yz) polarization geometry. This is unexpected, since the effective mass is isotropic. The reason for this will be given below in our discussion of spin-density fluctuations. The $L^{\pm}$ peaks appear in the (zz) geometry expected for

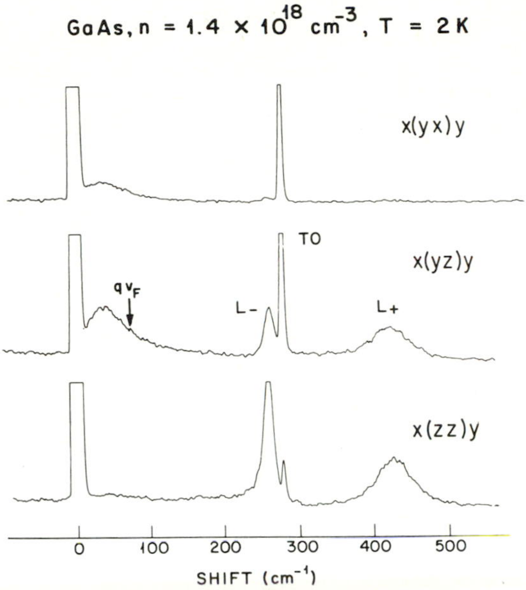

Fig. 4.2. Polarized Raman spectra from n-type GaAs at 2 K [4.12]

charge density fluctuation scattering in a cubic crystal, but they also appear in the (yz) geometry, which is allowed by the usual selection rules for the LO phonon [4.17].

4.1.4. Scattering by Coupled LO Phonon-Plasmon Modes

The LO phonon and plasmon interact because each produces a longitudinal electric field (or scalar potential) that interacts with the charge density of the other [4.14]. The resulting coupled mode frequencies are roots of the equation $\mathrm{Re}\{\varepsilon\}=0$, where for small q the dielectric constant is given by

$$\varepsilon=\varepsilon_s\left(\frac{\omega_l^2-\omega^2}{\omega_t^2-\omega^2-i\omega 0^+}\right)-\frac{\varepsilon_s\omega_p^2}{\omega^2+i\omega\Gamma}. \tag{4.20}$$

Here ε_s is the bound electron contribution, the first term is the contribution of bound electrons and phonons, and the second term is due to the plasmon. The equation $\mathrm{Re}\{\varepsilon\}=0$ is quadratic in ω^2; the two solutions obey the inequalities $\omega^+>\omega_l$, ω_p and $\omega^-<\omega_t$, ω_p. A comparison of experiment and the theoretical solutions for $\omega^\pm$ in GaAs is given in Fig. 4.3.

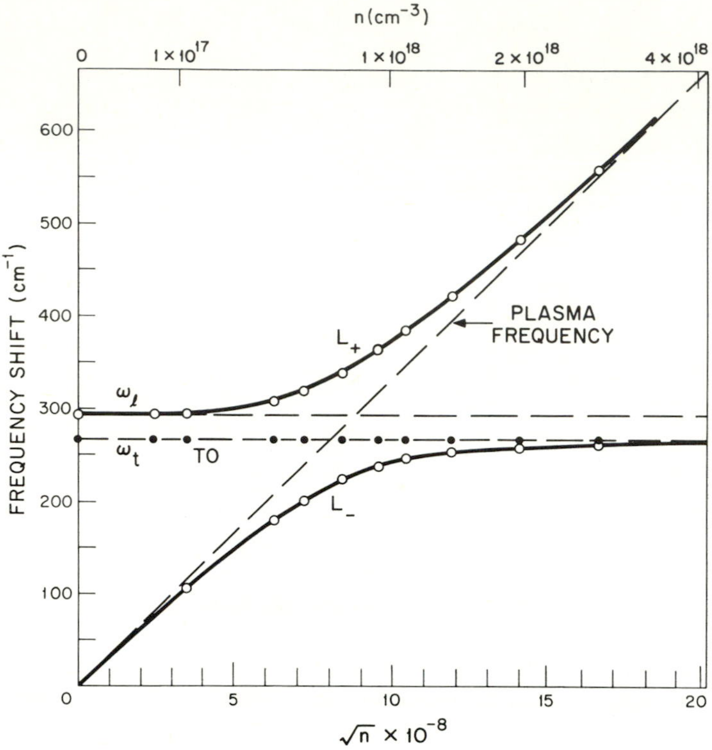

Fig. 4.3. Frequency shift of Raman scattered light due to $L^{\pm}$ phonon-plasmon mixed modes in GaAs at room temperature [4.11]. The solid curves are the calculated roots of the equation $\mathrm{Re}\{\varepsilon(q\to 0, \omega)\} = 0$

There are three mechanisms responsible for light scattering from the coupled modes; they will be denoted ϱ, u, and $\boldsymbol{E}$ mechanisms. In the ϱ mechanism the key elements as given in (4.17) are $r_0 A_1 A_2^+$ times the electronic density fluctuation ϱ_q. Scattering by the ionic density fluctuation is negligible since it involves the ionic Thompson radius $(mr_0/M) \ll r_0$. In the u mechanism scattering occurs via modulation of the bound electronic susceptibility $\overleftrightarrow{\chi}$ by the phonon coordinate u. This is often called the "deformation potential" mechanism discussed in detail in Chapters 2 and 3. The corresponding scattering Hamiltonian may be obtained by considering modulation of the polarization energy of the bound electrons in the presence of the optical field $\boldsymbol{E}_0$

$$-\tfrac{1}{2}\int \boldsymbol{E}_0(R)\cdot\overleftrightarrow{\chi}(R)\cdot\boldsymbol{E}_0(R)\,d^3R\,.$$

The resulting cross-term of interest is

$$\mathscr{H}_u' = (\mathrm{i}\omega_1 A_1/c)(\mathrm{i}\omega_2 A_2/c)^\dagger \exp(-\mathrm{i}\omega t)(\partial\chi/\partial u)\,U_q + \mathrm{H.C.}, \tag{4.21a}$$

where

$$U_q = \int u(\boldsymbol{R}) \exp(\mathrm{i}\boldsymbol{q}\cdot\boldsymbol{R})\, d^3R\,, \tag{4.21b}$$

and

$$\partial\chi/\partial u = \sum_{abc} (\partial\chi_{ab}/\partial u_c)\, e_{1a} e_{2b} q_c/|\boldsymbol{q}|\,. \tag{4.21c}$$

In the $\boldsymbol{E}$ mechanism the modulation of the polarization energy is by the total longitudinal field $\boldsymbol{E}$. This is often called the "electro-optic" mechanism and leads to a scattering Hamiltonian

$$\mathscr{H}_{\boldsymbol{E}}' = -(\mathrm{i}\omega_1 A_1/c)(\mathrm{i}\omega_2 A_2/c)^\dagger \exp(-\mathrm{i}\omega t)(\partial\chi/\partial \mathrm{E})\mathrm{E}_q + \mathrm{H.C.}, \tag{4.22}$$

where E_q and $\partial\chi/\partial\mathrm{E}$ are analogous to U_q and $\partial\chi/\partial u$[1]. The LO phonon part of the coupled modes has both a displacement u and a contribution to $\boldsymbol{E}$; it scatters light via both (4.24) and (4.22). The plasmon part of the coupled modes scatters light via (4.17) and (4.22). These mechanisms for the plasmon are 90° out of phase, since from Poisson's equation ϱ_q is 90° out of phase from the plasmon's contribution to E_q, and they do not interfere; the scattered intensities simply add.

The efficiency of light scattering by the coupled modes due to the electron density fluctuations (4.17) is given by [4.18]

$$\begin{aligned}&(d^2R/d\omega\, d\Omega)_\varrho \\ &= R_{21}^2 (n_\omega+1)(\omega_2/\omega_1)^2 (\mu r_0 \omega_\mathrm{p} q)^2 (4\pi^2 e^2)^{-1} \hbar\varepsilon_\mathrm{s}\omega\, \Gamma (\omega_l^2-\omega^2)^2/\Delta\,,\end{aligned} \tag{4.23}$$

where

$$\Delta = [\omega^2(\omega_l^2-\omega^2) - \omega_\mathrm{p}^2(\omega_\mathrm{t}^2-\omega^2)]^2 + \omega^2\Gamma^2(\omega_l^2-\omega^2)^2\,. \tag{4.24}$$

The first term [in the square bracket in (4.24)] is zero at $\omega=\omega^\pm$. Note the presence of the term $(\omega^2-\omega_l^2)^2$ in the numerator of (4.23). This produces a zero in the scattering at ω_l for all values of the damping Γ.

[1] We have set the components of $\boldsymbol{E}$ in Roman type in order to distinguish them from the eigenvalues of the energy.

The result of scattering of light by the coupled modes via deformation potential and electro-optic mechanisms is [4.11, 18, 19]

$$(d^2R/d\omega\, d\Omega)_{u,\boldsymbol{E}} \tag{4.25}$$
$$=(n_\omega+1)(4\pi\varepsilon_s^{-1})^2(\partial\chi/\partial \mathrm{E})^2(4\pi^2)^{-1}(\omega_2/c)^4\hbar\omega_p^2\varepsilon_s\omega\Gamma(\omega_0^2-\omega^2)^2/\Delta\,,$$

where

$$\omega_0^2=\omega_t^2(1+C) \tag{4.26}$$

with [Ref. 8.1, Eq. (2.93)]

$$C=\frac{e^*(\partial\chi/\partial u)}{M\omega_t^2\,\partial\chi/\partial \mathrm{E}}\,, \tag{4.27}$$

e^* is the effective charge associated with the optical phonons, and M is their reduced mass.

Apparently [4.18] is the only reference to derive (4.23). Blum and Mooradian [4.20] stated it correctly, but they refered to Mooradian and McWhorter for the derivation [4.11]. These latter authors gave a formula which reduces to the correct one only for ω near ω^+ or ω^-. Large plasmon damping Γ yields appreciable scattering away from ω^+ and ω^-; in such a case their result will give an incorrect spectrum.

The parameter C is called the "Faust-Henry coefficient" [4.21]. It is denoted by C_1 in [4.11] and by $1/\gamma$ in [4.18] and [4.19]. It determines the ratio of scattering by LO and TO phonons in an undoped crystal

$$\frac{(dR/d\omega)(\mathrm{LO})}{(dR/d\omega)(\mathrm{TO})}=\frac{(\omega_1+\omega_l)^4\omega_t}{(\omega_1+\omega_t)^4\omega_l}\left(1+\frac{\omega_t^2-\omega_l^2}{C_1\omega_t^2}\right)^2. \tag{4.28}$$

C is an important parameter also in the theory of light scattering from polaritons [4.23].

Equation (4.25) says that the scattering is zero at $\omega=\omega_0$ for the $(u,\boldsymbol{E})$ mechanism, whereas (4.23) say it is zero at $\omega=\omega_l$ for the ϱ mechanism. In both cases, the spectra are not simply two peaks at $\omega=\omega^+$ and $\omega=\omega^-$. This is amply illustrated by experimental data on n-type CdS [4.24], GaP [4.19], and SiC [4.18]. In all three cases the $u,\boldsymbol{E}$ mechanism dominates. A detailed discussion of the theoretical response functions and spectra due to the $u,\boldsymbol{E}$ mechanism has been given by Hon and Faust [4.19]. They showed that for large Γ the L^- branch can be so severly distorted and smeared out that it cannot be observed in practice. Figure 4.4 shows their fits in the L^+ region. Note that the plasmon is over-damped ($\Gamma\gtrsim\omega_p$). The parameter C is -0.53 for GaP. This gives a value of $\omega_0=250\,\mathrm{cm}^{-1}$. The scattering in this frequency region was too weak for one to notice the zero at $\omega=\omega_0$.

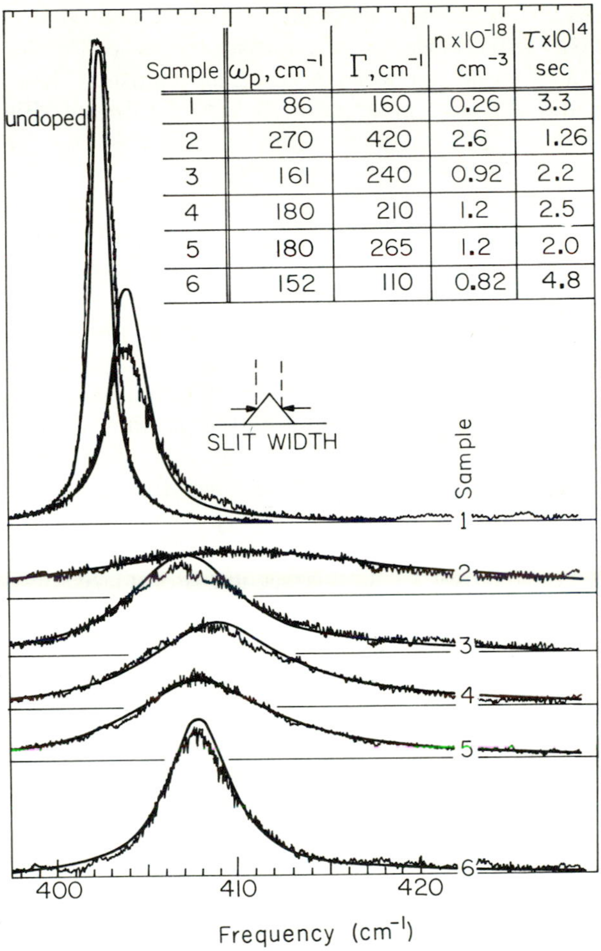

Sample	ω_p, cm^{-1}	Γ, cm^{-1}	n x 10^{-18} cm^{-3}	τ x 10^{14} sec
1	86	160	0.26	3.3
2	270	420	2.6	1.26
3	161	240	0.92	2.2
4	180	210	1.2	2.5
5	180	265	1.2	2.0
6	152	110	0.82	4.8

Fig. 4.4. Raman scattering from pure and n-type GaP [4.19]. The solid lines are theoretical fits using parameters shown in the table

A direct comparison of theoretical spectra due to $(u, \boldsymbol{E})$ and ϱ mechanisms has been given for SiC: N by KLEIN et al. [4.18]. There Γ was approximately $3\omega_p$. The zero in scattered intensity at $\omega = \omega_0$ was observed and an experimental value of $C = +0.39$ determined. From it and from a previous value of $\partial\chi_{xx}/\partial E_z$ a value for $\partial\chi_{xx}/\partial u_z$ was determined.

In n-type CdS SCOTT et al. [4.22] observed a zero at $\omega = \omega_0$ and found C to be $+0.50$.

In GaAs the damping Γ is relatively small, and MOORADIAN and MCWHORTER [4.10, 11] were able to measure cross-sections for scattering from L^+ and L^- modes via both ϱ and $u, \boldsymbol{E}$ mechanisms. The mechanisms can be separated by the polarization selection rules (see Fig. 4.2): The $x(yz)y$ spectrum (∥, ⊥ in their original notation) shows the results of the $u, \boldsymbol{E}$ mechanism; the $x(zz)y$ spectrum (⊥, ⊥ in their notation) shows the results of the ϱ mechanism. Figure 4.2 reveals that the two mechanisms produce comparable scattering for 1.06 μm light in GaAs, and MOORADIAN and MCWHORTER obtained excellent agreement between their observed intensities and calculations from expressions equivalent to our (4.23) and (4.25) for various values of the electron concentration.

In addition to the systems discussed above, scattering from coupled plasmon-phonon modes has been observed with 1.06 μm laser light in CdTe and InP [4.12], and scattering by the plasmon in InSb has been observed with 10.6 μm light [4.25]. A study of the coupled phonon-plasmon modes as a function of electron concentration has been made in InSb with the use of a 5.39 μm laser [4.20].

There is a trend in the experimental results on coupled plasmon-phonon modes referred to in this section. For GaAs observed with 1.06 μm light the $(u, \boldsymbol{E})$ and ϱ mechanisms are comparable in intensity. for smaller band-gap materials studied with longer wavelength lasers the ϱ mechanism dominates. For larger band-gap materials studied with visible lasers the $(u, \boldsymbol{E})$ mechanism dominates. This trend can be explained by examining how the various quantities in (4.23) and (4.25), or sum rules derived from these equations [4.28], scale with the band-gap. In making this comparison one uses laser frequencies which are approximately proportional to the corresponding band-gap energy.

4.2. Raman Scattering from Bound Electrons and Holes

Electrons on donor atoms or holes on acceptor atoms scatter light very efficiently while undergoing transitions among bound states. These are somewhat like the well-known infrared excitations [4.26, 27]. We shall show that unlike the scattering of light by plasmon density fluctuations, the cross-section is independent of q and is of order $(r_0 m/m^*)^2$ multiplied be a resonance enhancement factor, where relevant. Per electron or hole, this is greater than that for plasmons.

Complete calculations of the scattering intensity start from the Hamiltonian of (4.1) [4.28, 29]. The dominant term can be easily obtained using the effective mass approximation, as shown below.

Measurements of the positions, selection rules, and intensities of the scattered light from donors and acceptors can, in principle, yield detailed information about some of the impurity wave functions and about the band structure of the host crystal. We first discuss the existing measurements and then discuss the effective-mass theory of the effect.

4.2.1. Experimental Results—Acceptors

The first experiments of this type were done by HENRY et al. [4.30] on GaP containing Zn and Mg acceptors. They saw five lines or bands: A, A′, B, C, and D, four of which are indicated schematically in Fig. 4.5. These lines can be explained as follows. The acceptor ground state is pulled up and derived from the top of the four-fold degerate $p_{\frac{3}{2}}$, or Γ_8 valence band. Within the effective mass approximation, it is made up of Bloch functions from these bands multiplied by (predominantly) 1S hydrogenic envelope functions. This level splits into two Kramers doublets under stress. Line A was observed at practically zero energy shift and was attributed to transitions between such doublets split by residual stresses in the crystal. Band A′ was attributed to transition A accompanied by the emission of an acoustical phonon. As shown in Fig. 4.5 line A shifts upwards in energy when a stress is applied. Line B occurs at 33.1 meV for Zn acceptors and 34.4 meV for Mg acceptors. It is found to shift half as much with stress as line A at low temperatures and split into two components at higher temperatures [4.31]. Its final state must therefore be a Kramers doublet, Γ_6 or Γ_7. Band C was attributed to unresolved transitions to higher bound states, presumably

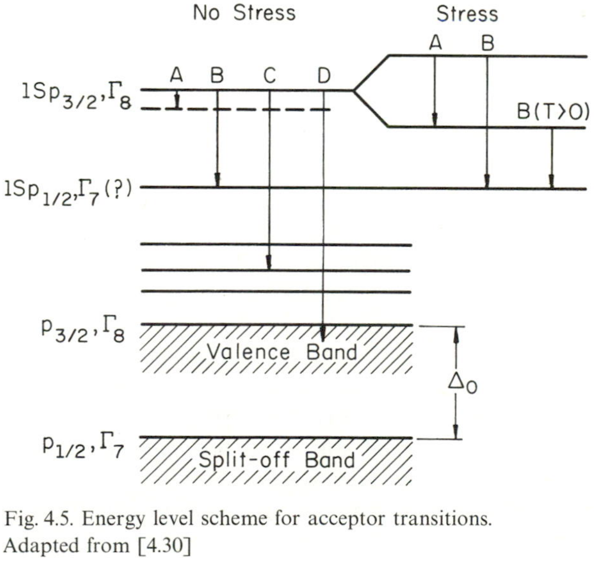

Fig. 4.5. Energy level scheme for acceptor transitions. Adapted from [4.30]

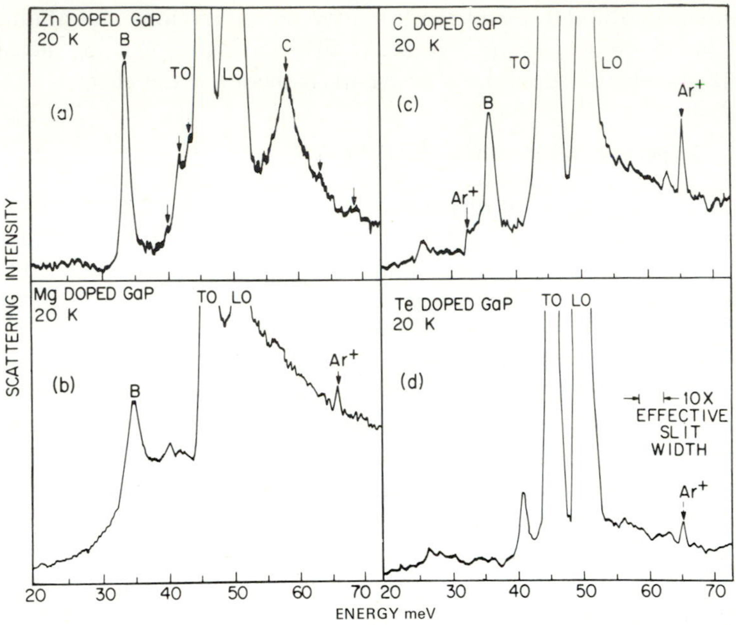

Fig. 4.6a–d. Raman spectra due to acceptors (a)–(e) and Te donors (d) in GaP [4.31]. Dopings are; (a) 5×10^{17} cm^{-3} Zn and (d) 2×10^{18} cm^{-3} Te. The geometry is $[1\bar{1}1]$ ($[1\bar{1}0]$, unanalyzed) $[1\bar{1}0]$

derived from the Γ_8 valence band, and Band D was attributed to transitions to the valence band.

A more detailed study of acceptors in GaP was performed by Manchon and Dean [4.31]. They found that the energy of the B transition decreases somewhat as the binding energy of the hole to the acceptor increases. Their spectra for three acceptors are shown in Fig. 4.6. They also studied the splitting under stress of the B line for the Zn doped sample.

Wright and Mooradian [4.28] investigated light scattering from Boron acceptors in Si. A sharp line of "B" type was found at 23.4 meV, depolarized like the optical phonon line. In a later paper, they found that this line splits with stress like the B line in GaP [4.32].

The splitting of the B line due to the Boron acceptor in Si under stress and in a magnetic field was studied in detail by Cherlow et al. [4.33]. They determined the shear deformation potentials due to the splitting of the $1S\Gamma_8$ acceptor ground state and the spin Hamiltonian parameters for the same state. The excited state remained unsplit in both cases. Their spin Hamiltonian parameters disagreed with those determined from magnetic resonance measurements.

Table 4.1. Lowest observed acceptor Raman transitions

Host	Δ_0 [MeV]	Acceptor	Transition energy [meV]	Binding energy [meV]	Type
Si	44[a]	B	23.4[b], 22.7[c]	≧44.32[d]	B
GaP	100[e]	C	36.5[f]	48[g]	B
		Mg	34.7[f]	53.5[g]	B
		Zn	33.3[f]	64[g]	B
Ge	300[e]	Ga	6.5–10.5[h]	11[i]	C
GaAs	330[e]	Cd	25.4[j]	21[i]	C
		Zn	21[j]	24[i]	C

[a] [4.37].
[b] [4.28].
[c] [4.33].
[d] [4.38].
[e] [4.39].
[f] [4.31].
[g] [4.40].
[h] [4.35].
[i] [4.41].
[j] [4.32].

WRIGHT and MOORADIAN observed a single line due to Cd, Zn, Mn, and Mg acceptors in GaAs [4.32, 34]. In the first two cases, the line was at about 25 meV. No stress measurements were performed.

DOEHLER et al. have studied the Ga acceptor in Ge using the 2.1 μm line of an ABC-YAG laser [4.35]. They did not see a B line, but observed a band of unresolved lines of the C type in the 6.5–10.5 meV range in a crystal contaning $2.5 \times 10^{16}\,\mathrm{cm}^{-3}$ acceptors. Some resolved lines appeared at lower doping [4.36].

The acceptor transitions are broader in the heteropolar GaP and GaAs lattices than in the homopolar Si lattice. This is attributed to stronger coupling to phonons and hence a shorter lifetime in the heteropolar case.

Some information concerning the lowest observed acceptor transitions has been assembled in Table 4.1.

4.2.2. Experimental Results—Donors

The first donor transition to be seen by light scattering was observed at 13.1 meV in Si(P) by WRIGHT and MOORADIAN [4.28]. It was polarized and presumably had group theoretical E-symmetry. Its position agreed with the known energy of the $1S(\mathrm{A}_1) \rightarrow 1S(\mathrm{E})$ valley-orbit transition. WRIGHT and MOORADIAN worked out the theory of the scattering and showed that such valley-orbit transitions should be the strongest donor transitions. Later WRIGHT and MOORADIAN reported observing the same transition on As and Sb donors in Si and on Te donors in AlSb [4.34].

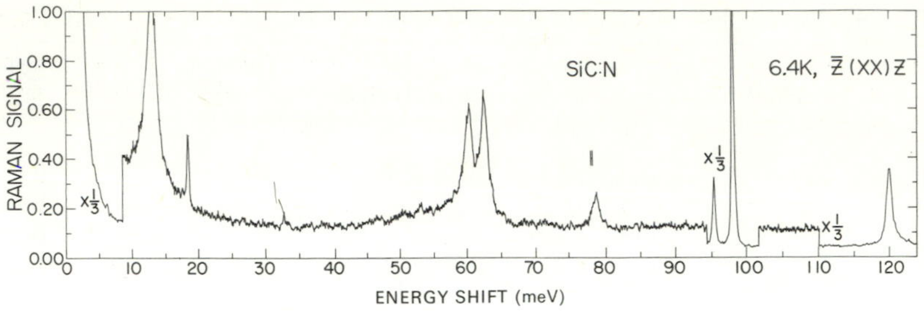

Fig. 4.7. The Raman spectrum of the 6H polytype of SiC containing N donors [4.29]. This (XX) polarization geometry couples to excitations of A_1 and E_2 symmetry. The valley-orbit donor peaks are at 13.0, 60.3, and 62.6 meV. The peak at 78.8 meV is thought to be a localized phonon gap mode. The A_1(LO) phonon is at 119.5 cm^{-1}. There are E_2 phonon lives at 18.0, 18.5, 32.5, 95.0, and 97.7 meV

Manchon and Dean observed $1S(\mathrm{A}_1) \rightarrow 1S(\mathrm{E})$ transitions due to Se, S, and Te donors in GaP [4.31]. The line split symmetrically under stress, a fact consistent with this assignment.

Colwell and Klein observed lines of E_2 symmetry at 13.0, 60.3, and 62.6 meV in the 6H polytype of SiC doped with N donors [4.29] (Fig. 4.7). They were attributed to $1S(\mathrm{A}_1) \rightarrow 1S(\mathrm{E}_2)$valley-orbit transitions of donors occupying the three inequivalent carbon sites in that lattice. In the 15R polytype four and possibly five lines were seen; this was consistent with the existence of five inequivalent C sites in that lattice. For the 6H polytype, the fact that precisely three lines of E_2 symmetry were observed locates the conduction band minima along the line ML at the edge of the Brillouin zone [4.29, 42]. The donor lines were also seen in 6H SiC by Dean and Hartman [4.43].

Doehler et al. observed the $1S(\mathrm{A}_1) \rightarrow 1S(\mathrm{T}_2)$ valley-orbit transition at 34.6 cm^{-1} (4.3 meV) due to As donors in Ge [4.35]. An experimental measurement gave a scattering cross-section within a factor of three of the theoretical prediction.

As mentioned above, the strongest donor or acceptor transitions that originate on the ground state should be $1S \rightarrow 1S$. The reason the transition is seen at finite energy is that degeneracy in the bands is split, by the valley-orbit interaction for the donor case and by the spin-orbit interaction for the acceptor case. For donors in GaP [4.31] and SiC [4.29] the light scattering data have yielded the first direct measurements of the $1S(\mathrm{A}) - 1S(\mathrm{E})$ splittings. Other transitions, while possibly allowed by symmetry, are much weaker and may be difficult to observe.

An exception occurs near resonance, for then cross-sections can be considerably enhanced. There are no valley-orbit transitions in CdS, since the conduction band minima are at $k=0$, but in that material doped with Cl donors HENRY and NASSAU were able to observe light scattering from the $1S \rightarrow 2S$ and $1S \rightarrow 2P_z$ donor transitions [4.44]. This was done with 4880 Å laser excitation under near resonance conditions.

4.2.3. Effective-Mass Theory of Donor Raman Transitions

The wave functions for the electron on a donor (hole on an acceptor) are derived from the Bloch functions $\phi_j(r)$ at the lowest conduction band minima (highest valence band maxima). These are assumed to be degenerate. The donor or acceptor wave functions can be written in the form [4.45]

$$\psi^n(r) = \sum_j F_j^n(r)\,\phi_j(r)\,, \tag{4.29}$$

where the F_j^n are "envelope functions" that are slowly varying on the scale of one lattice constant. The effective mass wave equation is

$$\sum_{j'} \left\{ \Lambda_{jj'} + \sum_{\alpha\beta} D_{jj'}^{\alpha\beta} p_\alpha p_\beta + [U(r) - E]\,\delta_{jj'} \right\} F_{j'}^n(r) = 0\,. \tag{4.30}$$

Here the $\Lambda_{jj'}$ matrix represents the spin-orbit interaction. This is important for valence band states having p-character. The coefficients D are generalized reciprocal effective mass parameters resulting from interband transitions to states l outside the set j

$$D_{jj'}^{\alpha\beta} = m^{-2} \sum_l \langle j|p_\alpha|l\rangle \langle l|p_\beta|j'\rangle (E_j - E_l)^{-1} + \delta_{jj'}\,\delta_{\alpha\beta}(2m)^{-1}\,. \tag{4.31}$$

$U(r) = e^2/(\varepsilon_s r)$ is the Coulomb potential of the donor or acceptor. The eigenvalue E is the energy relative to the band extremum. An effective mass Hamiltonian for the scattering of radiation may be obtained from (4.30) by replacing $\boldsymbol{p}$ by $\boldsymbol{p} + e\boldsymbol{A}/c$ and collecting the A^2 terms. Such a treatment is valid for photon wavelengths much longer than a lattice constant and for photon energies well below the band gap. The matrix element for the transition from n to n' is

$$\begin{aligned} &\langle n'|\mathscr{H}'|n\rangle \\ &= 2e^2 c^{-2} A_1 A_2^\dagger \left[\sum_{jj'\alpha\beta} D_{jj'}^{\alpha\beta} \tfrac{1}{2}(e_{1\alpha} e_{2\beta} + e_{1\beta} e_{2\alpha}) \int F_j^{n'*} F_{j'}^n\, d^3r \right]. \end{aligned} \tag{4.32}$$

The contribution of this transition to the power scattering cross-section is

$$d\sigma(n, n')/d\Omega = (\omega_2/\omega_1)^2\, r_0^2 R_{12}^2 m^2 |[\ \ ]|^2\,, \tag{4.33}$$

where the empty square bracket denotes the factor in square brackets in (4.32). We have inserted the factor R_{12}^2, defined by (4.18), to approximate the effect of near-resonant conditions [4.28, 29].

For donors in a crystal having a many-valley conduction band, the matrix D is diagonal in the valley index j and is just $\mu_{\alpha\beta}^{(j)}(2m)^{-1}$ where $\mu_{\alpha\beta}^{(j)} = (m/m^*)_{\alpha\beta}$ for valley j. Within the effective mass approximation the F_j^n are written $F_j^n = \alpha_j^n H_j^{h(n)}$, where the $H_j^h(r)$ form a complete orthonormal set of hydrogenic wave-functions that, in principle, take into account the mass anisotropy of valley j. The α_j^n are numerical coefficients determined from symmetry considerations [4.45]. The cross-section becomes [4.29]

$$d\sigma(n, n')/d\Omega = (\omega_2/\omega_1)^2 r_0^2 R_{12}^2 \cdot \left\| \left[\sum_{\alpha\beta j} e_{1\alpha} e_{2\beta} \mu_{\alpha\beta}^{(j)} \alpha_j^{n'*} \alpha_j^n \int H_j^{h(n)*} H_j^{h(n')} d^3 r \right] \right\|^2 . \tag{4.34}$$

The overlap integral gives the selection rule that $h(n) = h(n')$: The hydrogenic quantum numbers must be the same. Thus if n refers to the ground state, only transitions within the $1S$ manifold are allowed by this effective mass theory. Calculations starting with virtual interband transitions due to $\boldsymbol{p} \cdot \boldsymbol{A}$ terms in the Hamiltonian show that other transitions that are allowed by symmetry are indeed much weaker than those considered here [4.29].

The term in square brackets in (4.34) is readily evaluated. For Si and GaP, the valleys are in $\langle 100 \rangle$ directions, and there will be parallel and perpendicular components of the μ-tensor with respect to these directions. We find for E-polarization geometry—$e_1 \| [1\bar{1}0]$ and $e_2 \| [110]$—that the square bracket is equal to $(\mu_\| - \mu_\perp)/\sqrt{6}$. For Ge, there are four valleys along $\langle 111 \rangle$ directions. The valley-orbit transition has T_2 symmetry. For $e_1 \| x$ and $e_2 \| y$, one obtains $(\mu_\| - \mu_\perp)/3$ for the square bracket [4.33]. For 6H SiC, there are effectively three valleys along the lines ML in the center of the rectangular faces of the hexagonal Brillouin zone. For E_2 polarization geometry [(xx), (yy), or (xy)], one finds $(\mu_{xx} - \mu_{yy})/\sqrt{8}$ for the square bracket. Here z is along the hexagonal axis, x is perpendicular to the zone face in question, and y is parallel to that face. Thus

$$d\sigma(1S\mathrm{A}, 1S\mathrm{E})/d\Omega = (\omega_2/\omega_1)^2 r_0^2 R_{12}^2 (\mu_\| - \mu_\perp)^2/6 , \quad \text{for Si} , \tag{4.35a}$$

$$d\sigma((1S\mathrm{A}, 1S T_2)/d\Omega = (\omega_2/\omega_1)^2 r_0^2 R_{12}^2 (\mu_\| - \mu_\perp)^2/9, \quad \text{for Ge} , \tag{4.35b}$$

$$d\sigma(1S\mathrm{A}, 1S\mathrm{E}_2)/d\Omega = (\omega_2/\omega_1)^2 r_0^2 R_{12}^2 (\mu_{xx} - \mu_{yy})^2/8, \quad \text{for 6H SiC} . \tag{4.35c}$$

In addition, there will be elastic (Rayleigh) scattering with a cross-section

$$d\sigma/d\Omega = (\omega_2/\omega_1)^2 r_0^2 R_{12}^2 \bar{\mu}^2 , \tag{4.35d}$$

where $\bar{\mu} = \boldsymbol{e}_1 \cdot [A v_j \overleftrightarrow{\boldsymbol{\mu}}^{(j)}] \cdot \boldsymbol{e}_2$. DOEHLER et al. calculated $d\sigma/d\Omega$ from (4.35b) to be 1.4×10^{-24} cm^2/sr for the scattering of 2.1 μm laser light by the valley-orbit transition in Ge [4.35]

4.2.4. Effective-Mass Theory of Acceptor Raman Transitions

Apart from a comment by WRIGHT and MOORADIAN [4.32], there are apparently no discussions in the literature of the theory of light scattering by acceptor transitions. We sketch such a theory here.

We start with the matrix elements $\langle m|\mathscr{H}'|n\rangle$ given by (4.32). The D coefficients are relatively simple if the states j transform like p-orbitals times spin $\frac{1}{2}$ spin functions. The spin-orbit interaction is diagonalized in the $J m_J$ representation. In it the matrix

$$D_{jj'} = \sum_{\alpha\beta} D_{jj'}^{\alpha\beta} \tfrac{1}{2}(e_{1\alpha} e_{2\beta} + e_{1\beta} e_{2\alpha}) \tag{4.36}$$

has the form of the matrix in DRESSELHAUS et al. (Ref. [4.46], Eq. (62)), if one replaces the matrix $k_\alpha k_\beta$ in the latter reference by $S_{\alpha\beta} \equiv \frac{1}{2}(e_{1\alpha} e_{2\beta} + e_{1\beta} e_{2\alpha})$. The detailed nature of $D_{jj'}$ thus depends on the photon polarization vectors $\boldsymbol{e}_1$ and $\boldsymbol{e}_2$, but in general there will be non-zero entries both within and between the blocks representing the $J = \frac{3}{2}$ and $J = \frac{1}{2}$ manifolds.

This last statement holds only for the traceless part of $S_{\alpha\beta}$. Fully-symmetric A_1 (or Γ_1) scattering will not take place at finite frequency shift. This may be seen by replacing $S_{\alpha\beta}$ by $\delta_{\alpha\beta}$. Then $D_{jj'}$ takes the form $D_0 \delta_{jj'}$. When this is inserted in (4.32) and use is made of (4.39), given below, one finds that $\langle n'|\mathscr{H}'|n\rangle$ is proportional to $\delta_{nn'}$.

The point group is $\bar{T}_d$, the double tetrahedral group [4.47]. Let us consider a $1Sp_{\frac{3}{2}}(\Gamma_8)$ to $1Sp_{\frac{1}{2}}(\Gamma_7)$ acceptor transition (Fig. 4.5). The symmetry of the Raman transition will be either $\Gamma_3(\mathrm{E})$ with, say, $\boldsymbol{e}_1 \| [1\bar{1}0]$ and $\boldsymbol{e}_2 \| [110]$ or $\Gamma_4(\mathrm{T}_2)$, with, say $\boldsymbol{e}_1 \| x$ and $\boldsymbol{e}_2 \| y$. The $F_j^m(r)$ envelope functions in (4.32) have $1S$, $3D$, etc., components, all belonging to Γ_8^+ or Γ_7^+, in the notation of SCHECHTER [4.47]. To estimate cross-sections, we assume that all the F_j^m are the same function. This assumption will yield rough estimates for the numerical values of the cross-sections, but the ratios $\sigma_\mathrm{B}(\mathrm{E})/\sigma_\mathrm{B}(\mathrm{T}_2)$ and $\sigma_\mathrm{A}(\mathrm{E})/\sigma_\mathrm{A}(\mathrm{T}_2)$, obtainable from (4.37a) and (4.38a) and (4.38b) will be correct within the effective mass approximation. With $D_{jj'}$ determined, as described above, it is possible to calculate the cross-sections for the E and T_2 components of the B and A transitions in Fig. 4.5. For the B transition, assuming it is $1Sp_{\frac{3}{2}}(\Gamma_8^+) \rightarrow 1Sp_{\frac{1}{2}}(\Gamma_7^+)$,

$$d\sigma/d\Omega_B(E) = r_0^2 R_{12}^2 [2m(L-M)/\hbar^2]^2/3, \tag{4.37a}$$

$$d\sigma/d\Omega_B(T_2) = r_0^2 R_{12}^2 [2mN/\hbar^2]^2/3. \tag{4.37b}$$

The dimensionless quantities $2m|L-M|/\hbar^2$ and $2m|N|/\hbar^2$ were defined by DRESSELHAUS et al. [4.46] and equal 2.25 and 9.36, respectively, for Si [4.48] and 25.9 and 33.7, respectively, for Ge [4.49]. The A transition results from splitting under stress of the $m_J = \pm\frac{3}{2}$ levels from the $m_J = \pm\frac{1}{2}$ levels in the $J=\frac{3}{2}$ manifold. The cross-sections for them are

$$d\sigma/d\Omega_A(E) = r_0^2 R_{12}^2 [2m(L-M)/\hbar^2]^2/6, \tag{4.38a}$$

$$d\sigma/d\Omega_A(T_2) = r_0^2 R_{12}^2 [2mN/\hbar^2]^2/6. \tag{4.38b}$$

Band C, observed in GaP [4.30, 31], has been attributed to unresolved transitions from the ground state to higher bound states [4.30]. If there were a single envelope function for each of these states, it would have to be orthogonal to the envelope function of the ground state. According to (4.29), there is a set $F_j^n(r)$ of envelope functions for each state n; the orthogonality condition is

$$\sum_j \int F_j^{n'*} F_j^n \, d^3r = \delta_{nn'}. \tag{4.39}$$

Thus, in general, a sum derived from (4.32) and (4.36) of the form

$$[\ldots] \equiv \sum_{jj'} D_{jj'} \int F_j^{n'*} F_{j'}^n \, d^3r \tag{4.40}$$

will not be zero for $n \neq n'$. The actual value of this quantity depends upon the details of the envelope functions [4.47, 50–52]. A general theory of the intensities of the C transitions cannot be given without detailed calculations for each crystal.

4.2.5. Discussion

The general features of the experimental valley-orbit donor spectra seem to be well-understood. The acceptor transitions present some difficulties. The interpretation of transition B given here imples certain features of some relevant data. If line B involves a "1S" to "1S" transition, its energy should equal the spin orbit splitting of the valence band, Δ_0, plus a correction due to differences in effective-mass binding energies in the two states, plus the difference in central-cell corrections. MENDELSON and SCHULTZ have performed an effective mass calculation of the energies of various acceptor levels for Si [4.51]. They predict a ground state binding of 37.1 meV and a B transition energy of 30.2 meV. From Table 4.1, we thus conclude that the central cell correction for the ground state would be $44.3 - 37.1 = 7.2$ meV; the difference in central cell corrections would be $30.2 - 22.7 = 7.5$ meV; the central cell correction for the $S_{\frac{1}{2}}(\Gamma_7^+)$ state would then be $7.5 + 7.2 = 14.7$ meV. The GaP data may also imply that the central cell correction for the upper level is greater than that for the ground state and increases more rapidly than that for the ground state as the acceptor binding increases.

Some alternate explanations for line B have been advanced that would not require such a large difference in central cell corrections. MANCHON and DEAN have suggested that the spin orbit splitting may be reduced by the dynamic Jahn-Teller effect [4.31], and MORGAN has suggested that line B is a Jahn-Teller phonon sideband of line A [4.53]. The experimental evidence is insufficient to decide among these explanations. There is a need for additional theoretical and experimental work on the acceptor transitions.

4.3. Raman Scattering from Coupled Electron-Phonon Excitations

We discussed scattering from coupled LO phonon-plasmon excitations in Section 4.1. Other types of coupled excitation will be considered in this section.

4.3.1. Theoretical Introduction

The experimental data to be presented below will be interpreted using various modifications of the simple coupled two excited level system shown in Fig. 4.8. The unperturbed system has a ground state $|g\rangle$, an excited electronic state $|e\rangle$ with excitation energy E_e, and a one-phonon excited state $|p\rangle$ with energy $E_0 = \hbar\omega_0$. Higher phonon oscillator states will ordinarily be neglected. Transitions to the states $|e\rangle$ and $|p\rangle$ are Raman-active; the Raman matrix elements connecting them to the ground state are T_e and T_p. Electron-phonon coupling is assumed to be present in the form of a matrix element V between $|e\rangle$ and $|p\rangle$. This produces a mutual repulsion of the levels E_e and E_0 to positions $E_\pm$

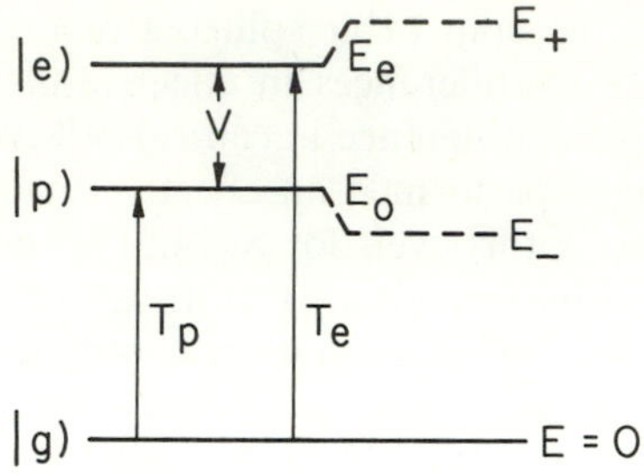

Fig. 4.8. System with two excited states coupled by a matrix element V

that are the roots of the secular equation

$$(E_0 - E)(E_e - E) - V^2 = 0\,. \tag{4.41}$$

(For simplicity V is assumed to be real.)

Rather than deal with the proper eigen-states $|\pm)$ to calculate the intensities, it is more convenient to use the Green's function operator [4.54]

$$G(z) = (\mathscr{H} - z)^{-1} = \begin{pmatrix} E_0 - z & V \\ V & E_e - z \end{pmatrix}^{-1}. \tag{4.42}$$

The Raman spectrum is proportional to

$$\begin{aligned} I(E) &= \pi \sum_{a=\pm} |(g|T|a)|^2\, \delta(E_a - E) \\ &= \mathrm{Im}\left\{ \sum_{i,j=e,p} T_i (i|G(z)|j)\, T_j \right\} \\ z &= E + i0^+\,. \end{aligned} \tag{4.43}$$

(For simplicity T_e and T_p are assumed to be real.) The matrix inversion in (4.42) is easily performed, and (4.43) gives

$$I(E) = \mathrm{Im}\left\{ \frac{T_e^2 (E_0 - z) - 2V\, T_e T_p + T_p^2 (E_e - z)}{(E_0 - z)(E_e - z) - V^2} \right\}. \tag{4.44}$$

The poles in the denominator of (4.44) at $E = E_\pm$ lead to delta-functions. The spectrum can be written in various ways, one of which is

$$\begin{aligned} I(E) &= \pi (E_+ - E_-)^{-1} [\delta(E - E_+) + \delta(E - E_-)] \\ &\quad \cdot |E_0 - E|\, [T_e + T_p (E - E_0)^{-1} V]^2\,. \end{aligned} \tag{4.45}$$

The last factor in (4.45) shows that an interference exists between two Raman amplitudes for the $g \to e$ transition, namely between the direct, first-order, transition with amplitude T_e and the indirect, second-order, transition $g \to p \to e$ with amplitude $T_p(E-E_0)^{-1} V$.

We now discuss a case considered first by FANO [4.55] and then by many other authors in various versions (see NITZAN for a listing [5.54])—the interaction of a sharp (phonon) level p with $a(n)$ (electronic) continuum. We assume for simplicity that each level e in the continuum couples to the ground state with a constant Raman matrix element T_e and to the excited phonon state with a constant matrix element V. Then (4.44) will still hold if we make the replacement

$$(E_\mathrm{e}-z) \to 1/g(z)$$

where

$$g(z)=\sum_\mathrm{e}(E_\mathrm{e}-z)^{-1} \tag{4.46}$$

is the unperturbed Green's function for the electronic continuum. The quantity

$$\varrho(E)=\pi^{-1}\,\mathrm{Im}\{g(E+\mathrm{i}0^+)\} \tag{4.47a}$$

is the number of electronic excitations per unit energy interval. Its Hilbert transform is

$$\begin{aligned} R(E) &= -\mathrm{Re}\{g(E+\mathrm{i}0^+)\} = \mathrm{P}\sum_\mathrm{e}(E-E_\mathrm{e})^{-1} \\ &= \mathrm{P}\int \varrho(E')(E-E')^{-1}\,dE', \end{aligned} \tag{4.47b}$$

where P stands for "principal value".

When these changes are made in (4.44) one takes the imaginary part and obtains

$$I(E)=\frac{\pi\varrho(E)\,T_e^2(E_0-E-V\,T_p/T_e)^2}{[E_0-E+V^2R(E)]^2+\pi^2V^4\varrho(E)^2}\,. \tag{4.48}$$

The denominator of (4.48) says that the interaction V has produced a width $2\pi V^2\varrho(E)$ to the phonon excitation, which has been shifted due to the interaction and now has maximum response at $E=E_0+V^2R(E)$. Equation (4.48) can be written in a form equivalent to that given by FANO, namely

$$I(E)=\pi\varrho(\varepsilon)\,T_e^2(q+\varepsilon)^2/(1+\varepsilon^2)\,, \tag{4.49a}$$

where

$$\varepsilon = [E - E_0 - V^2 R(\varepsilon)]/[\pi V^2 \varrho(E)] , \tag{4.49b}$$

and

$$q = [V T_p/T_e + V^2 R(E)]/[\pi V^2 \varrho(E)] . \tag{4.49c}$$

Equations (4.48) and (4.49) differ from those in FANO's paper [4.55]. We have included the density of state $\varrho(E)$, since our (quasi-) continuum wave functions $|e)$ are normalized to unity, whereas his are given a delta-function energy normalization [4.56]. The line shape given in (4.49) is particularly useful when the continuum is broad so that the energy dependence of ϱ and R can be neglected. Then q is a constant, and ε is a scaled energy variable.

If $\varrho(E)$ is non-zero when $E = E_a \equiv E_0 - V T_p/T_e$, or, equivalently, when $\varepsilon = -q$, the spectral function $I(E)$ will reveal an "antiresonance" at $E = E_a$; there is then an exact cancellation between the Raman amplitudes T_e and $T_p(E - E_0)^{-1} V$.

4.3.2. Phonon Coupled to Interband Hole Transition in Silicon

In heavily doped p-type silicon the Raman-active LO phonon line is strongly broadened and distorted by interactions with free holes [4.57]. CERDEIRA et al. studied this effect in detail [4.58, 56] and noted that there is scattering present from a continuum and a Fano-type interference between the LO phonon line and the continuum. An example of their results is shown in Fig. 4.9. The points represent experimental data. The solid lines were calculated using (4.49a) with a constant $R(E)$ and with a constant $\Gamma \equiv \pi V^2 \varrho = 8.24\ \mathrm{cm}^{-1}$ and with their parameter $\omega - \Omega$ the same as our $E - E_0 - V^2 R$. The observed frequency shift for maximum phonon response was $\delta\Omega = +V^2 R = -4.9\ \mathrm{cm}^{-1}$ for all the curves.

The curves in Fig. 4.9 have different shapes because the parameter q is dependent on the laser wavelength—it varies from 7.0 at 4545 Å to 2.0 at 6471 Å, CERDEIRA et al. found the dependence of q on the laser photon energy E_L to be approximated well by $q \propto (3.3\ \mathrm{eV} - E_L)^{-1}$, except for the 6471 Å value. They suggested that this dependence is due to a resonant behavior of the ratio $(T_p/T_e) \propto (3.3 - E_L)^{-1}$ for photon energies near the E_0' critical point at 3.3 eV.

According to (4.49c) a more precise application of this line of reasoning suggests that $q - R(\pi\varrho)^{-1} = (T_p/T_e)(\pi V \varrho)^{-1}$ should vary as $(3.3 - E_L)^{-1}$. The correction, $-R(\pi\varrho)^{-1}$, numerically amounts to $+0.59$. I have replotted the data of CERDEIRA et al. [4.58] and find that, as expected, the fit is indeed better with the correction.

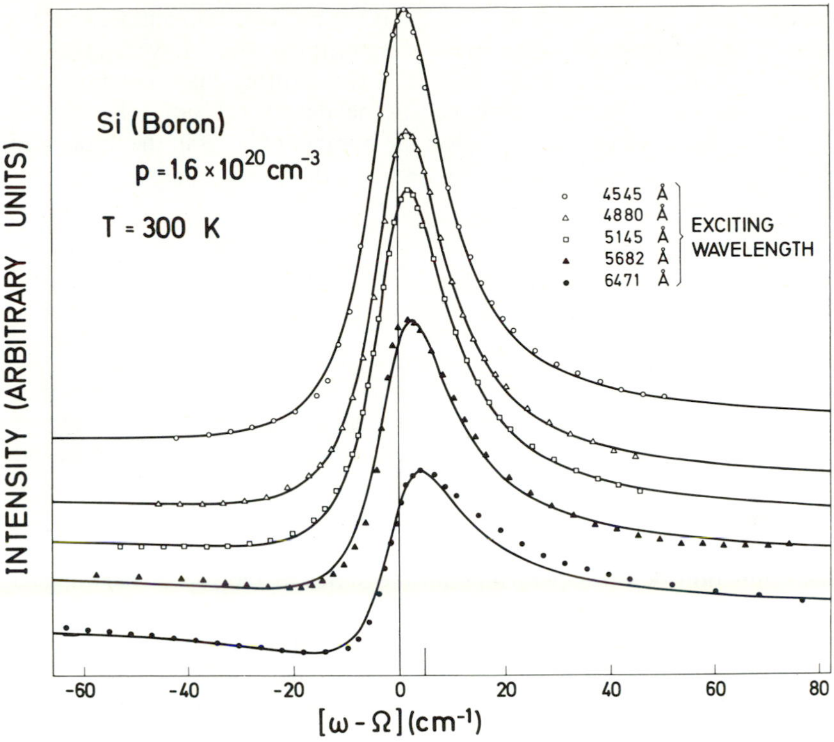

Fig. 4.9. Room temperature Raman spectra for p-type Si [4.56, 58]. For easier viewing the curves have been shifted vertically with respect to one another. The points are experimental data; the lines are calculated curves using (4.49a) with constant ϱ, R

The nature of the electronic continuum has not been studied experimentally in detail, but theoretical papers by Mills et al. [4.59] and by Wright and Balkanski [4.60] may be relevant. The continuum is thought to be caused by nearly vertical transitions between occupied light-hole valence band states and unoccupied heavy-hole states [4.56]. Given a Fermi level about 100 MeV below the top of the valence band, appropriate for 10^{20} holes/cm^3, and with the large known directional anisotropy in the separation between the two bands, one can explain the existence of a very broad electronic continuum from almost zero energy almost to the Fermi energy [4.56].

Cerdeira et al. have given a rough derivation of the width parameter $\Gamma = \pi V^2 \varrho$ appearing in (4.48) and (4.49) [4.56]. We provide here an order of magnitude estimate done in the same spirit. The matrix element $V = (e|H_{\text{el-ph}}|p)$ is approximately $d_0(g|u|p)/a$, where $d_0 \sim 29$ eV is the valence band deformation potential, u is the phonon coordinate, and a is the lattice constant. At low temperatures we have $|(g|u|p)|^2 = \hbar(2MN\omega_l)^{-1}$, where $\hbar\omega_l = E_0$, M is the reduced mass per unit cell, and N is the number of unit cells. The density of states is roughly $\varrho \sim (N_h - N_l)/\mu$, where N_h and N_l are the number of holes in the upper and lower valence bands, respectively, and μ is the Fermi energy. Thus

$$\Gamma = \pi V^2 \varrho(E) \sim \pi \frac{d_0^2}{\mu} \left(\frac{\hbar}{2M\omega_l a^2}\right) \left(\frac{N_h - N_l}{N}\right). \tag{4.50}$$

For the sample of Fig. 4.9 we guess $(N_h - N_l)/N = 1.5 \times 10^{-3}$, use $\mu = 0.1$ eV and find $\Gamma \sim 3$ meV $= 24\,\mathrm{cm}^{-1}$, which is three times higher than the experimental value. The more precise, but still rough, estimate of CERDEIRA et al. was five times greater than the experimental result [4.56].

CERDEIRA et al. have performed additional measurements as a function of concentration at 77 and 300 K and have also studied the spectrum as a function of stress (which splits the states at the top of the valence band) [4.56]. More recently they have observed a Fano-type of interference between the electronic continuum and the localized optical phonon modes of the boron impurity atoms in Si [4.61, 97, 98].

4.3.3. Optical Phonon Modes Bound to Neutral Donors in GaP

By Raman and photo-luminescent techniques DEAN et al. observed LO phonon localized modes in GaP containing neutral donors [4.62]. Their Raman data are reproduced in Fig. 4.10. The localized modes denoted by the arrows lie 0.8 to 1.6 meV lower in energy than the unperturbed phonon (LO_Γ) at $E_0 = 50.2$ meV. The localized modes are due to the bound state that forms when the "free", i.e. plane wave, phonon states couple to a spatially localized electronic excitation of slightly higher energy. For the Raman data in Fig. 4.10, the latter transitions are assumed to be $1S(\mathrm{A}_1)$ to $2P$ donor transitions [4.62]. According to our discussion in Subsection 4.1.3, the bare $1S \to 2P$ donor transition is expected to give relatively weak Raman scattering. The Raman parameter T_e is approximately zero for this case. The general problem of electron-phonon and exciton-phonon bound states has recently been reviewed by LEVINSON and RASHBA [4.63].

We now sketch the theory of the bound localized phonon modes in n-type GaP. We have modified somewhat the original discussion of DEAN et al. [4.62] to better fit the notation of this section.

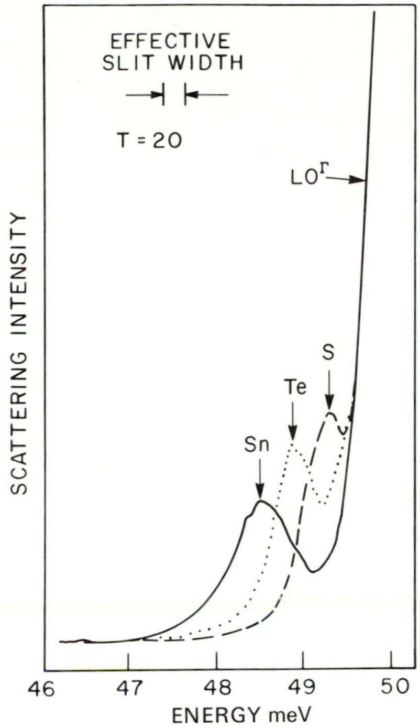

Fig. 4.10. Local phonon modes due to neutral Sn, Te, and S donors in GaP [4.62]. Donor concentration: about $10^{18}/\mathrm{cm}^3$. Laser wavelength: 5145 Å

The electron-phonon coupling is due to the Fröhlich interaction. The interaction Hamiltonian can be written

$$\mathscr{H}_{\text{el-ph}} = -\frac{4\pi e e^*}{\varepsilon_\infty \sqrt{N}\, v_0^2} \sum_k \left[\int d^3r\, \psi_f^*(r)\, e^{i\boldsymbol{k}\cdot\boldsymbol{r}} \psi_i(r)\right] U_k/ik\,. \tag{4.51}$$

Here $e^* = \omega_0 [M v_0 \varepsilon_\infty (\varepsilon_s - \varepsilon_\infty)/(4\pi\varepsilon_s)]^{\frac{1}{2}}$ is the effective charge of the phonon [already used in (4.27)], v_0 is the volume of a unit cell, ψ_i and ψ_f are initial and final donor electron envelope functions, and U_k is defined in (4.21b). The excited phonon state denoted by $|p)$ in Subsections 4.3.1 and 4.3.2 now becomes a set of (nearly) degenerate states $|k)$. The matrix element of (4.51) between a one phonon excited state of wave-vector k, denoted by $|k)$, and the excited states of the $1S \to 2P_j$ donor transition ($j = x, y$, or z), denoted by $|ej)$, is

$$(ej|\mathscr{H}_{\text{el-ph}}|k) = \frac{4\pi e e^*}{\varepsilon_\infty v_0} \sqrt{\frac{\hbar}{2MN\omega_l}}\, \frac{F_j(k)}{ik}\,, \tag{4.52}$$

where $F_j(k)$ is the integral in the brackets in (4.51) and obeys

$$\frac{F_j(k)}{ik} = \frac{3\sqrt{8}k_j}{a_0^5 k(k^2+k_0^2)^3}, \qquad (4.53)$$
$$k_0 = 3/(2a_0).$$

Here a_0 is the donor Bohr radius, and k_j is the jth Cartesian component of $\boldsymbol{k}$.

We write (4.52) in the form

$$(ej|\mathscr{H}_{\text{el-ph}}|k) = V\,\psi_j(k) \qquad (4.54a)$$

where $\psi_j(k)$ is normalized to unity and is given by

$$\psi_j(k) = \eta^{-1}(k_j/k)(k^2+k_0^2)^{-3}, \quad \text{with} \quad \eta^2 = 7Nv_0 a_0^9/(2\pi 3^{10}). \qquad (4.54b)$$

Equations (4.54a) and (4.54b) have a simple interpretation if the phonon dispersion is neglected. Then the states $|k)$ are degenerate. One particular linear combination of them, namely $|pj) \equiv \sum_k \psi_j(k)|k)$ couples to the $1S \to 2P_j$ electronic transition $|ej)$:

$$(ej|\mathscr{H}_{\text{el-ph}}|pj) = V. \qquad (4.55)$$

All other linear combinations, orthogonal to $|pj)$ will be uncoupled to $|ej)$. Their energy remains $E_0 = \hbar\omega_l$. Thus we return to an effective coupled 2-level system as in Fig. 4.8. $\psi_j(k)$ is the phonon wave function in k-space. Equation (4.41) holds for the perturbed energies, namely

$$(E_0 - E)(E_e - E) = V^2 \qquad (4.41)$$

with $E_0 = \hbar\omega_l$ and $E_e = E_{2p} - E_{1s}$. The parameter V^2 is given by

$$V^2 = \frac{56}{6561}\,\frac{e^2}{a_0}\left(\frac{1}{\varepsilon_\infty} - \frac{1}{\varepsilon_s}\right)\hbar\omega_l. \qquad (4.56)$$

Now E_0 is less than E_e, and except for the Sn donor in GaP, E_0 differs enough from E_e so that the second-order perturbation result

$$E_B = E_0 - E_- = V^2/(E_e - E_0) \qquad (4.57)$$

is a good approximation to the binding energy for the lower energy solution of (4.41). The eigen-function will be mostly that of the phonon state $|pj)$. Actually, (4.57) is somewhat inaccurate, as the operator U_k in (4.51) can both create and annihilate a phonon. A better second-order expression for E_B adds the result of a virtual transition to a two-phonon excited electronic state of energy $2E_0+E_e$ to (4.57), giving [4.62]

$$E_B = 2E_e V^2/(E_e^2 - E_0^2)\,. \tag{4.58}$$

DEAN et al. have compared the experimental values of E_B for S, Te, and Si donors with the prediction of (4.58). For the Sn donor case, they had to use the coupled-mode result (4.41) corrected by perturbation theory for the two-phonon electronic excited state at $2E_0+E_e$. The calculations underestimate E_B. This is not surprising, since one should really have a sum over electronic excitations e in (4.58).

The Raman scattering Hamiltonian for both the regular LO phonon and for the bound state $|pj)$ may be obtained from (4.21) and (4.22) with slight modifications. Consider an (xy) polarization Raman scattering geometry. Then for LO phonons in the zincblende structure the only non-zero susceptibility derivatives are $\partial\chi_{xy}/\partial u_z$ and $\partial\chi_{xy}/\partial \mathrm{E}_z$. The longitudinal electric field and the displacement of the LO phonon are related as follows

$$\mathrm{E}_q = -4\pi e^* U_q(\varepsilon_\infty v_0)^{-1}\,. \tag{4.59}$$

The scattering Hamiltonian is

$$\mathscr{H}_u' + \mathscr{H}_{\mathrm{E}}' = (\mathrm{i}\omega_1 A_1/c)(\mathrm{i}\omega_2 A_2/c)^\dagger \exp(-\mathrm{i}\omega t)\,\chi' U_q q_z/q\,, \tag{4.60}$$

where

$$\chi' = \partial\chi_{xy}/\partial u_z - 4\pi e^*(\varepsilon_\infty v_0)^{-1}\,\partial\chi_{xy}/\partial \mathrm{E}_z\,.$$

The matrix element of U_q from the phonon ground state to an unperturbed LO phonon state of wave-vector q is

$$(q|U_q|g) = v_0[\hbar N/(2M\omega_l)]^{\frac{1}{2}} \tag{4.61}$$

and that to the bound phonon state $|pj) = \sum_k \psi_j(k)|k)$ is

$$(pj|U_q|g) = (q|U_q|g)\,\psi_j(q)^*\,. \tag{4.62}$$

Thus the ratio of Raman intensities for $N_d = n_d N v_0$ total donors is

$$\begin{aligned}\frac{I_{\mathrm{bound}}}{I_{\mathrm{LO}}} &= \frac{N_d \sum_j |(pj|U_q|g)|^2}{|(q|U_q|g)|^2} = N_d \sum_j |\psi_j(q)|^2 \\ &= 8192\pi a_0^3 n_d/63\,.\end{aligned} \tag{4.63}$$

We have gone to the $q=0$ limit of $\psi_j(q)^2$ after performing the sum over j. This is appropriate, since $q \ll a_0^{-1}$.

Equation (4.63) is $\frac{32}{7}$ times larger than a result given by DEAN et al. [4.62]. The ratio in (4.63] has been independently verified as correct [4.64].

BARKER has shown that most of the properties of these localized LO phonon modes can be understood using a phenomenological approach [4.65]. The donor atom is replaced by a dielectric sphere embedded in a polar medium. The resonance frequency of the dielectric is close to that of the medium. They interact via the electric field of the phonon, which polarizes the dielectric. As a result a bound phonon mode is formed. In essence, BARKER's picture would be consistent with the microscopic theory of DEAN et al. [4.62] if one could approximate the transition charge density $-e\psi_f(r)^* \psi_i(r)$ in (4.51) as a constant for r less than the radius of the sphere, which is of order a_0, and zero for r greater than this radius.

4.3.4. Coupled Valley-Orbit and E_2 Phonon Excitations in SiC

In Subsection 4.2.2 there was a discussion of Raman scattering from the E_2 valley-orbit transition in nitrogen-doped 6H SiC. At a donor concentration of $4 \times 10^{18}/cm^3$ the 13 meV E_2 valley-orbit line has a half width at half maximum of 1 meV [4.29]. The high energy tail of this line shows an anti-resonant interaction with a phonon of E_2 symmetry at 18 meV. This is barely visible as a sharp high energy drop in that phonon peak in Fig. 4.7, but it shows up clearly in higher resolution data published by COLWELL and KLEIN [4.29]. A similar effect was observed between a 22 meV phonon line and a broadened valley-orbit transition in the 15R polytype of SiC [4.29].

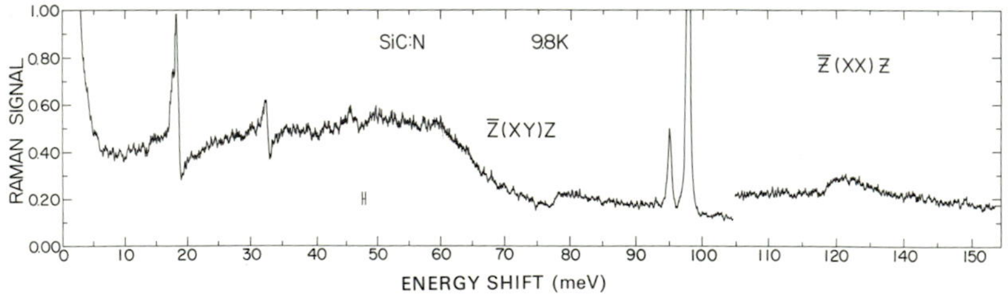

Fig. 4.11. Raman spectra of 6H SiC : N, $6 \times 10^{19}/cm^3$ at 9.8 K [4.29]. The $\bar{z}(xy)z$ geometry couples to E_2 excitations only. The $\bar{z}(xx)z$ geometry couples to A_1 and E_2 excitations. This spectrum reveals Fano-type interferences between a broad electronic continuum and E_2 phonon modes at 18 and 33 meV, and a coupled LO phonon-plasmon mode near 125 meV. Laser wavelength: 5145 Å

Figure 4.11 shows Raman data from 6H SiC doped with 6×10^{19} N/cm^3 [4.29]. Such a sample shows metallic conductivity and, as already discussed in Section 4.1, has a broad LO phonon-plasmon coupled mode at about 125 meV. Fano-type interferences are readily seen near 18 and 33 meV between E_2 phonon modes and a greatly broadened electronic continuum.

COLWELL and KLEIN fit their data from 10–25 meV on the 4×10^{18}/cm^3 sample using (4.48) with $\varrho(E)$ assumed to be a Lorentzian centered at 13 meV [4.29]. Use of this equation in this context is an oversimplification because it considers neither the localized nature of the coupled modes nor the simultaneous presence of Raman scattering from unperturbed phonon modes.

A more correct theory will now be discussed. It essentially synthesizes the treatments of Subsections 4.3.1 and 4.3.3. The details of the electron-phonon coupling are specific to 6H SiC, but the remaining discussion has more general applicability.

As mentioned above in the discussion following (4.34), there are effectively three valleys in the conduction band of 6H SiC. Let them be at k_1, k_2, and k_3. These will be assumed to couple to an E_2 phonon coordinate u and thereby change their energies by the amounts

$$\begin{aligned} &\delta E(k_2) = -\delta E(k_3) = \delta E \\ &\delta E(k_1) = 0\,, \end{aligned} \tag{4.64}$$

where $\delta E = ud/a$, and d is an appropriate deformation potential.

If the envelope functions of $\phi_j(r)$ in (4.29) are assumed to be independent of the valley index j (and of n), the effective-mass valley-orbit functions are for the $1S$(A) level

$$3^{-\frac{1}{2}}(\phi_1 + \phi_2 + \phi_3)\, f(r) \tag{4.65a}$$

and for the $1S(E_2)$ level

$$2^{-\frac{1}{2}}(\phi_2 - \phi_3)\, f(r)\,. \tag{4.65b}$$

Now $f(r)$ is slowly varying on a distance of order a (at least within the effective mass approximation). For phonon displacements $u(r)$ that are also slowly varying we have from (4.64) and (4.65).

$$(E_2|\mathscr{H}_{\text{el-ph}}|A) = 2d\, 6^{-\frac{1}{2}} a^{-1} \int |f(r)|^2 u(r)\, d^3 r\,.$$

Now introduce U_k from (4.21 b) via

$$u(\boldsymbol{r}) = v_0^{-1} N^{-1} \sum_k U_k \,\mathrm{e}^{-i\boldsymbol{k}\cdot\boldsymbol{r}}$$

and find

$$(E_2|\mathscr{H}_{\text{el-h}}|A) = 2d\, N^{-1} v_0^{-1} a^{-1} 6^{-\frac{1}{2}} \sum_k U_k F(k)\,, \tag{4.66a}$$

where

$$F(k) = \int |f(r)|^2 e^{-ik\cdot r} d^3 r = (1 + q^2 a_0^2/4)^{-2} \tag{4.66b}$$

for a $1S$ function $f(r)$.

Using (4.61) we find for the coupling between an E_2 phonon state of wave-vector k and the excited E_2 valley-orbit state

$$(e|\mathscr{H}_{\text{el-ph}}|k) = \frac{2d}{a(6N)^{\frac{1}{2}}} \left(\frac{\hbar}{2M\omega_k}\right)^{\frac{1}{2}} F(k).$$

We write this in a form similar to (4.54):

$$(e|\mathscr{H}_{\text{el-ph}}|k) = V H(k), \tag{4.67}$$

where

$$\sum_k H(k)^2 = 1.$$

If we approximate ω_k by $\omega_0 = \omega_{k=0}$, we have

$$H(k) = \left(\frac{16\pi a_0^3}{N v_0}\right)^{\frac{1}{2}} (1 + q^2 a_0^2/4)^{-2} \tag{4.68a}$$

and

$$V^2 = \frac{v_0}{a_0^3} \frac{d^2}{48\pi} \frac{\hbar}{M\omega_0 a^2}. \tag{4.68b}$$

The coupled modes can be treated using a generalization of the Green's function of (4.42). Again we let $G(z) = (\mathscr{H} - z)^{-1}$, where now $(e|\mathscr{H}|e) = E_e$, $(k|\mathscr{H}|k) = E_k = \hbar\omega_k$, and $(e|\mathscr{H}|k) = (k|\mathscr{H}|e) = VH(k)$. We find that

$$g_e = (e|G|e) = \left[E_e - z - V^2 \sum_k H(k)^2 (E_k - z)^{-1}\right]^{-1} \tag{4.69a}$$

$$(k|G|e) = (e|G|k) = -VH(k)\, g_e (E_k - z)^{-1}, \tag{4.69b}$$

$$\begin{aligned}(k|G|k') = (k'|G|k) &= \delta_{kk'}(E_k - z)^{-1} \\ &+ V^2 H(k) H(k')\, g_e (E_k - z)^{-1} (E_{k'} - z)^{-1}.\end{aligned} \tag{4.69c}$$

The Raman amplitude T_p will be the appropriate matrix element of $\mathscr{H}_u' \propto \chi' U_q$ for $q \approx 0$. Thus the Raman spectrum will be given by this modification of (4.43)

$$\begin{aligned}I(E) &= \text{Im}\left\{\sum_{i,j=e,q=0} T_i (i|G|j)\, T_j\right\} \\ &= \text{Im}\{T_0^2 (E_0 - z)^{-1} + [T_e - VH(0)\, T_0 (E_0 - z)^{-1}]^2 g_e\}.\end{aligned} \tag{4.70}$$

This result holds for a single electronic center. If there are N_d non-interacting centers, we must multiply the second term in (4.70) by N_d.

A simplification results when we can neglect phonon dispersion and set $E_k = E_0$. Then for N_d impurities (4.70) becomes

$$I(E) \tag{4.71}$$

$$= \mathrm{Im}\left\{\frac{T_0^2[1-N_d H(0)^2]}{E_0 - z} + \frac{N_d[(E_0-z)T_e^2 - 2V H(0) T_0 T_e + H(0)^2 T_0^2(E_e - z)]}{(E_0-z)(E_e-z)-V^2}\right\}.$$

The first term in (4.71) gives scattering at the unperturbed phonon energy E_0 with an intensity $[1-N_d H(0)^2]$ times that of an undoped crystal. This factor corrects for the phonon scattering strength that has been transferred to the coupled modes via the second term.

The second term in (4.71) gives N_d multiplied by our previous result (4.44) with T_p replaced by $H(0)\, T_0$. With these changes the manipulations we performed on (4.44) apply to this term also. In particular, if the electronic level E_e is homogeneously broadened and has a density of states $\varrho(E)$, then the appropriately modified version of (4.48) will replace the second term in (4.71), giving

$$I(E) \tag{4.72}$$

$$= \pi T_0^2[1-N_d H(0)^2]\,\delta(E-E_0) + \pi N_d T_e^2 \frac{\varrho(E)[E_0 - E - V H(0) T_0/T_e]^2}{[E_0 - E + V^2 R(E)]^2 + \pi^2 V^4 \varrho(E)^2}.$$

For donors at finite concentration (below that of the metal-insulator transition) the donor transition energy E_e is inhomogeneously broadened due to concentration-produced fluctuations in the potential. If $N_d\, \varrho(E_e)\, dE_e$ represents the number of donors with excitation energy between E_e and $E_e + dE_e$, then (4.71) becomes

$$I(E) = \pi T_0^2[1-N_d H(0)^2]\,\delta(E-E_0) \tag{4.73a}$$

$$+ N_d \,\mathrm{Im}\left\{\int \varrho(E_e)\, dE_e \frac{[(E_0-z)T_e^2 - 2V H(0) T_0 T_e + H(0)^2 T_0^2 (E_e - z)]}{(E_0-z)(E_e-z) - V^2}\right\}$$

or

$$I(E) = \pi T_0^2[1-N_d H(0)^2]\,\delta(E-E_0) \tag{4.73b}$$

$$+ \pi N_d T_e^2 [E_0 - E - V H(0) T_0/T_e]^2 (E_0-E)^{-2}\, \varrho[E + V^2(E_0-E)^{-1}].$$

Equation (4.73b) was obtained by evaluating (4.73a) at its poles and by assuming $\varrho(E_e)$ is sufficiently well-behaved.

Except with a Lorentzian $\varrho(E)$ function, (4.72) and (4.73) give different results. COLWELL and KLEIN fitted their low concentration SiC:N data by using the second term of (4.72) with a Lorentzian $\varrho(E)$ and by ignoring the first term [4.29]. It would be valuable to take more extensive data and repeat the computer calculations for varying concentrations on both sides of the metal-insulator transition in an effort to determine experimentally whether the electronic transition is homogeneously or inhomogeneously broadened.

4.4. Single Particle Spectra

4.4.1. Introduction

In the limit of small q an electron moving with velocity $\boldsymbol{v}$ will scatter a photon with a frequency shift $\boldsymbol{q}\cdot\boldsymbol{v}$ due to the Doppler effect. For a non-interacting free electron gas the light scattering spectrum would thus be a superposition of Doppler shifted lines. The result, as given in (4.8) and (4.10) and the following discussion is

$$\begin{aligned} N(d^2\sigma/d\omega\, d\Omega)/V = d^2 R/d\omega\, d\Omega &= (\omega_2/\omega_1)^2\, r_0^2(1+n_\omega) \\ &\quad \cdot \hbar^{-1}\pi\, \mathrm{Im}\{-F(q,\omega)\} \qquad (4.74) \\ &= (\omega_2/\omega_1)^2\, r_0^2(1+n_\omega)\sum_k\, [n(\boldsymbol{k}+\boldsymbol{q})-n(\boldsymbol{k})]\,\delta(\omega-\boldsymbol{q}\cdot\boldsymbol{v}_k)\,. \end{aligned}$$

The response of a simple electron gas to a $1\to2$ scattering process can be treated in terms of the scattering Hamiltonian of the form $(\boldsymbol{e}_1\cdot\boldsymbol{e}_2)\cdot r_0 A_1 A_2^\dagger \varrho_q \exp(-i\omega t)$ given in (4.5). A free electron gas would respond to this perturbation to give the cross section in (4.74), but when electron-electron Coulomb interactions are taken into account within the RPA the longitudinal electric field associated with ϱ_q suppresses the strength of the response. The result is that (4.74) must be multiplied by $|\varepsilon(q,\omega)|^{-2}$, which is much less than unity for q less than the screening wave-vector q_s.

4.4.2. Scattering from Spin Density Fluctuations [Ref. 8.7, Sect. 2.3.2]

In a solid it is possible to take advantage of several band structure effects to drive an electron gas by the external radiation fields A_1 and $A_2^\dagger$ in ways other than via the charge density fluctuation ϱ_q. It is often possible to drive the electron system via the spin density fluctuation operator

$$\boldsymbol{\sigma}(\boldsymbol{q}) = \sum_j \boldsymbol{\sigma}_j\, \mathrm{e}^{i\boldsymbol{q}\cdot\boldsymbol{r}_j} \qquad (4.75a)$$

which is essentially the Fourier transform of the spin density operator $\boldsymbol{\sigma}(\boldsymbol{r})$. A connection with ϱ_q can be established as follows. If the z-direction is chosen as the spin quantization axis, and if we take the trace with respect to spin variable of $\boldsymbol{\sigma}(\boldsymbol{q})$, we find

$$\mathrm{Tr}\,\boldsymbol{\sigma}(\boldsymbol{q}) = \hat{z}(\varrho_{q\uparrow}-\varrho_{q\downarrow})\,, \qquad (4.75b)$$

where $\varrho_{q\uparrow}$ and $\varrho_{q\downarrow}$ are the density fluctuation operators for spin-up and spin-down electrons. Since the total electron density fluctuation ϱ_q

$=\varrho_{q\uparrow}+\varrho_{q\downarrow}$ remains unexcited during this process, a coupling via (4.75a) or (4.75b) will not be screened by the long range part of the Coulomb interactions.

The detailed theory of this light scattering process was first given by HAMILTON and MCWHORTER [4.66]. The mechanism for coupling to $\boldsymbol{\sigma}(\boldsymbol{q})$ is the same as that used by YAFET [4.67] in his theory of spin-flip Raman scattering in a magnetic field, but the dynamics are different in the two cases. A simple band structure was assumed in these theories [4.68]. The effect relies on mixing by spin-orbit coupling of the three p-like states at the top of the valence band to produce states of mixed orbital character. Spin flips can then occur when the optical radiation fields induce suitable virtual transitions between the conduction and valence bands. The effect is present for both electrons and holes, but we shall limit the discussion to electrons. Physical ideas will be emphasized here. For detailed discussions of the theory, and for detailed comparisons with experiment, the reader is referred to the literature [4.8, 66, 11, 13].

To understand the spin-density fluctuation mechanism, it is instructive to treat the spin-orbit coupling

$$\mathscr{H}_{\mathrm{so}}=\frac{\hbar\boldsymbol{p}\cdot(\boldsymbol{\sigma}\times\nabla V)}{4m^2c^2} \tag{4.76}$$

in perturbation theory. Consider the scattering of an electron in the conduction band with wavevector $\boldsymbol{k}$ and spin state α to a conduction band state with wavevector $\boldsymbol{k}+\boldsymbol{q}$ and spin β in third-order perturbation theory with the $\boldsymbol{p}\cdot\boldsymbol{A}$ terms of $\mathscr{H}_{\text{el-rad}}$ in (4.1) used twice and $\mathscr{H}_{\mathrm{so}}$ used once. Let the incident and scattered polarization vectors $\boldsymbol{e}_1$ and $\boldsymbol{e}_2$ be along x and y axes. Let the conduction band states be s-like and be denoted by $|s\alpha\rangle$ and $|s\beta\rangle$; let the valence band states transform as x, y, and z; and let holes in the valence band be denoted by bars: $|\overline{x\alpha}\rangle$, etc. Ignore the wavevector dependence of the interband matrix elements of the momentum $\boldsymbol{p}$ and, for now, of the energy gap E_{G}. $\mathscr{H}_{\mathrm{so}}$ causes transitions among hole states [4.69]. The matrix element $\gamma_{\alpha\beta}$ as used by BLUM [4.9] and by HAMILTON and MCWHORTER [4.66] becomes

$$\begin{aligned}\gamma_{\alpha\beta}=&\frac{1}{m}\\ &\cdot\frac{\langle s\alpha|p_x|\overline{x\beta},s\alpha,s\beta\rangle\langle\overline{x\beta},s\alpha,s\beta|\mathscr{H}_{\mathrm{so}}|\overline{y\alpha},s\alpha,s\beta\rangle\langle\overline{y\alpha},s\alpha,s\beta|p_y|s\beta\rangle}{(E_{\mathrm{G}}-\hbar\omega_1)^2}\\ &+\frac{1}{m}\\ &\cdot\frac{\langle s\alpha|p_y|\overline{y\beta},s\alpha,s\beta\rangle\langle\overline{y\beta},s\alpha,s\beta|\mathscr{H}_{\mathrm{so}}|\overline{x\alpha},s\alpha,s\beta\rangle\langle\overline{x\alpha},s\alpha,s\beta|p_x|s\beta\rangle}{(E_{\mathrm{G}}+\hbar\omega_2)^2}.\end{aligned} \tag{4.77}$$

We use the approximation $\omega_2 = \omega_1$ and the relation

$$\langle \overline{x\beta} | \mathscr{H}_{\text{so}} | \overline{y\alpha} \rangle = -\langle \overline{y\beta} | \mathscr{H}_{\text{so}} | \overline{x\alpha} \rangle = -i \langle \alpha | \sigma_z | \beta \rangle \, \Delta_0/3 \,, \tag{4.78}$$

where Δ_0 is the usual spin-orbit parameter (Fig. 4.5). In addition, we let $P \equiv i\langle x | p_x | s \rangle = -i\langle s | p_y | y \rangle$, etc. Then

$$\gamma_{\alpha\beta} = -\frac{P^2 4\hbar\omega_1 E_G i \Delta_0 \langle \alpha | \sigma_z | \beta \rangle}{3m(E_G^2 - \hbar^2 \omega_1^2)^2} \,. \tag{4.79}$$

Equation (4.79) is correct in the limit $\Delta_0 \ll E_G$.

The general result for the KANE band model [4.68] has been derived by HAMILTON and MCWHORTER [4.66] for the spin-dependent scattering:

$$\gamma_{\alpha\beta} = -i \boldsymbol{e}_1 \times \boldsymbol{e}_2 \cdot \langle \alpha | \boldsymbol{\sigma} | \beta \rangle \, B \tag{4.80a}$$

with

$$B = \frac{\hbar P^2 \omega_1}{3m} \left(\frac{1}{E_1^2 - \hbar^2 \omega_1^2} + \frac{1}{E_2^2 - \hbar^2 \omega_1^2} - \frac{2}{E_3^2 - \hbar^2 \omega_1^2} \right) \tag{4.80b}$$

plus a term anisotropic in $\boldsymbol{k}$ that will be neglected. The energy E_i represents the vertical gap between the conduction band and the ith valence band at wavevector $\boldsymbol{k}$. The heavy hole band is denoted by $i = 1$, the light hole band by $i = 2$, and the split-off band by $i = 3$.

The factor of i in (4.80a) is not present in the analogous expression for the spin-independent part of $\gamma_{\alpha\beta}$ [see (4.89) below]. Its presence means that the two types of scattering cannot interfere.

If we set $E_2 = E_1 = E_G$ and $E_3 = E_G + \Delta_0$, (4.80) becomes

$$\gamma_{\alpha\beta} = -\frac{iP^2}{3m} \frac{2\hbar\omega_1 (2E_G \Delta_0 + \Delta_0^2) \, [(\boldsymbol{e}_1 \times \boldsymbol{e}_2) \cdot \langle \alpha | \boldsymbol{\sigma} | \beta \rangle]}{(E_G^2 - \hbar^2 \omega_1^2) \, [(E_G + \Delta_0)^2 - \hbar^2 \omega_1^2]} \,. \tag{4.81}$$

Equation (4.81) reduces to (4.79) for small Δ_0 and for the appropriate polarization geometry.

If we write (4.79), (4.80), or (4.81) in the form

$$\gamma_{\alpha\beta} = i\gamma_0 |\boldsymbol{e}_1 \times \boldsymbol{e}_2| \, \langle \alpha | \sigma_z | \beta \rangle \,, \tag{4.82a}$$

where the z axis is defined to be along $\boldsymbol{e}_1 \times \boldsymbol{e}_2$, and if we introduce electron creation and annihilation operators $C^{\dagger}_{k+q,\alpha}$ and $C_{k,\beta}$, we find that the transition operator M, whose matrix element is $\gamma_{\alpha\beta}$, can be written as

$$M = i \sum_{k,\alpha,\beta} C^{\dagger}_{k+q,\alpha} C_{k,\beta} \langle\alpha|\sigma_z|\beta\rangle \, |e_1 \times e_2| \, \gamma_0 \,. \tag{4.82b}$$

Using an obvious extension of the definition (4.4) of ϱ_q, we find

$$M = i(\varrho_{q\uparrow} - \varrho_{q\downarrow}) \, \gamma_0 \, |e_1 \times e_2| \,. \tag{4.82c}$$

The scattering Hamiltonian consistent with (4.82) is

$$\mathscr{H}' = i\gamma_0 |e_1 \times e_2| \, A_1 A_2^{\dagger} \, e^{-i\omega t} (\varrho_{q\uparrow} - \varrho_{q\downarrow}) + \text{H.C.} \tag{4.83}$$

The power scattering cross-section per unit volume is [4.9]

$$\frac{d^2 R}{d\omega \, d\Omega} = (\omega_2/\omega_1)^2 \, Av_i \left[\sum_f |\langle f | M | i \rangle|^2 \right] \delta(\omega_{if} - \omega) \,. \tag{4.84}$$

This should be compared with (4.6) and (4.7).

When (4.84) is evaluated using the fluctuation-dissipation theorem, within the RPA the Coulomb interactions have no effect, since the electrostatic potential induced by $-\varrho_{q\downarrow}$ cancels that induced by $\varrho_{q\uparrow}$. The result is

$$\frac{d^2 R}{d\omega \, d\Omega} = (\omega_2/\omega_1)^2 \, |e_1 \times e_2|^2 \, \gamma_0^2 (1 + n_\omega) \, \hbar \pi^{-1} (-\operatorname{Im}\{F\}) \,, \tag{4.85}$$

where F was defined in (4.8). To (4.85) we must add the contribution of the spin-independent processes (4.19), see (4.104) below.

When the k-dependence of the energy gaps cannot be neglected, $\gamma_0^2 F$ in (4.85) is replaced by a more complicated sum over k [4.66]. This will affect the shape of the spectrum under resonance conditions.

Since F is a response function for the non-interacting electron gas, $\operatorname{Im}\{F\}$ obeys a sum rule [4.3, 4], which leads to the following low temperature result that is similar to (4.14)

$$\int_0^\infty \omega (d^2 R/d\omega \, d\Omega) \, d\omega = (\omega_2/\omega_1)^2 \, r_0^2 |e_1 \times e_2|^2 \, \gamma_0^2 \hbar q^2 N/(2mV) \,. \tag{4.86}$$

For a degenerate electron gas at small q, the integral on the left side of (4.86) is $2qv_{\mathrm{f}} \int_0^\infty (d^2 R/d\omega \, d\Omega) \, d\omega/3$. Thus the integrated scattering efficiency is

$$\frac{dR}{d\Omega}=\frac{3}{4}\left(\frac{\omega_2}{\omega_1}\right)^2 r_0^2\,|\boldsymbol{e}_1\times\boldsymbol{e}_2|^2\,\gamma_0^2(N/V)(\hbar q/m v_{\mathrm{f}})\,. \tag{4.87}$$

The last factor $(\hbar q/m v_{\mathrm{f}})$ represents the fraction of the electrons allowed by the Pauli principle to participate in the scattering. Below resonance order of magnitude estimates are

$$d\sigma/d\Omega\sim r_0^2(m/m^*)^2\,(\hbar\omega_1\,\Delta_0/E_{\mathrm{G}}^2)^2\,(\hbar q/m v_{\mathrm{f}}) \tag{4.88a}$$

in the small Δ_0 limit [4.8] and

$$d\sigma/d\Omega\sim r_0^2(m/m^*)^2\,(\hbar\omega_1/E_{\mathrm{G}})^2\,(\hbar q/m v_{\mathrm{f}}) \tag{4.88b}$$

in the large Δ_0 limit.

MOORADIAN [4.12] has compared calculated cross-sections due to spin and charge density fluctuations with measured intensities in GaAs for various polarization geometries and carrier densities. At low densities and high temperatures where $q>q_{\mathrm{s}}$, there is no screening, and the charge density fluctuations dominate. When $q<q_{\mathrm{s}}$, the spin density mechanism dominates, except at high temperatures, where the energy density fluctuation mechanism gives a comparable effect.

4.4.3. Light Scattering by Energy Density Fluctuations

This mechanism, first proposed by WOLFF [4.70, 5] is related to the non-parabolic nature of the energy bands of real semiconductors. Using the simplified energy bands of KANE [4.68], HAMILTON and MCWHORTER [4.66] derived the following expression for the spin-independent part of the matrix element for photon scattering accompanied by the scattering of an electron from $\boldsymbol{k}$ to $\boldsymbol{k}+\boldsymbol{q}$

$$\gamma_{\alpha\beta}=\gamma\,\delta_{\alpha\beta}\,, \tag{4.89a}$$

where

$$\gamma=\boldsymbol{e}_1\cdot\overset{\leftrightarrow}{\boldsymbol{A}}\cdot\boldsymbol{e}_2\,, \tag{4.89b}$$

where, except for conditions of extreme resonance to be discussed below, $\overset{\leftrightarrow}{\boldsymbol{A}}$ depends upon $\boldsymbol{k}$, and not upon $\boldsymbol{q}$ in the small q limit

$$\begin{aligned}\overset{\leftrightarrow}{\boldsymbol{A}}=\overset{\leftrightarrow}{\boldsymbol{I}}\left[1+\frac{2P^2}{3m}\left(\frac{E_1}{E_1^2-\hbar^2\omega_1^2}+\frac{E_2}{E_2^2-\hbar^2\omega_1^2}+\frac{E_3}{E_3^2-\hbar^2\omega_1^2}\right)\right]\\ -(\hat{k}\hat{k}-\overset{\leftrightarrow}{\boldsymbol{I}}/3)\,\frac{P^2}{m}\left(\frac{E_1}{E_1^2-\hbar^2\omega_1^2}-\frac{E_2}{E_2^2-\hbar^2\omega_1^2}\right).\end{aligned} \tag{4.89c}$$

The notation is the same as that in (4.80); E_1, E_2 and E_3 are the vertical gaps at $\boldsymbol{k}$ between the top three valence bands and the conduction band.

In the small ω_1 limit, the tensor $\overleftrightarrow{A}$ in (4.89c) becomes the (m/m^*) tensor of the electrons. When the $\boldsymbol{k}$-dependence of $\overleftrightarrow{A}$ is neglected, the resulting perturbation $\mathscr{H}'$ on the electronic system is proportional to the number density fluctuation ϱ_q. As was mentioned in the discussion between (4.7) and (4.8), this perturbation takes the form

$$\mathscr{H}' = -e\,\varphi_{\text{ext}}\varrho_q = -e\,\varphi_{\text{ext}}\sum_k C_{k+q}^{\dagger} C_q\,,$$

and within the RPA the total effective Hamiltonian is $\mathscr{H}_{\text{eff}} = \mathscr{H}' - e\,\varphi_{\text{ind}}\varrho_q$. In the small q limit the potential caused by the induced charge density obeys $\varphi_{\text{ind}} = -\varphi_{\text{ext}}$ so that as q tends to zero $\mathscr{H}_{\text{eff}}$ tends to zero; the density fluctuations are screened.

When the $\boldsymbol{k}$ dependence of $\overleftrightarrow{A}$ and hence of γ in (4.89a) and (4.89b) cannot be neglected, the perturbing Hamiltonian takes the form

$$\mathscr{H}' \propto \sum_k \gamma(\boldsymbol{k})\, C_{k+q}^{\dagger} C_q\,. \tag{4.90}$$

For a degenerate Fermi gas off resonance, the first term in (4.89c) is isotropic. Its contribution to $\mathscr{H}'$ will still be screened since we can write

$$\gamma_{\text{is}}(\boldsymbol{k}) = \gamma_{\text{is}}(k) \approx \gamma_{\text{is}}(k_f) = \text{constant}\,.$$

The dominant effect then comes from an anisotropic second term in (4.89c) [4.70]. If we expand E_1 and E_2 about their values at $k=0$ (denoted by E_{10} and E_{20}), we find for $\boldsymbol{e}_1 \parallel \boldsymbol{e}_2$ that γ_{an}, the anisotropic part of γ, is given by

$$\gamma_{\text{an}} = \frac{4P^4(E_{10}^2 + \hbar^2\omega_1^2)}{3m^2 E_{10}(E_{10}^2 - \hbar^2\omega_1^2)^2}\;\frac{\hbar^2 k^2}{2m}\left(\cos^2\beta_k - \frac{1}{3}\right) \tag{4.91}$$

where β_k is the angle between $\boldsymbol{k}$ and $\boldsymbol{e}_1$. Here we have used the relation $E_2 - E_1 = 2P^2\hbar^2k^2/(3m^2E_1)$ [4.68].

The perturbation on the electronic system as given by (4.90) and (4.91) can be described as a direction-dependent energy density fluctuation. It will not be screened. The resulting calculation of the cross-section is complicated, but an estimate of the average of the cross-section over polarization directions has been given by Wolff [4.70]:

$$\frac{d\sigma}{d\Omega} \sim 2r_0^2(\omega_2/\omega_1)^2\,(E_{\text{F}}/E_{\text{G}})^2\,(\hbar q/m v_{\text{f}})\,. \tag{4.92}$$

In most cases this will be less than the estimates in (4.88), and this process is not likely to be important.

At finite temperature another type of energy density fluctuation is possible. We write

$$E_j \cong E_{j0} + (1 + m_e/m_j)\, K\,, \tag{4.93}$$

where $K = \hbar^2 k^2/2m_e$ is the electron's kinetic energy, m_e is the electron mass, and m_j is the hole mass in band j. If we then expand the isotropic part of A in (4.89c) about $k=0$, we find

$$\begin{aligned} A_{\rm is} = &\left[1 + \frac{2P^2}{3m} \sum_j \left(\frac{E_{j0}}{E_{j0}^2 - \hbar^2\omega_1^2}\right)\right] \\ &- \frac{2P^2}{3m} \sum_j \frac{(\hbar^2\omega_1^2 + E_{j0}^2)(1 + m_e/m_j)}{(E_{j0}^2 - \omega_1^2)} K\,. \end{aligned} \tag{4.94}$$

When the fluctuation-dissipation theorem is used with the RPA to calculate the scattering efficiency [4.9, 66, 70], the result in the small q limit is

$$\begin{aligned} &d^2R/d\Omega\, d\omega \\ &= -r_0^2 \hbar \pi^{-1} (\omega_2/\omega_1)^2 (n_\omega + 1)\, \mathrm{Im}\{(L_2/L_0) - (L_1/L_0)^2\}\,, \end{aligned} \tag{4.95}$$

where L_0 equals F as given in (4.8) and where

$$\hbar L_j = \sum_k \gamma^j [n(\boldsymbol{k}) - n(\boldsymbol{k}+\boldsymbol{q})]\,[\omega + \mathrm{i}0^+ + \omega(\boldsymbol{k}) - \omega(\boldsymbol{k}+\boldsymbol{q})]^{-1}\,, \tag{4.96}$$

where now $\gamma = (\boldsymbol{e}_1 \cdot \boldsymbol{e}_2)\, A_{\rm is}$. It is instructive to interpret

$$\langle \gamma^j \rangle \equiv L_j/L_0 \tag{4.97}$$

as a weighted average [in the sense of (4.96)] of γ^j. Then we have

$$\begin{aligned} &d^2R/d\Omega\, d\omega \\ &= -r_0^2 \hbar \pi^{-1} (\omega_2/\omega_1)^2 (n_\omega + 1)\, \mathrm{Im}\{F \langle (\gamma - \langle\gamma\rangle)^2 \rangle\}\,. \end{aligned} \tag{4.98}$$

One can show that at high temperatures

$$-\hbar\pi^{-1} \int_{-\infty}^{\infty} (n_\omega + 1)\, \mathrm{Im}\{F\}\, d\omega = N/V. \tag{4.99}$$

The constant (first) term in (4.94) drops out, since it has no fluctuation. If we neglect the weak ω-dependence of $\langle\gamma\rangle$ and $\langle\gamma^2\rangle$, we find that the integrated cross-section is proportional to $\langle(K-\langle K\rangle)^2\rangle$, the mean square fluctuation in the kinetic energy K; this is essentially $(k_B T)^2$.

The integrals L_j have been calculated by Wolff [4.70] for a Maxwellian electron distribution. He, in fact, used both isotropic and anisotropic parts of γ. The result for the integrated cross section after averaging over polarization directions is

$$d\sigma/d\Omega \sim 8(m r_0/m^*)^2 (\omega_2/\omega_1)^2 (k_B T/E_G)^2 . \tag{4.100}$$

For a degenerate electron gas, this estimate should be multiplied by $\hbar q/m v_f$.

Wolff found the frequency-dependence of the cross section to be slightly broadened from a Gaussian of the form [4.70]

$$\exp[-\omega^2/(q v_{th})^2]$$

with

$$v_{th}^2 = 2 k_B T/m_e .$$

The energy density fluctuation mechanism just described explains the strong polarized light scattering observed at room temperature in GaAs doped in the $q \ll q_s$ region [4.12, 13]. The polarized and depolarized intensities are nearly equal at a carrier concentration of $1.4 \times 10^{18}\ \mathrm{cm}^{-3}$.

4.4.4. Polarized Single Particle Scattering under Conditions of Extreme Resonance

Equations (4.92) and (4.100) have shown that off-resonance light is scattered very weakly by the energy density fluctuations of a degenerate electron gas. Under conditions of "extreme resonance" this changes dramatically. In n-GaAs Pinczuk et al. [4.71] observed a strong resonance enhancement of the polarized spectra at 10 K with 6471 Å (1.92 eV) laser light on a sample with a $1.3 \times 10^{18}\ \mathrm{cm}^{-1}$ doping and with 6328 Å (1.96 eV) light on a sample with a $4.8 \times 10^{18}\ \mathrm{cm}^{-1}$ doping. Their data for the former case are shown in Fig. 4.12. One can see that the polarized scattering is almost as strong as the depolarized scattering.

The resonant band gap in the experiments of Pinczuk et al. [4.71] was the one involving the split-off valence band, denoted above by E_3. For a vertical transition at wavevector k_f from this band to the top of the

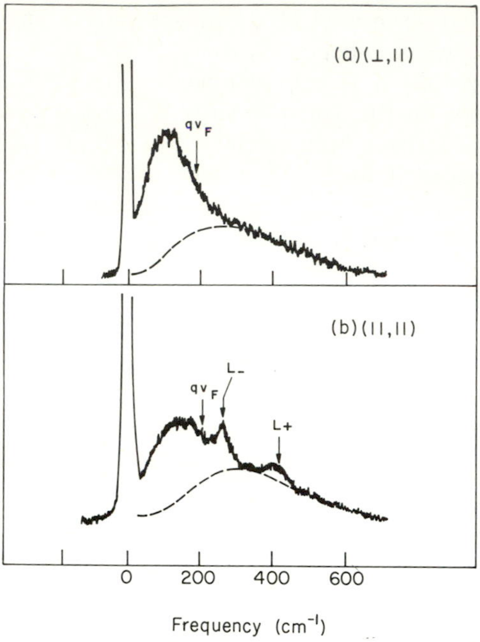

Fig. 4.12a and b. Single particle spectra and coupled LO phonon-plasmon modes ($L\pm$) for n-GaAs, $n=1.3\times10^{18}$ cm^{-3}. Temperature: 10 K. Excitation: 6471 Å. The interband transition energy from the split-off valence band to the Fermi level is very close to the laser photon energy. Estimated luminescence background is shown by the dashed line. After PINCZUK et al. [4.71]

Fermi surface the gap is $E_3=E_{30}+\eta E_{\mathrm{f}}$, where $\eta=(1+m_{\mathrm{e}}/m_{\mathrm{h}})$, with m_{e} the electron mass and m_{h} the mass in the third valence band. "Extreme resonance" means that the laser energy $\hbar\omega_1$ is within about qv_{f} of E_3. Then the non-vertical nature of the transition from the valence to conduction bands becomes very important. Consider the interband processes that take an electron from $\boldsymbol{k}$ below the Fermi surface to $\boldsymbol{k}+\boldsymbol{q}$ above it. An incident photon of wavevector $\boldsymbol{k}_1$ excites an electron of wavevector $\boldsymbol{k}+\boldsymbol{q}-\boldsymbol{k}_1$ in the filled valence band to $\boldsymbol{k}+\boldsymbol{q}$ in the conduction band. The electron at $\boldsymbol{k}$ below the Fermi surface then falls into the valence band hole at $\boldsymbol{k}+\boldsymbol{q}-\boldsymbol{k}_1=\boldsymbol{k}-\boldsymbol{k}_2$ and emits a photon of wavevector $\boldsymbol{k}_2$. The energy denominator is

$$D(\boldsymbol{k})=E_{30}+K_{\mathrm{e}}(\boldsymbol{k}+\boldsymbol{q})+K_{\mathrm{h}}(\boldsymbol{k}+\boldsymbol{q}-\boldsymbol{k}_1)-\hbar\omega_1\,. \tag{4.101}$$

HAMILTON and MCWHORTER's expression for the matrix element for the transition is [4.66]

$$\gamma_{\alpha\beta}=\boldsymbol{e}_1\cdot\overleftrightarrow{A}\cdot\boldsymbol{e}_2\,\delta_{\alpha\beta}-i(\boldsymbol{e}_1\times\boldsymbol{e}_2)\cdot(\alpha|\boldsymbol{\sigma}|\beta)\,B\,, \tag{4.102}$$

where B and $\overleftrightarrow{A}$ are defined in (4.80b) and (4.89b), respectively. The terms in $\gamma_{\alpha\beta}$ most resonant with the E_3 gap give

$$\gamma_{\alpha\beta} \approx P^2[\boldsymbol{e}_1 \cdot \boldsymbol{e}_2\, \delta_{\alpha\beta} + \mathrm{i}(\boldsymbol{e}_1 \times \boldsymbol{e}_2) \cdot (\alpha|\boldsymbol{\sigma}|\beta)]/(3mD)\,. \tag{4.103}$$

The scattering efficiency is [4.66]

$$\frac{d^2R}{d\Omega\, d\omega} = -r_0^2 \hbar \pi^{-1} (\omega_2/\omega_1)^2 (n_\omega + 1) \tag{4.104a}$$

$$\cdot \{|\boldsymbol{e}_1 \times \boldsymbol{e}_2|^2 \operatorname{Im}\{K_2\} + (\boldsymbol{e}_1 \cdot \boldsymbol{e}_2)^2 \operatorname{Im}\{L_2 - \Phi(L_2 L_0 - L_1^2)(1 - \Phi L_0)^{-1}\},$$

where

$$\Phi = 4\pi e^2/q^2 \tag{4.104b}$$

and where

$$\hbar K_j = \sum_k B^j [n(\boldsymbol{k}) - n(\boldsymbol{k}+\boldsymbol{q})] \tag{4.104c}$$

$$\cdot [\omega + \mathrm{i}0^+ + \omega(\boldsymbol{k}) - \omega(\boldsymbol{k}+\boldsymbol{q})]^{-1}\,,$$

$$\hbar L_j = \sum_k A^j [n(\boldsymbol{k}) - n(\boldsymbol{k}+\boldsymbol{q})] \tag{4.104d}$$

$$\cdot [\omega + \mathrm{i}0^+ + \omega(\boldsymbol{k}) - \omega(\boldsymbol{k}+\boldsymbol{q})]^{-1}\,.$$

With the approximation of (4.103), we have

$$\overleftrightarrow{A} = A\overleftrightarrow{I}\,, \tag{4.105a}$$

$$B = -A = -P^2/(3mD)\,, \tag{4.105b}$$

$$K_2 = L_2\,. \tag{4.105c}$$

We set up (4.104c) for evaluation at $T=0$. Let $\boldsymbol{k}_\mathrm{f}$ be a wavevector on the Fermi surface. The $[n(\boldsymbol{k}) - n(\boldsymbol{k}+\boldsymbol{q})]$ factor says that $\boldsymbol{k}+\boldsymbol{q}$ can range from $\boldsymbol{k}_\mathrm{f}$ to $\boldsymbol{k}_\mathrm{f}+\boldsymbol{q}$. We therefore define a dimensionless variable u, $0 < u \leqq 1$, so that $\boldsymbol{k}+\boldsymbol{q} = \boldsymbol{k}_\mathrm{f} + u\boldsymbol{q}$. Let θ be the angle between $\boldsymbol{k}_\mathrm{f}$ and $\boldsymbol{q}$. Then (4.104c) becomes

$$\hbar L_j = \frac{1}{4\pi^3} \int_{\theta=0}^{\pi} \int_0^1 du\, d\cos\theta \, \frac{2\pi k_\mathrm{f}^2 q \cos\theta}{(q v_\mathrm{f} \cos\theta - \omega - \mathrm{i}0^+)} \left(\frac{P^2}{3mD}\right)^j. \tag{4.106}$$

If we assume a backscattering geometry, where $\boldsymbol{k}_1 = \boldsymbol{q}/2$, then

$$D = (E_{30} + \eta E_{\mathrm{f}} - \hbar\omega_1) + \hbar q v_{\mathrm{f}} \cos\theta(\eta u - m_{\mathrm{e}}/2m_{\mathrm{h}})\,. \tag{4.107}$$

Now $P^2/3m$ is of order E_{G}. Under extreme resonance conditions $P^2/(3mD)$ will be large and rapidly varying as a function of u and θ. For the polarized $(\boldsymbol{e}_1 \cdot \boldsymbol{e}_2)^2$ term the fluctuation $\langle (D^{-1} - \langle D^{-1}\rangle)^2\rangle$ is likely to be dominated by $\langle D^{-2}\rangle$, giving a contribution almost as big as that of the $|\boldsymbol{e}_1 \times \boldsymbol{e}_2|^2$ term, which results from resonantly enhanced spin density fluctuations.

4.4.5. Electron Velocity Distributions

At room temperature GaAs at low electron densities ($\sim 10^{15}\ \mathrm{cm}^{-3}$) is in the $q \gg q_{\mathrm{s}}$ limit, and the charge density fluctuations are not screened. The dominant term in the scattering efficiency as given by (4.104) is

$$d^2R/d\omega\, d\Omega = -r_0^2 \hbar \pi^{-1} (\omega_2/\omega_1)^2 (\boldsymbol{e}_1 \cdot \boldsymbol{e}_2)^2 (n_\omega + 1)\, \mathrm{Im}\{L_2\}\,. \tag{4.108a}$$

If the k-dependence of A can be neglected, we have

$$-(1 + n_\omega)\, \mathrm{Im}\{L_2\} = -A^2(1 + n_\omega)\, \mathrm{Im}\{F\}\,. \tag{4.108b}$$

For a Maxwellian distribution [4.70]

$$\begin{aligned} &-(1 + n_\omega)\, L_2 \\ &= A^2 \omega^{-1} \int (\boldsymbol{k} \cdot \boldsymbol{q}/m_{\mathrm{e}}) [\omega - (\boldsymbol{k} \cdot \boldsymbol{q}/m_{\mathrm{e}})]^{-1} N(E)\, d^3k\,, \end{aligned} \tag{4.108c}$$

where $N(E)$ is the Maxwellian distribution function. This leads to

$$-(1 + n_\omega)\, \mathrm{Im}\{L_2\} \propto \exp(-\omega^2/q^2 v_{\mathrm{th}}^2)\,, \tag{4.109}$$

with $v_{\mathrm{th}}^2 = 2k_{\mathrm{B}} T/\mathrm{m}^*$.

Figure 4.13 shows a single particle spectrum for n-GaAs measured with 1.06 μm laser excitation [4.13]. The theoretical curve took into account the weak, but not entirely negligible, k-dependence of A.

With A considered to be constant, the scattering efficiency is proportional to (4.74), which says that the scattered spectrum gives the distribution of electrons having values of velocity in the direction of $\boldsymbol{q}$ equal to ω/q. For anisotropic crystals, this will depend on the direction of $\boldsymbol{q}$ [4.72].

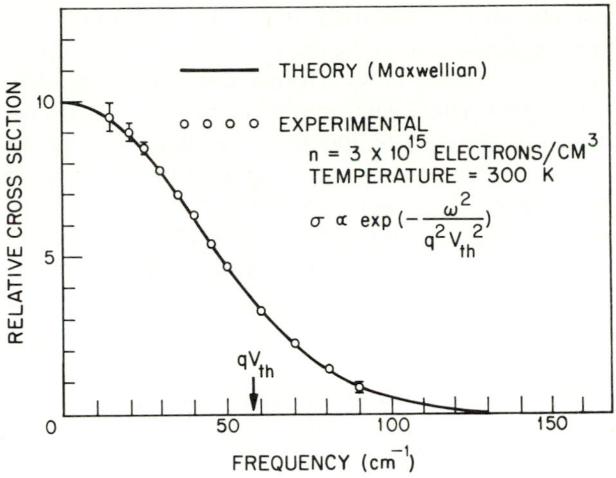

Fig. 4.13. Polarized single particle light scattering spectrum for n-GaAs at room temperature [4.13]. Theoretical curve from (4.108)

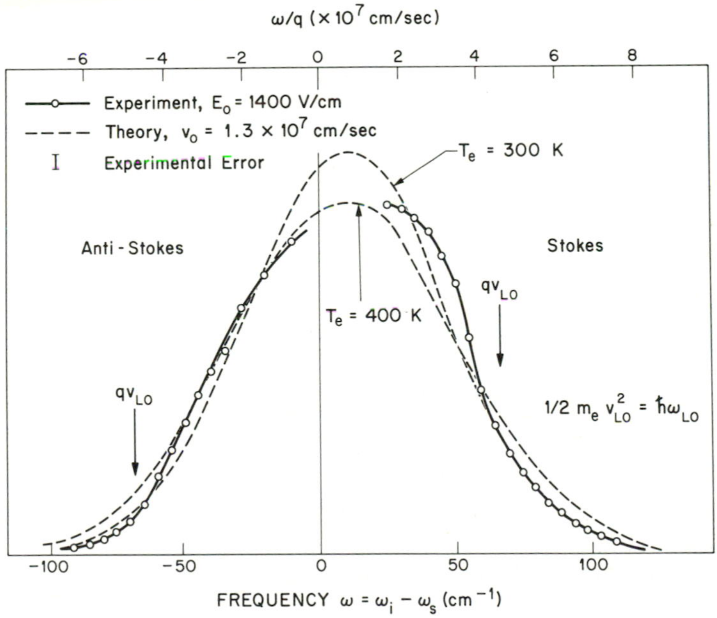

Fig. 4.14. Single particle light scattering spectrum of n-GaAs in the presence of an external electric field. The photon scattering angle was 90°. The dashed lines were calculated from (4.108) using a displaced Maxwellian distribution function [4.13]

In the presence of a DC electric field $\boldsymbol{E}_0$, the electron velocity distribution will have its mean value at the drift velocity $\boldsymbol{v}_0 = \mu \boldsymbol{E}_0$, where now μ is the mobility. Figure 4.14 shows MOORADIAN's data on such a system [4.73, 13]. The theoretical curves are essentially Gaussians for two different values of electron temperature displaced so that they center about $\omega = \boldsymbol{q} \cdot \boldsymbol{v}_0$. The agreement is only fair and suggests that the velocity distribution is distorted as well as shifted. Note the drop of intensity when $\omega = \pm q v_{\mathrm{LO}}$. At this point, the electrons have enough energy to excite LO phonons and thereby return more rapidly to equilibrium.

More detailed studies of light scattering from "hot" electron distributions would be valuable in understanding non-equilibrium plasmas in solids [4.74–76].

It would also be desirable to study experimentally the role of various damping processes on the shape of the single-particle spectrum [4.6, 77]. Strong damping should place the electron system in the "collision-controlled" regime, where hydrodynamics determines the response functions [4.1, 4, 75, 76]. Under appropriate circumstances, one might observe "motional narrowing" of the single-particle spectrum, in a way similar to what has been proposed for the spontaneous spin-flip Raman line [4.78–80].

4.5. Multicomponent Carrier Effects

There has been little experimental work on light scattering from semiconductors containing two or more types of mobile carriers, but there has been some very interesting theoretical work, starting with a paper by PLATZMAN [4.81]. Our emphasis in this section will be on carriers in a multivalley semiconductor, such as n-Si, n-Ge, or the lead salts. In such systems each of the components of the plasma is anisotropic. This brings new features to the light scattering spectra.

4.5.1. Introductory Considerations

Consider electrons in the jth valley, where the reciprocal effective mass tensor is $m^{-1}\overleftrightarrow{\boldsymbol{\mu}}_j$. Let their charge density fluctuation operator be ϱ_{qj}. Within the RPA, the induced response $\langle \varrho_{qj} \rangle$ to a total longitudinal perturbation having wavevector $\boldsymbol{q}$ is described by a function $F_j(\boldsymbol{q}, \omega)$, a generalization of the function $F(q, \omega)$ of (4.8) [4.81]

$$\begin{aligned} &\hbar F_j(\boldsymbol{q}, \omega) \\ &= \sum_{k} [n(\boldsymbol{k}) - n(\boldsymbol{k}+\boldsymbol{q})]\,[\omega + \mathrm{i}0^+ + \omega^{(j)}(\boldsymbol{k}) - \omega^{(j)}(\boldsymbol{k}+\boldsymbol{q})]^{-1}, \end{aligned} \tag{4.110a}$$

where

$$\omega^{(j)}(\boldsymbol{k}) = (\hbar/2m)\,\boldsymbol{k}\cdot\overleftrightarrow{\boldsymbol{\mu}}_j\cdot\boldsymbol{k}\,. \tag{4.110b}$$

F_j will depend on the orientation of $\boldsymbol{q}$ relative to the principal axes of $\overleftrightarrow{\boldsymbol{\mu}}_j$.

Coupling mechanisms for light scattering by the multicomponent gas of carriers include those discussed in Section 4.4. Now each of the density fluctuations ϱ_{qj} couples separately via the following generalization of (4.17) and (4.18)

$$\mathscr{H}'_\varrho = r_0 R_{12} A_1 A_2^\dagger \exp(-i\omega t) \sum_j \mu_j \varrho_{qj} + \text{H.C.} \tag{4.111}$$

where

$$\mu_j = \boldsymbol{e}_1\cdot\overleftrightarrow{\boldsymbol{\mu}}_j\cdot\boldsymbol{e}_2\,.$$

As the directions of the polarization vectors $\boldsymbol{e}_1$ and $\boldsymbol{e}_2$ vary, the external fields "drive" various linear combinations of the ϱ_{qj}. An important role is played by depolarized ($\boldsymbol{e}_1 \perp \boldsymbol{e}_2$) scattering, which involves solely relative fluctuations of the form $(\varrho_{qi}-\varrho_{qj})$, while leaving the total density fluctuation

$$\varrho_q = \sum_j \varrho_{qj} \tag{4.112}$$

unexcited. This can be seen by writing

$$\sum_j \overleftrightarrow{\boldsymbol{\mu}}_j \varrho_{qj} = \varrho_q \overleftrightarrow{\boldsymbol{\mu}}_{\text{ave}} + \sum_{i<j} (\overleftrightarrow{\boldsymbol{\mu}}_i - \overleftrightarrow{\boldsymbol{\mu}}_j)(\varrho_{qi}-\varrho_{qj})\,. \tag{4.113}$$

where $\overleftrightarrow{\boldsymbol{\mu}}_{\text{ave}}$ is the average of the $\overleftrightarrow{\boldsymbol{\mu}}_j$ over all the valleys. $\overleftrightarrow{\boldsymbol{\mu}}_{\text{ave}}$ will have the full symmetry of the crystal, and for a cubic crystal will be a multiple of the unit tensor. This lack of coupling to ϱ_q, while coupling to a fluctuation $(\varrho_{qj}-\varrho_{qi})$ is reminiscent of the situation for the spin density fluctuation mechanism. According to (4.83) that coupling is of the form $|\boldsymbol{e}_1\times\boldsymbol{e}_2|\,(\varrho_{q\uparrow}-\varrho_{q\downarrow})$.

If the valley j has axial symmetry about a direction represented by the unit vector $\boldsymbol{n}_j$, with principal values of $\overleftrightarrow{\boldsymbol{\mu}}$ parallel and perpendicular to $\boldsymbol{n}_j$ denoted by $\mu_\parallel$ and $\mu_\perp$, then

$$\overleftrightarrow{\boldsymbol{\mu}}_j = \mu_\perp \overleftrightarrow{\boldsymbol{I}} + (\mu_\parallel - \mu_\perp)\,\boldsymbol{n}_j\boldsymbol{n}_j \tag{4.114}$$

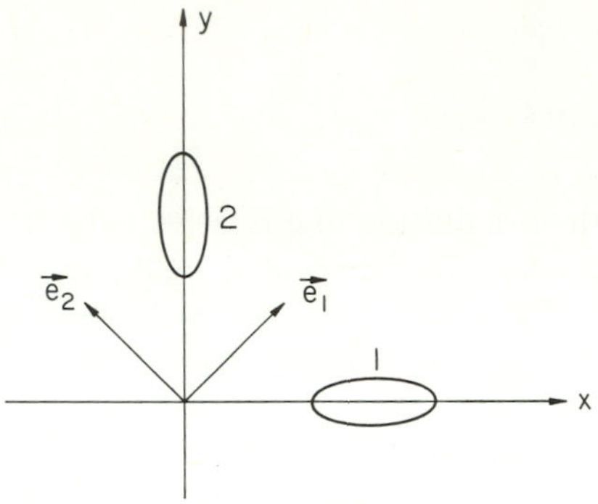

Fig. 4.15. Depolarized light scattering geometry for electrons in two valleys with axes at right angles

and for a cubic crystal

$$
\begin{aligned}
\sum_j \mu_j \varrho_{qj} = \varrho_q (\boldsymbol{e}_1 \cdot \boldsymbol{e}_2)(\mu_{\|} + 2\mu_{\perp})/3 \\
+ (\mu_{\|} - \mu_{\perp}) \sum_{i<j} [(\boldsymbol{n}_i \cdot \boldsymbol{e}_1)(\boldsymbol{n}_i \cdot \boldsymbol{e}_2) - (\boldsymbol{n}_j \cdot \boldsymbol{e}_1)(\boldsymbol{n}_j \cdot \boldsymbol{e}_2)] \\
\cdot (\varrho_{qi} - \varrho_{qj}) \,.
\end{aligned}
\tag{4.115}
$$

The second term in (4.115) could be rewritten further by using symmetry arguments. Only those linear combinations of the $(\varrho_{qi} - \varrho_{qj})$ appear that belong to irreducible representations that result from reducing the traceless tensors $(\boldsymbol{n}_i \boldsymbol{n}_i - \boldsymbol{n}_j \boldsymbol{n}_j)$.

As an example, we consider the application of (4.115) to the 2-valley system shown in Fig. 4.15. The result is

$$
\Sigma \, \mu_j \varrho_{qj} = (\mu_{\|} - \mu_{\perp})(\varrho_{q2} - \varrho_{q1})/2 \,. \tag{4.116}
$$

Figure 4.15 and (4.116) also apply to the $x-y$ plane of a cubic crystal with three $\langle 100 \rangle$ valleys. The representation involved is one component of E-symmetry—the one transforming as $(x^2 - y^2)$. The other traceless tensor representation for cubic symmetry is T_2, with tensor representations transforming as (xy, yz, and zx). It will not couple to $\langle 100 \rangle$ valleys. It will contribute, however, to light scattering via the $(\boldsymbol{e}_1 \times \boldsymbol{e}_2)$. $(\alpha | \boldsymbol{\sigma}_q | \beta)$ spin density fluctuation mechanism of (4.80). Thus with E polarization vector geometry one couples both to spin fluctuations and to "intervalley fluctuations"[2], whereas with T_2 geometry one couples only to spin fluctuations.

This situation is exactly reversed for $\langle 111 \rangle$ valleys in a cubic crystal. For then $(\boldsymbol{n}_i \cdot \boldsymbol{n}_i - \boldsymbol{n}_j \cdot \boldsymbol{n}_j)$ belongs to T_2, and so with T_2 polarization geometry one couples to both spin and intervalley fluctuations, whereas with E geometry the coupling is only to spin fluctuations.

4.5.2. Theory

PLATZMAN [4.81] has derived the relevant theoretical formulas for light scattering by density fluctuations in multivalley semiconductors. He has also computed some sample cross-section curves ($d^2\sigma/d\omega\, d\Omega$) for the two-ellipsoid case, as in Fig. 4.15, and he has discussed conditions for observation of "acoustic plasmons" (see below). Some of his work has been further discussed by PLATZMAN and WOLFF [4.1]. TZOAR and FOO [4.82] have calculated theoretical spectra for a system with four $\langle 111 \rangle$ valleys for various combinations of $\boldsymbol{e}_1$ and $\boldsymbol{e}_2$. The system parameters were appropriate for PbTe.

FOO and TZOAR [4.83] next added scattering via the spin density fluctuation density mechanism for each valley. They pointed out that for the E geometry $[\boldsymbol{e}_1 \| (110), \boldsymbol{e}_2 \| (\bar{1}10)]$ coupling would be to spin density fluctuations only. Their calculations were for PbTe, which is a direct semiconductor having both conduction band minima and valence band maxima at the zone boundary in $\langle 111 \rangle$ directions. They used appropriately defined A and B factors [Eqs. (4.80) and (4.89)] for that structure, as calculated numerically by JHA [4.84, 85]. These would not apply to Ge, although the valleys are $\langle 111 \rangle$ in both cases, because the rest of the band structure, especially the spin-orbit effect, is quite different.

PLATZMAN's results for the density fluctuation scattering is, in our notation, [4.81]

$$d^2R/d\omega\, d\Omega = -r_0^2 \hbar \pi^{-1} R_{12}^2 (1+n_\omega) \,\mathrm{Im}\left[\sum_{ij} (\mu_i^2 F_i \delta_{ij} + \Phi \mu_i \mu_j F_i F_j \varepsilon^{-1})\right], \tag{4.117a}$$

[2] It is tempting to use the term "intervalley transition" [4.75] or "intervalley fluctuation" to describe the response induced by $\varrho_{qi} - \varrho_{qj}$. The operator ϱ_{qi} induces a number density fluctuation having wave-vector $\boldsymbol{q}$ for the electrons in the ith valley; it does not change the number of electrons in that valley. Thus $\varrho_{qi} - \varrho_{qj}$ does not cause a transfer of electrons between valleys i and j; it induces density fluctuations of opposite sign in the valleys. It is in this sense that we use the term "intervalley fluctuation".

where $\Phi = 4\pi e^2/q^2$ and where the dielectric constant is now given by

$$\varepsilon = \varepsilon_s - \Phi \sum_j F_j . \tag{4.117b}$$

The term in square brackets can be written

$$[\ ] = \left[\sum_i \varepsilon_s \mu_i^2 F_i - \Phi \sum_{i<j} (\mu_i - \mu_j)^2 F_i F_j\right] \Big/ \varepsilon . \tag{4.118}$$

In the $q \to 0$ limit we have

$$[\ ] = \sum_{i<j} (\mu_i - \mu_j)^2 F_i F_j \Big/ \left(\sum_i F_i\right) . \tag{4.119}$$

Equation (4.119) shows that only the anisotropic part, $(\mu_\parallel - \mu_\perp)\, \boldsymbol{n}_j \boldsymbol{n}_j$, of $\boldsymbol{\mu}_j$ produces an unscreened scattering cross-section.

It is often possible to select the direction of $\boldsymbol{q}$ so that all the $F_j(\boldsymbol{q}, \omega)$ are equal. This is the case for $\boldsymbol{q} \| (100)$ for $\langle 111 \rangle$ valleys or $\boldsymbol{q} \| (111)$ for $\langle 100 \rangle$ valleys. Then (4.117a) and (4.119) give

$$d^2 R/d\omega\, d\Omega = -r_0^2 \hbar \pi^{-1} R_{12}^2 (1 + n_\omega) \sum_{i<j} (\mu_i - \mu_j)^2 \,\mathrm{Im}\{F_1(\boldsymbol{q}, \omega)/r\} , \tag{4.120}$$

where r is the number of valleys. Equation (4.120) equals the unscreened scattering efficiency from one valley times the factor $\sum_{i<j} (\mu_i - \mu_j)^2/r$.

In the high temperature limit, where (4.99) holds, (4.120) gives a cross-section per electron that is equal within a factor of order unity to that for the valley-orbit transitions given in (4.35). Examination of (4.81), (4.82a), and (4.84) shows that near resonance the spin density fluctuations mechanism and (4.120) can give comparable cross-sections. Below resonance (4.120) contains the factor $R_{12}^2 = E_G^4 (E_G^2 - \hbar^2 \omega_1^2)^{-2}$, which tends to unity, whereas the comparable factor in (4.81) for the spin density mechanism tends to $(\hbar\omega_1 \Delta_0/E_G^2)^2$ for $\Delta_0 \ll E_G$ and to $(\hbar\omega_1/E_G)^2$ for $\Delta_0 \gg E_G$ [see (4.88)]. This means that for volume scattering in an indirect semiconductor, where $\hbar\omega_1$ must be below the indirect band gap, $\hbar\omega_1$ will usually satisfy the inequality $(\hbar\omega_1)^2 \ll E_G^2$, where E_G is the direct band gap. In this case, the intervalley charge density scattering will be much stronger than the spin density fluctuation scattering.

PLATZMAN [4.81] has shown that $F_j(\boldsymbol{q}, \omega)$, the response function for the jth valley, may be related to $F(\boldsymbol{q}, \omega)$, as given in (4.8), an isotropic response function for the same number of electrons evaluated using the free electron mass m. For valleys with axial symmetry, PLATZMAN's result is

$$F_j(\boldsymbol{q}, \omega) = \mu_\perp^{-1} \mu_\parallel^{-\frac{1}{2}} F(\boldsymbol{q}^{(j)}, \omega) . \tag{4.121a}$$

where

$$\boldsymbol{q}^{(j)} = \mu_\perp^{\frac{1}{2}} \boldsymbol{q}_\perp^{(j)} + \mu_\parallel^{\frac{1}{2}} \boldsymbol{q}_\parallel^{(j)} . \tag{4.121b}$$

Here

$$\boldsymbol{q}_\parallel^{(j)} = (\boldsymbol{q} \cdot \boldsymbol{n}_j)\, \boldsymbol{n}_j \tag{4.121c}$$

and

$$\boldsymbol{q}_\perp^{(j)} = \boldsymbol{q} - \boldsymbol{q}_\parallel^{(j)} ,$$

i.e., $\boldsymbol{q}_\parallel$ and $\boldsymbol{q}_\perp$ are components of $\boldsymbol{q}$ parallel and perpendicular to the valley axis.

4.5.3. Acoustic Plasmons

We have seen that the general expression (4.119) takes a simple form when all the F_j are the same. The other extreme is of great interest, namely the case where the F_j are as different as possible. This can occur if the valleys are very anisotropic so that $\mu_\perp \gg \mu_\parallel$ and if the direction of $\boldsymbol{q}$ can be chosen so that for some of the valleys the ratio $|\boldsymbol{q}^{(j)}|/q$ is as large as possible, whereas for the others it is as small as possible. Then frequencies ω_a exist for which [4.81]

$$\mathrm{Re}\{\varepsilon(\boldsymbol{q}, \omega_a)\} = \varepsilon_s + (4\pi e^2/q^2)\, \mathrm{Re}\left\{\sum_j F_j(\boldsymbol{q}, \omega_a)\right\} = 0 \tag{4.122a}$$

and

$$\mathrm{Im}\{\varepsilon\} \ll 1 . \tag{4.122b}$$

A mode obeying (4.122) might be called an "acoustic plasmon". It is instructive to discuss such a collective mode first for a very simple case.

Consider a two-component plasma, one with light carriers having a plasma frequency ω_p and one with heavy carriers for which the plasma frequency is Ω_p. Select a small value of q and consider a frequency range where [4.86].

$$q^2 v_h^2 \ll \omega^2 \ll q^2 v_l^2 . \tag{4.123}$$

Here v_l and v_h are the r.m.s. or Fermi velocities of the light and heavy carriers. Then the complex dielectric constant can be written [4.1]

$$\varepsilon = \varepsilon_r + i\varepsilon_i , \tag{4.124a}$$

where

$$\varepsilon_r = (1 + q_s^2/q^2 - \Omega_p^2/\omega^2)\varepsilon_s \tag{4.124b}$$

and

$$\varepsilon_i \ll 1 .$$

Here $q_s^2 = 3\omega_p^2/v_l^2$ is the screening parameter for the light carriers. The acoustic plasmon frequency satisfies $\mathrm{Re}\{\varepsilon\} = 0$, or (for $q \ll q_s$)

$$\omega_a^2 = \Omega_p^2 q^2/q_s^2 = \Omega_p^2 v_l^2 q^2/(3\omega_p^2) \tag{4.125}$$

where ω_a^2 must still satisfy Eq. (4.123). It has recently been observed [4.99].

If the two components have opposite charges, they oscillate together, and the light charges screen almost completely the electric field due to the heavy charges. Thus the name "acoustic plasmon", in analogy with acoustic phonons. If the components have the same charge, they will oscillate 180° out of phase, and there will still be almost complete screening.

The formula of PLATZMAN [4.81], as given in (4.117) or (4.118), automatically allows for acoustic plasmons. When (4.122a) and (4.122b) are satisfied, there is essentially a pole in ε^{-1} with a large enhancement in the cross-section over what is given by a formula such as (4.120).

It will be difficult in practice for ω_a to satisfy inequalities of the type in (4.123) in a multicomponent electron plasma in solids. We can usually expect $\mathrm{Im}\{\varepsilon\}$ to be large enough to produce substantial Landau damping of the acoustical plasmon, i.e., decay of the plasmon into single particle excitations.

It is known that damping can be reduced and the cross-section increased if a DC electric field is applied to impart a net drift velocity v_0 to the carriers. The reason is that some of the energy of the drifting carriers is fed into the plasma wave [4.86]. This effect was verified by PLATZMAN's calculations for a degenerate 2-valley plasma [4.81]. He used ratios of $\mu_\perp/\mu_\parallel$ of 5 and 10. He found an acoustic plasmon-like peak in the calculated cross-section for $\omega \sim q v_{fh}$, where v_{fh} is the Fermi velocity of the heavy carriers. This peak was greatly enhanced whan a dc drift velocity was added. This effect anticipates the onset of the so-called "two-stream" instability in which the acoustic plasmon grows in time, receiving energy from the dc field [4.86].

4.5.4. The Metal-Semiconductor Transition

As the concentration of donors in a semiconductor increases, the wavefunctions on neighboring individual donors begin to overlap, and the donor levels broaden inhomogeneously. This is revealed in multi-valley n-type crystals as a broadening of the valley-orbit Raman line. At a sufficiently high donor concentration, n_m, the "Mott transition" occurs, in which the donor ground state broadens into a conducting impurity band, and the semiconductor becomes a metal [4.87]. Impurity band conduction is characterized by a short electron mean-free path of the order of the interdonor spacing [4.87]. At a higher concentration the impurity band passes into the conduction band. After this has occurred we expect to observe light scattering from a relatively free multicomponent plasma, as described in the previous subsection.

The data on n-SiC, shown in Fig. 4.11 [4.29] exhibit scattering from a sample doped into the impurity band region. The three E_2 valley-orbit lines shown at 13, 60, and 62 meV in the more lightly doped sample of Fig. 4.7, seem to have broadened into a continuum. Studies have recently been made on n-Ge [4.88] and on n-Si [4.89] at doping levels in and just below the impurity band region to determine to what extent the valley-orbit transition occurs in the metallic state and to try to study how the valley-orbit spectra change into spectra characteristic of a multicomponent plasma.

4.6. Concluding Remarks

4.6.1. Present Knowledge and Possible Future Trends

The mechanisms for light scattering by the electronic excitations discussed in this article are generally well understood. The nature of the excitations themselves is usually not in question, although this is not true for inter-bound state acceptor transitions and for the continuum observed in metallic SiC, Ge, and Si.

Future work in this field is likely to emphasize the shapes and widths of the various spectral lines. Much more detailed knowledge is needed on the effects of collisions on plasmon and plasmon-phonon coupled line shapes and on the shapes of single particle spectra. This is particularly true in the impurity-band concentration range, where the one-electron eigenstates are not well represented by Bloch waves among which collision-induced transitions take place.

In solids many more experiments need to be done on non-equilibrium plasmas. Studies on multicomponent plasmas are needed, and efforts should be made to produce plasma instabilities, or near-instabilities for study by light scattering techniques.

4.6.2. Remarks on Spin-Flip Raman Scattering

We have deliberately excluded scattering by electronic excitations in a magnetic field. This subject is too complex and important to be merely a part of an article such as the present one. Spontaneous Raman scattering by spin flips in a magnetic field is the most important of these omitted processes because it has led to the invention of the spin-flip Raman laser and also because it has involved some important new solid-state physics.

Chapter 7 by SHEN briefly discusses spontaneous spin-flip Raman scattering and gives several key references. To those I would add two recent conference review papers by LAX [4.90] and by PATEL [4.91]. Also worthy of mention are studies of spin-flip processes in II–VI compounds using visible laser light by THOMAS and HOPFIELD [4.92], FLEURY and SCOTT [4.93], SCOTT and DAMEN [4.94], SCOTT et al. [4.95], and HOLLIS et al. [4.96].

Acknowledgements

The author wishes to thank M. ALTARELLI for several useful discussions. Research support for the preparation of this article from the NSF under Contract GH 37757 and the ARPA under Contract US DAHC-15-73-G-10 is gratefully acknowledged.

References

4.1. P.M. PLATZMAN, P.A. WOLFF: *Waves and Interactions in Solid State Plasmas* (Academic Press, New York, 1973).
4.2. P.M. PLATZMAN, N. TZOAR: Phys. Rev. **136**, A11 (1964).
4.3. D. PINES: *Elementary Excitations in Solids* (Benjamin, New York, 1963).
4.4. D. PINES, P. NOZIERES: *The Theory of Quantum Liquids* (W.A. Benjamin, New York, 1966).
4.5. P.A. WOLFF: In *Light Scattering Spectra of Solids*, ed. by G.B. WRIGHT (Springer, New York, Heidelberg, Berlin, 1968), p. 273.
4.6. N.D. MERMIN: Phys. Rev. B **1**, 2362 (1970).
4.7. P.A. WOLFF: Phys. Rev. Letters **16**, 225 (1966).
4.8. S.S. JHA: Nuovo Cimento **58** B, 331 (1969).
4.9. F.A. BLUM: Phys. Rev. B **1**, 1125 (1970).
4.10. A. MOORADIAN, A.L. MCWHORTER: Phys. Rev. Letters **19**, 849 (1967).
4.11. A. MOORADIAN, A.L. MCWHORTER: In *Light Scattering Spectra of Solids*, ed. by G.B. WRIGHT (Springer, New York, Heidelberg, Berlin, 1968), p. 297.
4.12. A. MOORADIAN: In *Light Scattering Spectra of Solids*, ed. by G.B. WRIGHT (Springer, New York, Heidelberg, Berlin, 1968), p. 285.
4.13. A. MOORADIAN: In *Laser Handbook*, ed. by F.T. ARECCHI and E.O. SCHULZ-DUBOIS (North-Holland Publishing Co., Amsterdam, 1972), vol. II, p. 1309.
4.14. B.B. VARGA: Phys. Rev. **137**, A1896 (1965).
4.15. A. MOORADIAN, G.B. WRIGHT: Phys. Rev. Letters **16**, 999 (1966).
4.16. E. BURSTEIN, A. PINCZUK, S. IWASA: Phys. Rev. **157**, 611 (1967).
4.17. R. LOUDON: Advan. Phys. **13**, 423 (1964).

4.18. M.V. Klein, B.N. Ganguly, P.J. Colwell: Phys. Rev. B**6**, 2380 (1972).
4.19. D.T. Hon, W.L. Faust: Appl. Phys. **1**, 241 (1973).
4.20. F.A. Blum, A. Mooradian: In *Proc. 10th Intern. Conf. Physics of Semiconductors*, ed. by Keller, Hensel, and Stern (USAEC Division of Technical Information Extension, Oak Ridge, Tenn., 1970), p. 755.
4.21. W.L. Faust, C.H. Henry: Phys. Rev. Letters **17**, 1265 (1966).
4.22. J.F. Scott, T.C. Damen, J. Shah: Opt. Commun. **3**, 384 (1971).
4.23. E. Burstein, S. Ushioda, A. Pinczuk, J.F. Scott: In *Light Scattering Spectra of Solids*, ed. by G.B. Wright (Springer, New York, Heidelberg, Berlin, 1969), p. 43.
4.24. J.F. Scott, T.C. Daman, J. Ruvalds, A. Zawadowski: Phys. Rev. B**3**, 1295 (1971).
4.25. C.K.N. Patel, R.E. Slusher: Phys. Rev. **167**, 413 (1968).
4.26. P. Fisher, A.K. Ramdas: In *Physics of the Solid State*, ed. by S. Balakrishna, M. Krishnamurthi, and B. Ramachandra Rao (Academic Press, New York, 1969), p. 149.
4.27. H.R. Chandrasekhar, P. Fisher, A.K. Ramdas, S. Rodriguez: Phys. Rev. B**8**, 3836 (1973).
4.28. G.B. Wright, A. Mooradian: Phys. Rev. Letters **18**, 608 (1967).
4.29. P.J. Colwell, M.V. Klein: Phys. Rev. B**6**, 498 (1972).
4.30. C.H. Henry, J.J. Hopfield, L.C. Luther: Phys. Rev. Letters **17**, 1178 (1966).
4.31. D.D. Manchon, Jr., P.J. Dean: *Proc. 10th Intern. Conf. Physics of Semiconductors*, ed. by S.P. Keller, J.C. Hensel, and F. Stern (USAEC Division of Technical Information Extension, Oak Ridge, Tenn., 1970), p. 760.
4.32. G.B. Wright, A. Mooradian: In *Proc. 9th Intern. Conf. Physics of Semiconductors, Moscow* (Nauka Publishing House, Leningrad, 1968), p. 1067.
4.33. J.M. Cherlow, R.L. Aggarwal, B. Lax: Phys. Rev. B**7**, 4547 (1973); Erratum Phys. Rev. B**9**, 3633 (1974).
4.34. G.B. Wright, A. Mooradian: Bull. Am. Phys. Soc. **13**, 479 (1968).
4.35. J. Doehler, P.J. Colwell, S.A. Solin: Phys. Rev. B**9**, 636 (1974).
4.36. P.J. Colwell, J. Doehler, S. Solin: Bull. Am. Phys. Soc. **19**, 226 (1974).
4.37. S. Zwerdling, K.J. Button, B. Lax, L.M. Roth: Phys. Rev. Letters **4**, 173 (1960).
4.38. A. Onton, P. Fisher, A.K. Ramdas: Phys. Rev. **163**, 686 (1967).
4.39. M. Cardona, K.L. Shaklee, F.H. Pollak: Phys. Rev. **154**, 696 (1967).
4.40. P.J. Dean, E.G. Schönherr, R.B. Zetterstrom: J. Appl. Phys. **41**, 3475 (1970).
4.41. S.M. Sze, J.C. Irwin: Solid State Electron. **11**, 599 (1968).
4.42. L. Patrick: Phys. Rev. B**5**, 2198 (1972).
4.43. P.J. Dean, R.L. Hartman: Phys. Rev. B**5**, 4911 (1972).
4.44. C.H. Henry, K. Nassau: Phys. Rev. B**2**, 997 (1970).
4.45. W. Kohn: In *Solid State Physics*, ed. by F. Seitz and D. Turnbull (Academic Press, New York, 1957), vol. 5, p. 257.
4.46. G. Dresselhaus, A.F. Kip, C. Kittel: Phys. Rev. **98**, 368 (1955).
4.47. D. Schechter: J. Phys. Chem. Sol. **23**, 237 (1962).
4.48. J.C. Hensel, G. Feher: Phys. Rev. **129**, 1041 (1963).
4.49. B.W. Levinger, D.R. Frankl: J. Phys. Chem. Sol. **20**, 281 (1961).
4.50. K.S. Mendelson, H.M. James: J. Phys. Chem. Sol. **25**, 729 (1964).
4.51. K.S. Mendelson, D.R. Schutz: Phys. Stat. Sol. **31**, 59 (1969).
4.52. A. Baldereschi, N. Lipari: Phys. Rev. B**8**, 2697 (1973).
4.53. T.N. Morgan: Phys. Rev. Letters **24**, 887 (1970).
4.54. A. Nitzan: Mol. Phys. **27**, 65 (1974).
4.55. U. Fano: Phys. Rev. **124**, 1866 (1961).
4.56. F. Cerdeira, T.A. Fjeldly, M. Cardona: Phys. Rev. B**8**, 4734 (1973).
4.57. R. Beserman, M. Jouanne, M. Balkanski: In *Proc. 11th Intern. Conf. Physics of Semiconductors, Warsaw* 1972; (PWN-Polish Scientific, Warsaw, 1972), p. 1181.
4.58. F. Cerdeira, T.A. Fjeldly, M. Cardona: Sol. State Commun. **13**, 325 (1973).

4.59. D. L. MILLS, R. F. WALLIS, E. BURSTEIN: In *Light Scattering in Solids*, ed. by M. BALKANSKI (Flammarion Sciences, Paris, 1971), p. 107.
4.60. G. B. WRIGHT, M. BALKANSKI: Mat. Res. Bull. **6**, 1097 (1971).
4.61. F. CERDEIRA, T. A. FJELDLY, M. CARDONA: Phys. Rev. B**9**, 4344 (1974).
4.62. P. J. DEAN, D. D. MANCHON, J. J. HOPFIELD: Phys. Rev. Letters **25**, 1027 (1970).
4.63. Y. B. LEVINSON, E. I. RASHBA: Rep. Progr. Phys. **36**, 1499 (1973).
4.64. J. J. HOPFIELD: Private communication.
4.65. A. S. BARKER, JR.: Phys. Rev. B**7**, 2507 (1973).
4.66. D. C. HAMILTON, A. L. MCWHORTER: In *Light Scattering Spectra of Solids,* ed. by G. B. WRIGHT (Springer, New York, Heidelberg, Berlin, 1969), p. 309.
4.67. Y. YAFET: Phys. Rev. **152**, 858 (1966).
4.68. E. O. KANE: J. Phys. Chem. Sol. **1**, 249 (1957).
4.69. B. S. WHERRETT, P. G. HARPER: Phys. Rev. **183**, 692 (1969).
4.70. P. A. WOLFF: Phys. Rev. **171**, 436 (1968).
4.71. A. PINCZUK, L. BRILLSON, E. BURSTEIN, E. ANASTASSAKIS: Phys. Rev. Letters **27**, 317 (1971).
4.72. D. HEALEY, T. P. MCLEAN: Phys. Letters **29**A, 607 (1969).
4.73. A. MOORADIAN, A. L. MCWHORTER: In *Proc. 10th Intern. Conf. Physics of Semiconductors*, ed. by S. P. KELLER, J. C. HENSEL, and F. STERN (USAEC, Oak Ridge, Tenn., 1970), p. 380.
4.74. P. M. TOMCHUK, V. A. SHENDEROVSKII: Zh. Eksp. Teor. Fiz **62**, 1131 (1972); Sov. Phys. JETP **35**, 598 (1972).
4.75. S. V. GANTSEVICH, V. L. GUREVICH, V. D. KAGAN, R. KATILIUS: In *Light Scattering in Solids*, ed. by M. BALKANSKI (Flammarion, Paris, 1971), p. 94.
4.76. S. V. GANTSEVICH, V. L. GUREVICH, R. KATILIUS: Zh. Eksp. Teor. Fiz. **57**, 503 (1969); Soviet Phys. JETP **30**, 276 (1970).
4.77. S. H. LIU: Ann. Phys. **59**, 165 (1970).
4.78. R. W. DAVIES, F. A. BLUM: Phys. Rev. B**3**, 3321 (1971).
4.79. R. W. DAVIES: Phys. Rev. B**7**, 3731 (1973).
4.80. S. Y. YUEN, P. A. WOLFF, B. LAX: Phys. Rev. B**9**, 3394 (1974).
4.81. P. M. PLATZMAN: Phys. Rev. **139**A, 379 (1965).
4.82. N. TZOAR, E. N. FOO: In *Light Scattering in Solids*, ed. by M. BALKANSKI (Flammarion, Paris, 1971), p. 119.
4.83. E. N. FOO, H. TZOAR: Phys. Rev. B**6**, 4553 (1972).
4.84. S. S. JHA: Phys. Rev. **179**, 764 (1969).
4.85. S. S. JHA: Phys. Rev. **182**, 815 (1969).
4.86. D. PINES, J. R. SCHRIEFFER: Phys. Rev. **124**, 1387 (1961).
4.87. N. F. MOTT: Adv. Phys. **21**, 785 (1971).
4.88. J. DOEHLER, P. J. COLWELL, S. A. SOLIN: Phys. Rev. Letters **34**, 584 (1975).
4.89. K. JAIN, S. LAI, M. V. KLEIN: Phys. Rev. B**13**, 5448 (1976).
4.90. B. LAX: In *Proc. 11th Intern. Conf. Physics of Semiconductors, Warsaw* (Elsevier, Amsterdam, 1972), p. 1115.
4.91. C. K. N. PATEL: In *Fundamental and Applied Laser Physics*, ed. by M. S. FELD, A. JAVAN, and N. A. KURNIT (Wiley, New York, 1973).
4.92. D. G. THOMAS, J. J. HOPFIELD: Phys. Rev. **175**, 1021 (1968).
4.93. P. A. FLEURY, J. F. SCOTT: Phys. Rev. B**3**, 1979 (1971).
4.94. J. F. SCOTT, T. C. DAMEN: Phys. Rev. Letters **29**, 107 (1972).
4.95. J. F. SCOTT, T. C. DAMEN, P. A. FLEURY: Phys. Rev. B**6**, 3856 (1972).
4.96. R. L. HOLLIS, J. F. RYAN, D. J. TOMS, J. F. SCOTT: Phys. Rev. Letters **31**, 1004 (1973).
4.97. M. CHANDRASEKHAR, J. B. RENUCCI, M. CARDONA: Phys. Rev. B **17**, 1623 (1978).
4.98. S. C. SHEN, C. J. FANG, M. CARDONA, L. GENZEL: Phys. Rev. B **22**, 2913 (1980).
4.99. A. PINCZUK, J. SHAH, P. A. WOLFF: Phys. Rev. Lett. **47**, 1487 (1980).

5. Raman Scattering in Amorphous Semiconductors

M. H. BRODSKY

With 29 Figures

Much of the study of Raman scattering in crystalline materials focuses on the few allowed single phonon processes. The corresponding interest in amorphous materials is on the breakdown of selection rules and the broad range of allowed scattering processes. Because of the absence of a well defined crystal momentum, a larger wealth of information about vibrational spectra can be extracted from Raman scattering in amorphous materials. In this chapter we shall be concerned with the experimental and theoretical results obtained from the prototypical elemental and compound amorphous semiconductors and how those results have been used to elucidate the vibrational levels and structure of the amorphous state. The structural results are both supplementary and complementary to information obtainable by the traditional structural analysis techniques of X-ray, electron and neutron diffraction.

The amorphous state [5.1, 2] is characterized by what it is not. It is not a form of matter with long range order, it does not have large regions in which the atoms are arranged in a periodic array, there are no crystallites of a size large enough to give sharp Bragg diffraction lines or spots. Often in the literature, as here in this article, the terms amorphous, non-crystalline, glassy and vitreous are used interchangeably. There are some technical aspects of the glassy state which have not been established for many of the amorphous semiconductors, e.g., a glass transition temperature; but we shall not make any such fine distinctions. The term "disordered" is more general than "amorphous" and can refer to crystals as well as glasses. Crystal alloys, for example, can be disordered in terms of which constituent lies on a given lattice site. Liquids, of course, are the prime example of the amorphous state, but except in passing we shall not discuss liquid semiconductors.

The key to the definition of "amorphous" is the distinction between long and short range ordering. In practice, simple amorphous materials have strong short range ordering for distances out to three or four bond lengths [5.1]. Longer range correlations are not observable except in crystals. There does not appear to be a continuous gradation in the correlation length over all values from single atom spacings to the crystal dimensions. In other words, there appears to be a lower limit on crystallite size (about 30–40 Å for silicon) and an upper limit on the correlation length in amorphous semiconductors (about 12–15 Å for silicon). A practical measure of non-crystallinity has been the width of the first Bragg diffraction peak; it is either broad and characteristic of a glass, or narrow and characteristic of a crystal. On the scale of our interests, there are no intermediate cases except for some slight broadening for small crystallites. Corresponding examples of no intermediate cases exist in Raman spectroscopy for the locations and widths of the spectral lines of amorphous Si (a–Si) and related tetrahedrally bonded amorphous semiconductors.

One further point needs to be made about the nature of the disorder in the amorphous state. In general there is a topological disorder. No direct and natural mapping can be made between the atom sites in the glass and the lattice sites of a corresponding crystal. For example, an amorphous solid is not well represented by a "hot" crystal in which each atom is displaced by small random amounts from lattice sites [5.3].

There are many other amorphous materials other than semiconductors which can and have been studied by Raman spectroscopy. Many theoretical [5.4, 5] and experimental [5.6, 7] studies have been reported for the traditional glasses. Although there has been expanded interest recently in the Raman spectra of metallic [5.8] and magnetic [5.9] crystals because of improved experimental techniques, there does not appear to be any reports yet of Raman spectra from amorphous metal alloys or amorphous magnetic systems, both of which exist and are of considerable current interest [5.10–12]. In this article we shall concentrate on the vibrational spectra of amorphous semiconductors. Recent reviews on vibrational spectra in general have been given by LUCOVSKY [5.13] for amorphous semiconductors, by BÖTTGER [5.14] for non-crystalline solids, and by BELL and DEAN for oxide glasses [5.4, 5].

SHUKER and GAMON [5.15, 16] have outlined how to go from the concept of a disorder-induced breakdown of selection rules to a quantitative representation of the vibrational density of states, as measured by Raman scattering. Naturally any density of states so derived is only a "Raman effective density of states" which is a convolution of the true vibrational spectrum with mode dependent matrix elements. Various approaches that try to calculate the transition matrix elements will be given in the course of this chapter. One way of viewing the breakdown of the $\boldsymbol{q} = \Delta \boldsymbol{k} \approx 0$ selection rule, where $\boldsymbol{q}$ is the wavevector of the phonon and $\Delta \boldsymbol{k}$ the difference between incident and scattered photon wavevectors, is in terms of a correlation length Λ which characterizes the spatial extent of a normal mode vibrational state. Of course, in a perfect harmonic crystal $\Lambda \to \infty$ and a vibrational normal mode is a true phonon with a plane wave envelope to its wavefunction and hence a well defined wavevector. If one assumes that in an amorphous solid, it is more accurate to picture the vibrational modes as nearly localized [5.4, 5], then in terms of the correlation length, the eigenvector envelope may be represented by a plane-wave factor $\exp(i\boldsymbol{q}\cdot\boldsymbol{r})$ times a spatial damping factor $\exp(-r/\Lambda)$ [5.16]. No longer is $\boldsymbol{q}$ a good quantum number and, in principle, there is no longer any $\boldsymbol{q}\approx 0$ wave vector selection rule. The exponential damping mixes the formerly distinct $\boldsymbol{q}$ states. SHUKER and GAMON [5.16] then argued that the space-time correlation is proportional to

$$e^{i\boldsymbol{q}_j\cdot\boldsymbol{r}}\,e^{-r/\Lambda} \tag{5.1}$$

for the jth mode. The Fourier transform of (5.1) is proportional to the scattered Raman intensity of the jth mode. Summing over all modes and allowing all $\boldsymbol{q}_j$ to contribute to the sum, they find that the non-resonant stokes component at a frequency shift ω is given by

$$I_{\alpha\beta\gamma\lambda}(\omega)=\sum_b C_b^{\alpha\beta\gamma\lambda}\left(\frac{1}{\omega}\right)[1+n(\omega,T)]\,g_b(\omega) \tag{5.2}$$

and the anti-Stokes component is

$$I_{\alpha\beta\gamma\lambda}(\omega)=\sum_b C_b^{\alpha\beta\gamma\lambda}(1/\omega)\,n(\omega,T)\,g_b(\omega)\,, \tag{5.3}$$

where

$$n(\omega,T)=[\exp(\hbar\omega/kT)-1]^{-1} \tag{5.4}$$

is the Bose-Einstein distribution function at a temperature T. Each band of vibrational states b is assumed, for simplicity, to have a coupling constant $C_b^{\alpha\beta\gamma\lambda}$ with the superscripts $\alpha\beta$ and $\gamma\lambda$ describing the polarizations of the incident and scattered photons. In addition, there is the extra radiation factor of $(\omega_L\pm\omega)^4$, where ω_L is the incident photon frequency. Equations (5.2) and (5.3) are the principal results of SHUKER and GAMON's calculations and show how the measured Raman spectrum of an amorphous solid is composed of the Bose-Einstein factor, the $1/\omega$ harmonic oscillator factor, the transition probabilities, and finally the desired information, the vibrational density of states. If $I^{\mathrm{P}}(\omega)$ is the measured Stokes part of the Raman spectrum for a particular polarization geometry, then we define the reduced Raman spectrum as

$$I_{\mathrm{R}}^{\mathrm{P}}(\omega)=\omega(\omega_L-\omega)^{-4}[n(\omega)+1]^{-1}\,I^{\mathrm{P}}(\omega)\,. \tag{5.5}$$

We shall routinely use (5.5) throughout the discussion. The reduced Raman spectrum only involves the density of states and frequency dependent matrix element factors which we presume to be smooth, slowly varying and in some special cases calculable.

In Subsection 5.1.3 we shall give an alternate way of calculating the disorder-induced Raman scattering in terms of large unit cells with many atoms per unit cell and therefore many $\boldsymbol{q}=0$ phonons which are potentially Raman active. Finally, in Section 5.2 extensive use of molecular concepts and selection rules will be made in the interpretation of the Raman spectra from those materials which retain identifiable and distinct molecular units.

5.1. Tetrahedrally Bonded Amorphous Semiconductors

In this section we discuss the Raman spectra of a relatively simple class of amorphous semiconductors, namely those in which the chemical bonding dictates tetrahedral coordination. First we give a description of the nature of the material studied and then a summary of the basic experimental facts on amorphous Si and Ge. We follow with a description of the various theoretical approaches towards an understanding of the spectra in terms of structural models. Finally we mention the kind of information sought in the Raman studies of other tetrahedrally bonded amorphous semiconductors.

5.1.1. Elemental Group IV Films—Basic Concepts

Si and Ge have been studied in recent years as prototypes of amorphous semiconductors [5.17]. The motivation for much of this interest has been for two unique reasons. First, as amorphous materials, they possess a degree of compositional and structural simplicity unmatched by the conventional glasses. Secondly, because of their technological and physical importance the body of knowledge and understanding of their crystalline states is massive and widely disseminated. Many workers have hoped to use the crystal knowledge and the amorphous studies to bring forth some general understanding of amorphous semiconductors. This approach has not fulfilled expectation as far as a fundamental understanding of electronic energy states is concerned, but it has been eminently successful in elucidating the vibrational information of Raman, infrared, and neutron spectroscopy. The reasons for this dicotomy are the so far inseparable mixing of the relative importance of long range and short range electronic interactions and the comparatively simple situation of the dominance of short range forces in determining the vibrational properties of covalently bonded solids. This reasoning will form much of the basis for the discussion which follows.

Si and Ge can be prepared as thin amorphous films by several deposition techniques. The common feature of most of the successful deposition procedures is the quenching of dispersed atoms or small clusters of atoms (e.g. a vapor) onto a relatively cool substrate. Typically the substrate temperatures are $\leqslant 450\,^{\circ}\mathrm{C}$ for Si and $\leqslant 250\,^{\circ}\mathrm{C}$ for Ge although in practice the temperature limits may vary hundreds of degrees depending on the method of preparation, substrate material, rate of deposition, and presence of impurities. The commonly reported methods of preparation [5.1, 18] are evaporation followed by quenching of the vapor, dc or rf cathodic sputtering in an inert gas discharge, heavy-dose ion bombardment of a crystalline surface, pyrolytic or glow discharge decomposition of silane (SiH_4) or germane (GeH_4), and electrolytic plating. In general, amorphous Si or Ge cannot be prepared by the conventional glass making process of quenching from a melt to form a viscous glassy state. This is because the Si and Ge atoms are not covalently bonded in their liquid state but are arranged in a metallic closed pack configuration. On the other hand, the short range ordering of solid amorphous Si and Ge is more comparable to their covalently bonded diamond structure crystalline phases. Analysis of pair correlation functions deduced from diffraction data show that the nearest neighbor covalent bonds, i.e. the tetrahedral coordination and bond lengths, are essentially identical in crystalline and amorphous Si and Ge [5.1]. The remarkable retention of the covalent bond is clearly evident in the Raman scattering observations described in this chapter. Beyond the four nearest neighbors the distortions increase and it follows that the pair correlation function loses its distinct peaks for atomic separations beyond about 12–14 Å. The exact arrangements of the atoms within the correlated 12–14 Å region has resulted in much controversy over the years with arguments offered for either a microcrystalline or a random network interpretation [5.19]. A similar controversy has arisen from time to time concerning the structure of SiO_2 and related glasses. The Raman and infrared spectra of a–Si and a–Ge tend to favor a network point of view and may give some information on the parameters of the widely used dense random networks. However, it is important to point out that there is no such thing as a unique network or a unique microcrystalline model, but only various manifestations thereof. There is some hope that Raman scattering studies can help to check the various models against experiment.

Raman spectra are available for some other group IV materials which have been prepared as amorphous films, e.g., C, SiC, and Ge–Si and Ge–Sn alloys. All of these amorphous semiconductors are believed to be tetrahedrally coordinated with the exception of amorphous C(a–C). Raman spectra have been used to test these structural ideas [Ref. 8.1, Sect. 2.1.15].

5.1.2. Raman Scattering Spectra of Amorphous Si and Ge

There are two general and non-trivial physical implications that can be drawn from the experimental Raman spectra of either amorphous Si or Ge. These are:

1) The first-order Raman spectrum covers the entire vibrational energy range whereas for the corresponding crystal state only one zone center optical phonon is Raman active.

2) Not only does the observed Raman spectrum show the gross features of the entire amorphous vibrational density of states, but the amorphous density of states is very similar to that of the corresponding crystal.

SMITH and his co-workers [5.20] were the first to obtain Raman spectra of amorphous Si and Ge as well as a number of related III–V amorphous semiconductors. A direct trace of a typical low temperature (27 K) spectrum of amorphous Si is shown in Fig. 5.1. From (5.2) we see that the first-order Stokes Raman spectrum is proportional to the factor $n(\omega, T)+1$, where $n(\omega, T)$ is the Bose-Einstein distribution at temperature T for vibrational energy ω. The anti-Stokes spectrum is proportional simply to $n(\omega, T)$, and so vanishes as the distribution function goes to zero. At 27 K and $\omega > 100\,\mathrm{cm}^{-1}$, $n(\omega, T) \ll 1$ (as can be seen by the absence of the anti-Stokes spectrum), and the Stokes spectrum is independent of the distribution function. The spectrum shown therefore

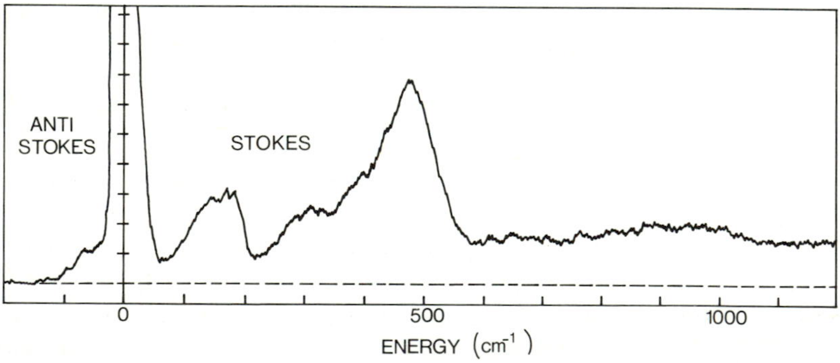

Fig. 5.1. Raman spectrum of amorphous Si at 27 K [5.20]. The a–Si was prepared by ion bombardment. The spectrum was obtained in a back scattering geometry with 0.25 W of a 488 μm laser line

differs from the vibrational density of states only by smoothly varying factors accounting for the dispersion in the Raman coupling. SMITH et al. observed that for energies below about 550 cm^{-1} this Raman spectrum bears a resemblance to the vibrational density of states of crystalline Si and suggests that 1) the dispersion in the Raman coupling is slight or slowly varying and 2) the vibrational spectra of the amorphous and crystalline forms of Si are very similar. The broad continuum above the first-order spectrum up to about 1050 cm^{-1} is probably second-order scattering; this is supported by the temperature dependence. The analysis of ALBEN et al. [5.21], which we review in Subsection 5.1.3, also supports the basic conclusions of a slowly varying Raman coupling which permits the observation of the vibrational density of states.

A common problem in the study of amorphous materials is the variability of results obtained on different samples [5.18], therefore, SMITH et al. [5.20] and WIHL et al. [5.22] performed a variety of experiments to establish that the reported spectra are characteristic of amorphous Ge and Si and independent of method and temperature of preparation. SMITH et al. [5.20] prepared amorphous Si on substrates held at temperatures from 300 K to 750 K. In addition, the methods of preparation included vapor condensation of Si or Ge, chemical vapor deposition of Si from SiH_4, rf sputtering of Si and Ge, and high-energy ion bombardment of single crystal Si or Ge surfaces. No measurable differences in the Raman spectra were observed. WIHL et al. [5.22] observed a 2 to 3% difference in the location of the high-frequency peak of amorphous Ge prepared by the electrolytic and other methods (sputtering or evaporation) methods and related this to density differences. However, it is not clear that such a small difference is of physical significance.

To our knowledge no structural or Raman studies have been performed on samples prepared and maintained below room temperature. This is a significant omission because it is possible that changes in short range order may occur for films quenched at very low temperatures.

The upper part of Fig. 5.2 shows the reduced Raman spectrum of amorphous Si [5.20]. The reduction corresponds to multiplication of the Stokes spectrum by the factor $\omega/[n(\omega, T)+1]$, see (5.5). SHUKER and GAMON have pointed out that in the absence of matrix element effects, the shape of this reduced spectrum would be that of the vibrational density of states. Figure 5.3 gives equivalent results for amorphous Ge [5.23]; the results of WIHL et al. [5.22] are in good agreement with those shown in Fig. 5.3. The lower parts of Figs. 5.2 and 5.3 show the crystalline densities of states [5.24] as well as broadened versions of them. It is clear that the general features of the reduced Raman spectra are

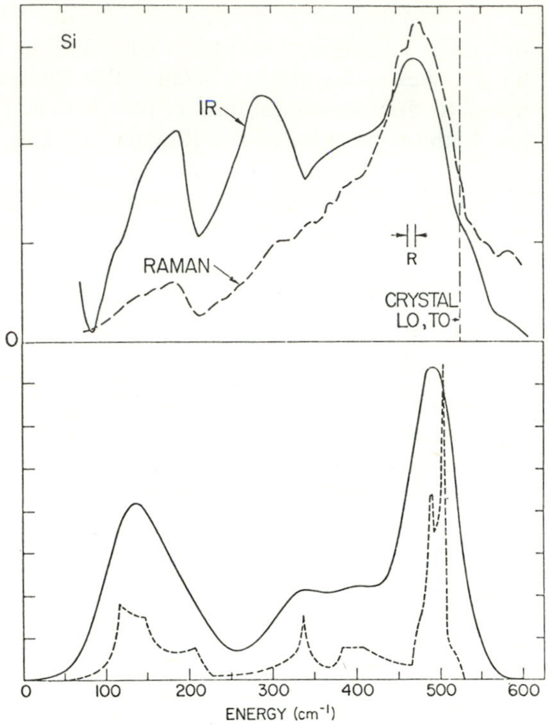

Fig. 5.2. Top: The reduced Raman spectrum (dashed line) of amorphous Si from room temperature data [5.20]. Also shown for comparison is the room temperature infrared absorption constant (solid line) versus energy [5.27]. The maximum absorption constant is 360 cm^{-1}. The vertical dashed line is the position of the single sharp line in the one phonon Raman spectrum of crystal Si. Bottom: The density of states of crystalline Si (dashed line) from a shell model fit to neutron scattering data [5.24]. The solid line is the crystalline density of states broadened by a convolution with a Gaussian factor of half-width 25 cm^{-1} [5.20]

given by the crystalline density of states, although the lower frequency modes of the reduced Raman spectra are relatively weak.

In addition to Raman scattering, similar information about the vibrational density of states of a–Si and a–Ge has been drawn from infrared spectroscopy [5.25–27]. More limited results have also been obtained from low-energy electron tunneling [5.28], high-energy electron energy loss [5.29], and neutron scattering [5.30].

LANNIN [5.31–33] has examined the frequency dependence of the coupling constants C_b in (5.2). For the spectral range where Debye behavior $g_b(\omega) \propto \omega^2$ is likely, $\omega \lesssim 65\ \mathrm{cm}^{-1}$ for Si and $\omega \lesssim 35\ \mathrm{cm}^{-1}$ for Ge, he found that $C_b(\omega) \propto \omega^2$ for both a–Si (Fig. 5.4) and a–Ge. Several

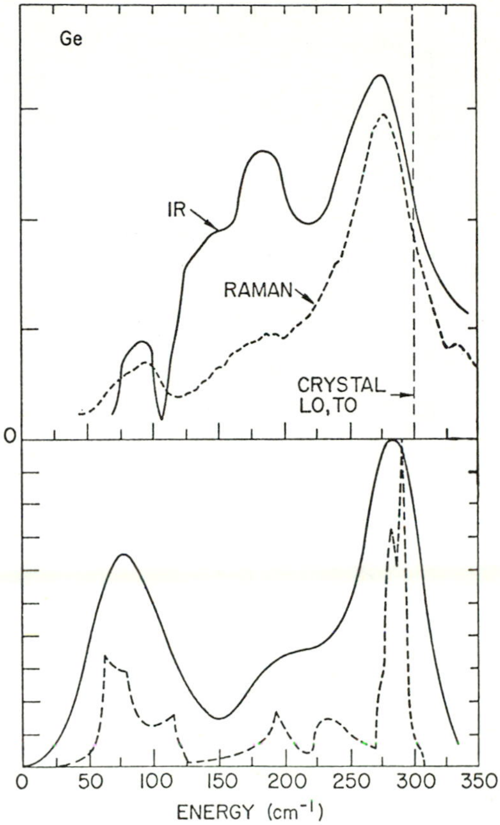

Fig. 5.3. Top: The reduced Raman spectrum (dashed line) of amorphous Ge from room temperature data [5.23]. Also shown for comparison is the room temperature infrared absorption constant versus energy [5.27]. The maximum absorption constant is 170 cm^{-1}. The vertical dashed line is the position of the single sharp line in the one phonon Raman spectrum of crystal Ge. Bottom: The density of states of crystalline Ge (dashed line) from a shell model fit to neutron scattering data [5.24]. The solid line is the crystalline density of states broadened by a convolution with a Gaussian factor of half-width 25 cm^{-1} [5.20]

authors [5.26, 34, 35] have attempted to calculate the spectral form of C_b and the results are in general agreement with LANNIN's observation. If $C_b(\omega)$ does go as ω^2 for small ω and if there is a similarly simple analytical form for the low frequency infrared absorption, then a quantitative comparison is possible between the Raman, infrared, and neutron scattering measured density of states. CONNELL [5.36] has assumed that $C_b(\omega) \propto \omega^2$, then from (5.2) it follows that

$$I(\omega) \propto \omega [n(\omega, T) + 1] \, g_R(\omega) \tag{5.6}$$

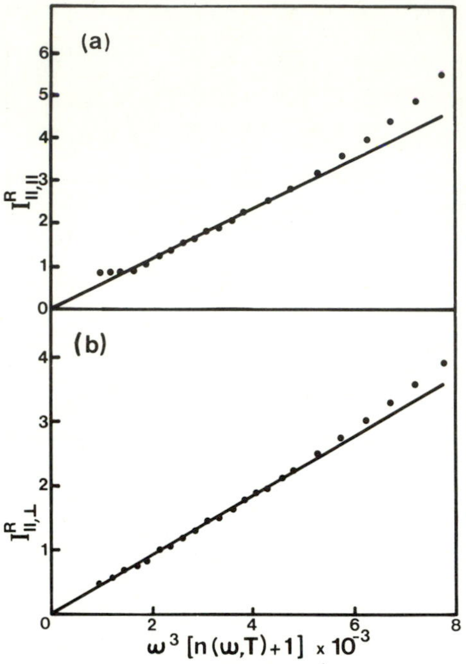

Fig. 5.4a and b. Raman scattering intensity (dots) of amorphous Si versus $\omega^3[n(\omega, T)+1]$ for low frequencies [5.32]. The solid lines are the theoretical predictions of (5.6) for Debye density of states. (a) Incident and scattered light are orthogonally polarized. (b) Incident and scattered light have parallel polarizations. The data was gathered in a back scattering geometry

for the Stokes scattered intensity $I(\omega)$ at sufficiently low frequencies. Here $g_R(\omega)$ in the Raman density of states after correcting for matrix element effects. Connell then also uses the arguments of Prettl et al. [5.26] which predict that the infrared absorption coefficient $\alpha(\omega)$ can be expressed in term of the infrared density of states as

$$g_{IR}(u) \propto \alpha(\omega)/\omega^4\,. \tag{5.7}$$

In Fig. 5.5 we reproduce the comparison [5.36] of $g_R(\omega)$ and $g_{IR}(\omega)$ with $g_N(\omega)$ the neutron scattering density of states at measured by Axe et al. [5.30] for amorphous Ge. All three measurements give a peak in $g(\omega)$ near 88 cm^{-1}. Connell then proceeded to use $g(\omega)$ to analyze the specific heat data of King et al. [5.37] and found that he can account for the decreased Debye temperature of amorphous Ge relative to crystal Ge.

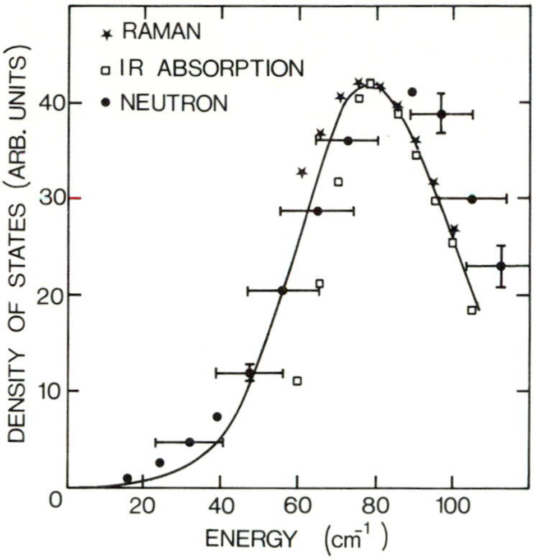

Fig. 5.5. Vibrational density of states of amorphous Ge as deduced [5.36] from Raman scattering [5.31], infrared absorption [5.25], and neutron scattering [5.30] data

The question of the form of the low-frequency Raman scattering spectrum is not unique to amorphous semiconductors. In fact SHUKER and GAMON originally proposed their analysis for vitreous SiO_2. HASS [5.7] had pointed out that the neglect of the Bose-Einstein population factor had lead to some misinterpretation of earlier high temperature, low frequency data [5.38]. Another example of the derivation and application of (5.2) was given by BARKER [5.39] for the case of low-frequency soft modes associated with ferroelectric phase transitions. BARKER also compares the frequency information derived from Raman and infrared spectroscopy. However, the specific heat results of SHUKER and GAMON [5.40] for a–SiO_2 [5.38] are in conflict with CONNELL's [5.36] analysis of the specific heat for a–Ge. SHUKER and GAMON assumed that the coupling constant C_b was independent of frequency while CONNELL assumed the square-law dependence reported by LANNIN [5.31]. Both analyses give good agreement with the specific heat data in the temperature range of 3 to 15 K although the SHUKER and GAMON results would be somewhat improved with a frequency dependent C_b. There is still room for more detailed analysis of the low frequency coupling constant in amorphous materials before we can draw firm conclusions.

Raman scattering provides a useful tool for the study of one of the more intriguing methods of preparation of amorphous Si or Ge, namely, ion bombardment [5.20, 23, 41–43]. For heavy doses of order 10^{15} ions/

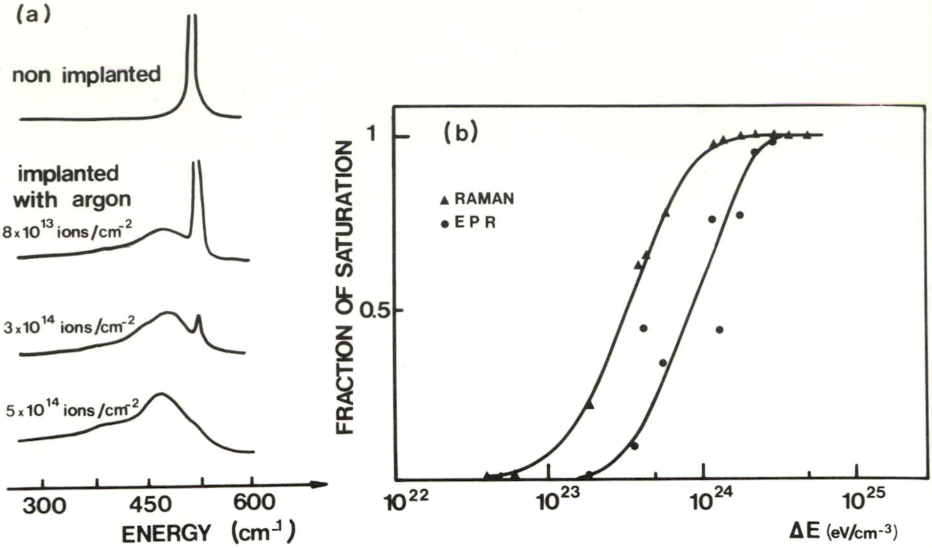

Fig. 5.6. (a) The Raman spectra from the surface of a single crystal of Si before and after increasing doses of 70 keV argon ions. The upper and lower curves are characteristic of the crystal and amorphous phases respectively. (b) The factor of the final maximum spin resonance (EPR) and Raman intensities as a function of total deposited ion energy [5.43]

cm^2 or greater, a continuous amorphous layer is formed on the surface of a target crystal. The layer thickness depends on the incident ion mass and ion energy, but typical numbers for Si are: an amorphous layer about 0.5 μm thick grows after a total flux $\gtrsim 10^{15}$ ions/cm^2 for Si^{28} ions accelerated to 300 keV. The exact nature of the layer formation is not known, e.g. does the amorphous layer grow from overlapping cylindrical damage regions about each ion track or does a thin buried amorphous layer form first which grows with planar symmetry towards the surface? Bourgoin et al. [5.43] have been probing the ion damaged region of silicon with Raman scattering in order to sort out the growth mechanism. The usefulness of the method comes from the ability to simultaneously see both the sharp crystalline peak with a frequency shift of 522 cm^{-1} as well as the broad features of the amorphous Si spectrum with its broad peak near 480 cm^{-1}. Bourgoin et al. [5.43] compared the ion doses required to saturate the amorphous Raman spectrum with the dose required to saturate an electron spin resonance (EPR) signal (see Fig. 5.6). The EPR signal, which arises from broken bonds [5.44, 45] requires a higher dose before the signal strength levels off. The Raman spectrum indicates a complete continuous amorphous layer is formed by the ion bombardment which then undergoes additional ion damage with continued irradiation. This conclusion is consistent with other ion damage studies of amorphous Si [5.46].

As exemplified above and in other studies [5.23] Raman scattering provides a unique characterization tool for testing an amorphous Si or Ge sample for traces of crystallinity. At the IBM laboratory we routinely examine evaporated Si samples by Raman scattering. Any presence of the sharp 522 cm^{-1} line in the Raman spectrum is conclusive evidence for crystallinity and cause for rejecting the sample as not being totally amorphous. The same results are obtainable by X-ray diffraction, but for thin Si films, the Raman spectrum over the range of interest is obtained in about 5 min compared with the several hours required for diffraction. A related use is the study of induced crystallization of amorphous materials, e.g. by annealing [5.47] or by the photo and thermal effects of the incident laser beam used for the scattering experiments [5.48]. An example will be given below in Subsection 5.2.5.

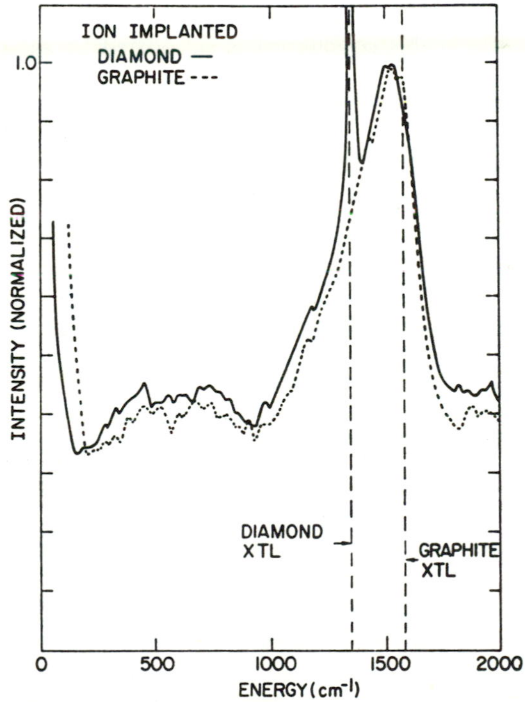

Fig. 5.7. Raman spectra of layers produced by high energy ion damage of diamond and graphite. The positions of first order Raman lines of the crystals are indicated. The data was obtained in a back scattering geometry [5.42]

Up to this point we have omitted any discussion of the Raman spectrum of amorphous carbon (a–C). Amorphous carbon is in many respects similar to a–Si and a–Ge but Raman studies [5.23, 41, 42, 49–51] have shown that a–C is not really tetrahedrally bonded with the same short range order as diamond. The Raman spectrum shows that the smooth density of states of a–C is more characteristic of a graphitically bonded solid than of diamond. Figure 5.7 shows the Raman spectra of two samples of a–C, one made by ion bombardment of diamond, the other by ion bombardment of graphite [5.42]. It is seen that both samples give essentially the same spectrum, except for the strong diamond crystal line showing through in the former sample because of the laser penetration through the amorphous layer. Also shown are the frequencies of the Raman active zone center phonons in diamond and graphite. These frequencies approximately mark the maximum phonon frequencies in these crystals. By comparison with the argument above (see Fig. 5.2) for a–Si, we see that the Raman spectrum of a–C is not going to be well represented by a smeared out version of the diamond density of states and we are led to the conclusion of a short range order that is more graphitic in nature. This conclusion is supported by the data of other studies on the Raman spectra of various forms of vitreous carbon [5.49–51]. (See [Ref. 8.1, Fig. 2.21].)

5.1.3. Numerical Theory

It is easy to understand intuitively that the vibrational spectra of covalently bonded amorphous solids are likely to be very similar to their crystalline counterparts because of the retention of short-range order. This follows from the strong short-range nature of covalent forces and the dominant role of the short range forces in determining vibrational spectra. For such a system, it is conceptually straightforward to use a limited number of short-range force constants and proceed to calculate vibrational spectra. In practice, much computational effort is required because it is necessary to use a model with a substantial number of atoms if the results are to be genuinely characteristic of an amorphous solid and not merely of a molecular-like cluster of atoms. BELL and DEAN [5.4, 5] have done extensive analyses of this kind for SiO_2 and more recently ALBEN and his collaborators [5.21, 52, 53] have computed the vibrational spectra for a variety of amorphous Si and Ge models. In addition, ALBEN et al. have computed the Raman, infrared, and inelastic neutron spectra for their models based on simple intuitive bond distribution concepts. We shall now discuss the assumption and results of their calculations.

For crystalline Si and Ge, one knows that any of several arbitrary representations of the nearest neighbor bond stretching and bond bending suffices to give most of the essential features of the vibrational dispersion curves. More distant interactions may be needed to get precisely correct dispersion curves, but the essential physical situation is well described by the short range forces. ALBEN et al. chose to use the KEATING form [5.54] of the short range potential energy

$$U = \tfrac{3}{4}\alpha \sum_{l\Delta} [(\boldsymbol{u}_l - \boldsymbol{u}_{l\Delta}) \cdot \boldsymbol{r}_\Delta(l)]^2 + \tfrac{3}{16}\beta \sum_{l(\Delta\Delta')} [(\boldsymbol{u}_l - \boldsymbol{u}_{l\Delta}) \cdot \boldsymbol{r}_{\Delta'}(l) + (\boldsymbol{u}_l - \boldsymbol{u}_{l\Delta'}) \cdot \boldsymbol{r}_\Delta(l)]^2 , \qquad (5.8)$$

where α and β are bond-stretching and bond bending force constants, respectively. Here $\boldsymbol{r}_\Delta(l)$ is the unit vector from the equilibrium position of atom l to its neighbor $l\Delta$; $\boldsymbol{u}_l$ and $\boldsymbol{u}_{l\Delta}$ are the displacements of the atom l and its neighbor $l\Delta$. With the above form of U and some models of the amorphous structure, which were chosen for both their physical reasonableness and computational tractability, ALBEN et al. [5.51] computed the vibrational eigenvalues and eigenvectors for long wavelength vibrations. The long wavelength vibrational frequencies give a satisfactory sampling of the entire vibrational spectrum, as can be seen for the schematic example of Fig. 5.8. In Fig. 5.8a, we schematically show the energy E versus wavevector q dispersion curves in some arbitrary q direction for a crystal, such as diamond structure Si, with $n=2$ atoms per unit cell. There are only $n \times 3 = 6$ eigenvalues in the long wavelength limit as $q \rightarrow 0$. However, if we, for example, distort the crystal somewhat so as to double the size of the unit cell, as might be the case of "wurtzite" structure Si [5.55], then the Brillouin zone dimensions are halved. Figure 5.8b shows how the dispersion curves are folded back into the first zone and naturally there are now $n \times 3 = 12$ zone center phonons. Finally, we exhibit in Fig. 5.8c how if the real space unit-cell grows so as to contain many atoms, then there are many zone center modes which together adequately represent the density of states of the total vibrational spectrum. ALBEN et al. [5.53] used several periodic models of the amorphous Si or Ge which had about 60 atoms in a unit cell; each atom had a slightly distorted tetrahedral environment. Although these models were periodic and therefore technically crystalline, the sixty-some atoms per unit cell serve the purposes of providing enough of a basis for giving the entire density of states, of being computationally manageable and of still possessing enough disorder within the unit cell to be a realistic amorphous model. While isolated clusters of about the same size as one of these unit cells are equally amenable to

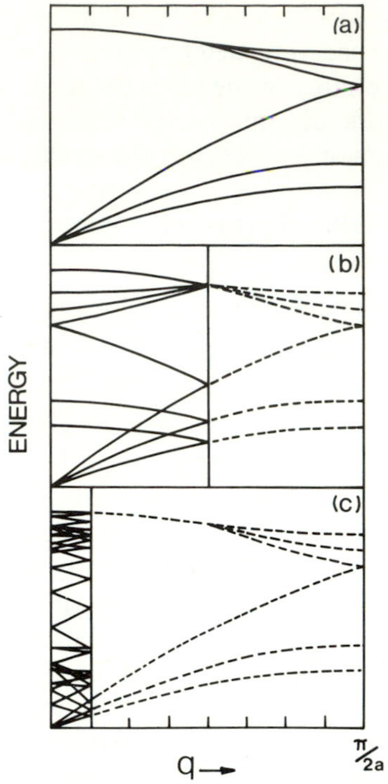

Fig. 5.8a–c. Schematic representation of the folding back of the energy versus wavevector of dispersion curves with an increasing number of atoms per unit cell. (a) The dispersion curves for two atoms per unit cell, e.g., Si in the diamond structuree. (b) The dispersion curves for four atoms per unit cell, e.g., Si in the wurtzite structure (solid lines). (c) The dispersion curves for sixteen atoms per unit cell, e.g., a sixteen atom periodic model of amorphous Si (solid lines). The dashed lines of (b) and (c) are the original two atom per unit cell dispersion curves

calculation they suffer from the disadvantage of not having as well defined and as realistic boundary conditions for its outside atoms in the framework of the periodic array. However, ALBEN et al. [5.53] did compute the density of states for a variety of cluster models also and they found essentially the same results, provided some account was made for the boundary effects. More recently, THORPE [5.56] has proposed a new method for handling the boundary atoms of small clusters (see Subsection 5.1.4).

Figure 5.9C shows the results of a density of states calculation for a periodic array of 61 atoms. The force constants α and β were chosen so as to be consistent with the silicon crystalline dispersion curves. Also shown in Fig. 5.9D and E are the results of Calculations of infrared absorption

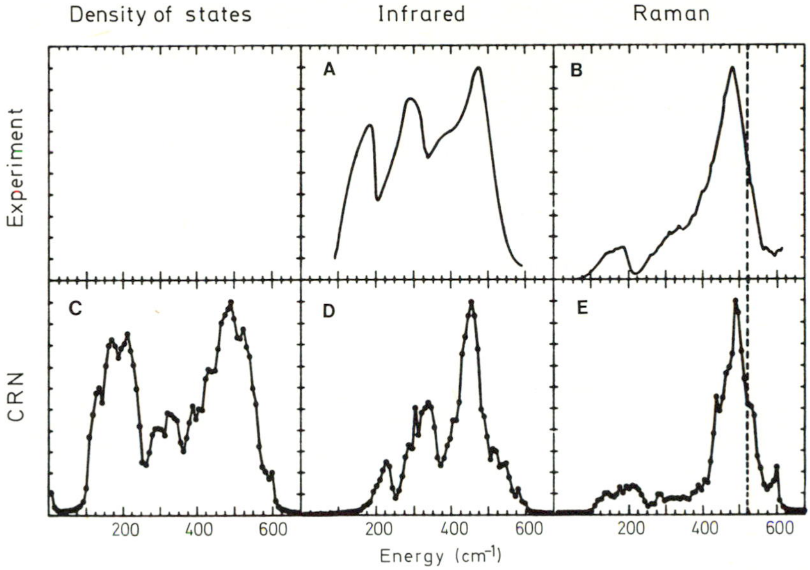

Fig. 5.9A–E. A comparison of the theory [5.21] for the periodic continuous random network (CRN) model of amorphous Si with the Raman and infrared data of Fig. 5.2. The theoretical spectra are formed by adding weighted Lorentzian contributions from $q=0$ modes. The weighting factors are given by (5.9)–(5.12), as described in the text

and Raman scattering spectra. ALBEN et al. arbitrarily chose as simple a set of mechanisms as they could for the infrared and Raman effects. Their infrared mechanism, as illustrated in Fig. 5.10, is entirely due to the dipole moment of pairs of bonds and is proportional to the difference in compression (or expansion) of adjacent bonds. The total infrared dipole strength is calculated at each frequency by summing over all pairs of bonds. Naturally, this sum nets out to zero for crystalline Si, as it must for all homopolar crystal structures with less than three atoms per unit cell [5.57]. The infrared absorption is proportional to

$$\begin{aligned} \boldsymbol{M} = & \sum_{l\{\Delta\Delta'\}} [\boldsymbol{r}_{\Delta'}(l) - \boldsymbol{r}_{\Delta}(l)] \\ & \cdot \{[\boldsymbol{u}_l - \boldsymbol{u}_{l\Delta}] \cdot \boldsymbol{r}_{\Delta}(l) - [\boldsymbol{u}_l - \boldsymbol{u}_{l\Delta'}] \cdot \boldsymbol{r}_{\Delta'}(l)\}, \end{aligned} \tag{5.9}$$

where here $\boldsymbol{M}$ is the net dipole moment due to the transferring of charge from extended to compressed bonds.

The Raman activity of a mode is given by the induced polarizability, which must be a second-order tensor that is linear in the displacements $\boldsymbol{u}_l$. ALBEN et al. [5.21, 52, 53] write this as a sum of contributions from

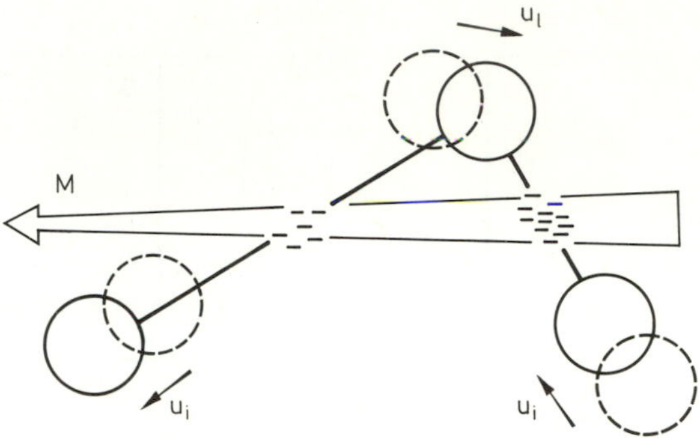

Fig. 5.10. Schematic illustration of the mechanism described by (5.9) for the infrared activity. During a vibration, some bond charge moves from extended to compressed bonds, resulting in a local electric dipole moment M. The local moments cancel for modes of the diamond cubic structure but they do not cancel in amorphous structures

individual bonds, imposing invariance with respect to translations and inversion. They further restricted the range of possibilities by the imposition of cylindrical symmetry about the bond, so that each bond is to be treated precisely as a homopolar diatomic molecule. Finally, they were left with three independent forms for the polarizability

$$\overleftrightarrow{\alpha}_1 = \sum_{l\Delta} [\overleftarrow{r}_\Delta(l)\, \vec{r}_\Delta(l) - \tfrac{1}{3}\overleftrightarrow{I}]\, \boldsymbol{u}_l \cdot \boldsymbol{r}_\Delta(l)\,, \tag{5.10}$$

$$\overleftrightarrow{\alpha}_2 = \sum_{l\Delta} \{\tfrac{1}{2}[\overleftarrow{r}_\Delta(l)\, \vec{u}_l + \overleftarrow{u}_l\, \vec{r}_\Delta(l)] - \tfrac{1}{3}\overleftrightarrow{I}\, \boldsymbol{u}_l \cdot \boldsymbol{r}_\Delta(l)\}\,, \tag{5.11}$$

$$\overleftrightarrow{\alpha}_3 = \sum_{l\Delta} \overleftrightarrow{I}\, \boldsymbol{u}_l \cdot \boldsymbol{r}_\Delta(l)\,. \tag{5.12}$$

Note that the first and third expressions depend only on the compression $(\boldsymbol{u}_l - \boldsymbol{u}_{l\Delta}) \cdot \boldsymbol{r}_\Delta(l)$ of each bond. There is an important distinction between the first expression and the other two. The latter vanish in the case of perfectly symmetric tetrahedral bonding, since in that case

$$\sum_\Delta \boldsymbol{r}_\Delta(l) = 0\,. \tag{5.13}$$

The first expression, on the other hand, has a quadratic dependence on the bond vectors $\boldsymbol{r}_\Delta(l)$ and hence does not vanish. If $\sum_\Delta \boldsymbol{r}_\Delta(l)$ were to be very small in a given structure, clearly $\overleftrightarrow{\alpha}_2$ and $\overleftrightarrow{\alpha}_3$ would be expected to contribute little to Raman activity. However, for structures of interest, the quantity $\sum_{\Delta} \boldsymbol{r}_{\Delta}(l)$ typically has magnitude 0.2 times the nearest neighbor distance and $\overleftrightarrow{\alpha}_2$ and $\overleftrightarrow{\alpha}_3$ cannot be neglected on grounds of approximate symmetry.

The weighting of the various contributions to the Raman activity was determined as follows. The $\overleftrightarrow{\alpha}_3$ mechanism is associated with a depolarization ratio of zero. The fact that the observed spectrum was found to have a similar shape in the HH (incident and scattered light both polarized parallel to each other) and HV (incident and scattered light polarized orthogonally) with a depolarization ratio of 0.8 ± 0.1 [5.20, 32]. See, however, [5.112] suggests that $\overleftrightarrow{\alpha}_3$ should be given zero weighting. This leaves (apart from the overall scale) only the ratio of the coefficients of $\overleftrightarrow{\alpha}_1$ and $\overleftrightarrow{\alpha}_2$ to be determined. By itself, $\overleftrightarrow{\alpha}_1$ provides a reasonable description of the upper half of the spectrum, but gives too little activity in the lower half. The inclusion of $\overleftrightarrow{\alpha}_2$ remedies this deficiency. The ratio 3:1 for the weighting of $\overleftrightarrow{\alpha}_1$ to $\overleftrightarrow{\alpha}_2$ was used in the calculations of Fig. 5.9e.

In summary, a numerical theory of the Raman scattering in amorphous Si and Ge works reasonably well with only one adjustable parameter. That adjustment is necessary because no theory is yet available to give the relative strength of the scattering from the two types of bond distortions that are consistent with the data.

5.1.4. Other Theoretical Approaches

THORPE [5.56] has proposed a method for establishing physically reasonable boundary conditions for the dangling bonds of surface atoms. He has then used small clusters [5.58] to calculate the vibrational spectra. In THORPE's approach, a small cluster is embedded in an infinite solid represented by an average potential. Each surface atom has an additional potential imposed so that the surface and bulk atoms have, on the average, the same mean-square displacements. Within the cluster, a Born model [5.59] with nearest neighbor forces is then used to calculate the vibrational density of states. To date the model has only been applied to a cluster with one interior atom, to a cluster with a hexagonal "chair" of 6 atoms, and to a cluster which is a five fold ring. Only the density of states has been calculated, but the model is capable of extension to give the Raman activity of the modes.

VON HEIMENDAHL [5.35] and MITRA et al. [5.60] have attempted to calculate information about the vibrational spectra via approaches which essentially broaden crystalline effects in accord with the broadening of the pair correlation function. VON HEIMENDAHL [5.35] used the eigen-

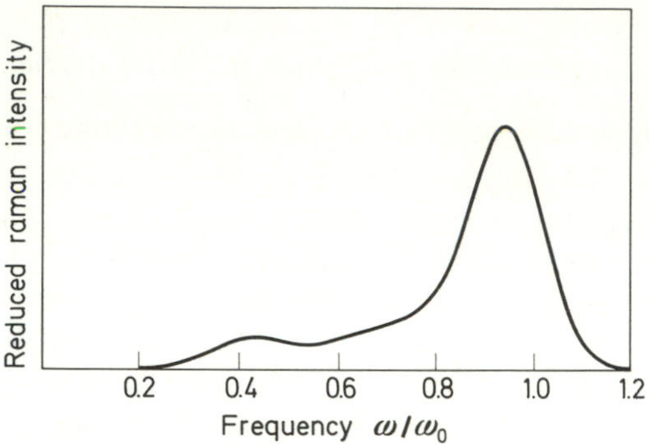

Fig. 5.11. Calculated Raman spectrum [5.35] for an array of tetrahedral units positionally and angularity disordered with respect to each other. Compare with Figs. 5.2 and 5.3. The crystal Raman frequency is ω_0

vector of a lattice whose sites are described by a Gaussian distribution function

$$p(\boldsymbol{r}-\boldsymbol{x}) = (\pi\alpha^2 x)^{-\frac{3}{2}} \exp[-(\boldsymbol{r}-\boldsymbol{x})^2/\alpha^2 x], \tag{5.14}$$

where $\boldsymbol{r}$ are real lattice vectors, $\boldsymbol{x}$ are the locations of atoms with respect to local coordinates around a site in an amorphous array, and α is a "positional disorder parameter". The tetrahedra centered at sites $\boldsymbol{x}$ are also angularly disordered with respect to their neighbors. A Keating model [5.54] was employed to calculate the eigenvectors. Finally the first-order Raman tensor is calculated from a single bond polarizability,

$$P_{\alpha\beta} = C \sum_{\Delta n} r_\Delta(n)_\alpha r_\Delta(n)_\beta \boldsymbol{r}_\Delta(n) \cdot \boldsymbol{u}(n) \tag{5.15}$$

where $r_\Delta(n)_\alpha$ are the Cartesian components of the vectors joining the equilibrium position of an atom with its nearest neighbors ($\Delta = 1, 2, 3, 4$). The displacement of an atom n is $\boldsymbol{u}(n)$. One of the main results of the calculation is that the ω^2 dependence of the reduced Raman spectra for low frequencies is reproduced analytically.

Figure 5.11 shows the result for a–Si or a–Ge with a reasonable choice of positional and angular disorder parameters. By comparison with Fig. 5.2 we see that the basic shape of the reduced Raman spectrum is reproduced.

Mitra et al. [5.60] have postulated that there is a broadened line in the Raman spectra of a–Si and a–Ge which arises from the effects of disorder on the sharp Raman line of the corresponding crystalline spectra.

In addition to this broad line at high frequency, they claim that there is a residual part of the spectrum that is purely disorder induced via the breakdown of the crystal momentum selection rules. An adjustable parameter method of subtracting out the broad main peak is used to find the residual disorder induced excess for comparison with the infrared spectrum, which is completely disorder induced for the case of homopolar a–Si and a–Ge. Although complete agreement is claimed, there appears to be certain flaws in the method. First, MITRA et al. [5.60] have to arbitrarily assume that 70% of the main broad peak in the Raman spectrum of a–Ge arises from the broadening mechanism and 30% from disordered induced Raman activity. Second, the disorder broadened Raman activity peaks at a considerably lower frequency than the zone-center Raman active modes of the crystal. Broadening alone cannot account for the shift in peak position [5.41].

WEAIRE and ALBEN [5.61] have presented a theorem which is useful for the understanding of the general features of the vibrational density of states of tetrahedrally bonded solids. Their theorem can be applied, for example, to the reduced Raman spectra of Figs. 5.2 and 5.3.

The WEAIRE-ALBEN theorem states that the phonon density of states of a tetrahedrally coordinated system subject to (5.8), the KEATING potential, should be as illustrated in the upper half of Fig. 5.12. The important features are three bands of states: 1) a band labeled TA which in the limit of $\beta/\alpha \to 0$ is a delta function at zero frequency; 2) a band labeled TO which in the same limit is a delta function centered at 8α; and 3) a continuous band labeled LA and LO which has the same form as electronic density of states of the following "one-band" Hamiltonian which has been widely used for tight binding calculations of electronic energy bands [5.62]

$$H^{(1)} = \alpha \left(4I - \sum_{l\Delta} |l\rangle \langle l\Delta| \right). \tag{5.16}$$

Here I is the unit operator and there is one state $|l\rangle$ associated with each site. When plotted as a function of ω^2, each band has a total of one state per atom.

The lower half of Fig. 5.12 shows, for comparison, the phonon density of states of crystal Si as determined by fitting neutron spectroscopy data by a parameterized shell model [5.24]. It is clear that even when $\beta/\alpha \neq 0$ there are identifiable bands corresponding to the TA and TO delta functions and the "one-band" Hamiltonian LA and LO states. There has been considerable discussion for the electronic case about what implications, if any, the observed disappearance in amor-

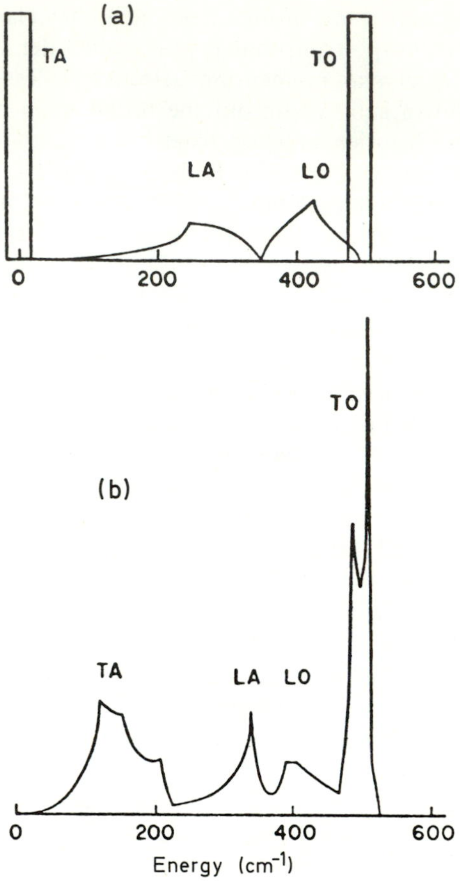

Fig. 5.12. (a) The phonon spectrum of cubic Si with diamand structure, as given by the Keating model (5.8) with $\alpha = 0.495 \times 10^5$ dyn/cm and $\beta = 0$. The contribution of the two δ functions have been spread over a small finite width so that their total weight is apparent [5.61]. (b) Phonon spectrum of diamond cubic Si as shell model fitted [5.24] to experimental neutron scattering data

phous Ge of the deep minimum between the two points of "one-band" Hamiltonian density of states. For example, some calculations [5.63] show that the minimum is filled in if five-fold rings of bonded atoms were present in addition to the six-fold rings of the crystal. However, more recent calculations for the electronic [5.64] and vibrational [5.53] cases do not indicate a washing out of the minimum. Alben [5.65] interprets the numerical theory of Subsection 5.1.3 and its match to Raman and infrared data for amorphous Si as follows. The three broad features of the amorphous spectra are: the low energy maximum corresponding to TA in Fig. 5.12; the high energy strong maximum corresponding to a mixing of the TO band and the LO half of the "one-band" Hamiltonian; and the middle maximum correspond to the LA half of the "one-band" Hamiltonian. This identification is substantiated by counting the states given by the model calculation used for comparison in Fig. 5.9; the minimum at around $350\,\mathrm{cm}^{-1}$ divides the total number of states in half. Such an identification seems to call into question the current structural interpretations of the electronic density of states as indicative of five-fold rings in amorphous Ge and of an absence of five-fold rings in amorphous GaAs [5.36, 66]. The interpretations depend upon an assumption of a minimum in the "one-band" Hamiltonian density of states without five-fold rings and a washing out of the minimum with five-fold rings. However, we have seen that the model interpretation of Fig. 5.9 indicates that a minimum can exist even with five-fold rings.

5.1.5. Amorphous III–V Compounds

Raman spectra of the amorphous III–V compounds have been reported by several groups [5.20, 22, 23, 26, 31, 41, 67]. To a certain extent, the basic results of the Si and Ge work carry over to the III–V's. That is, the amorphous vibrational densities of states apparently are, to first approximation, a broadened version of the crystalline vibrational density of states, and as for the group IV's the Raman spectra have structure characteristic of the entire vibrational spectra rather than merely the lines from Raman active zone center crystalline phonons. Figure 5.13 shows the Raman spectrum of amorphous GaAs compared with the crystalline density of states before and after broadening.

Figure 5.14 shows the spectrum for amorphous InP and here we see both an interesting and controversal new feature common to many III–V's. In addition to that part of the spectrum characteristic of the vibrational density of states, there is also an additional broad peak slightly above $400\,\mathrm{cm}^{-1}$. This peak is attributed to phosphorus-phosphorus vibrations. The controversy lies in whether to attribute the P–P bonds to inhomogeneities in the sample, e.g. phosphorus precipitates, or to say that there are isolated P–P bonds scattered through the amorphous InP network. The existence or non-existence of such wrong bonds is an important, unresolved point for those interested in the fundamental structural nature of tetrahedrally bonded amorphous semiconductors. It appears that the Raman data alone is incapable of resolving the controversy [5.36, 67].

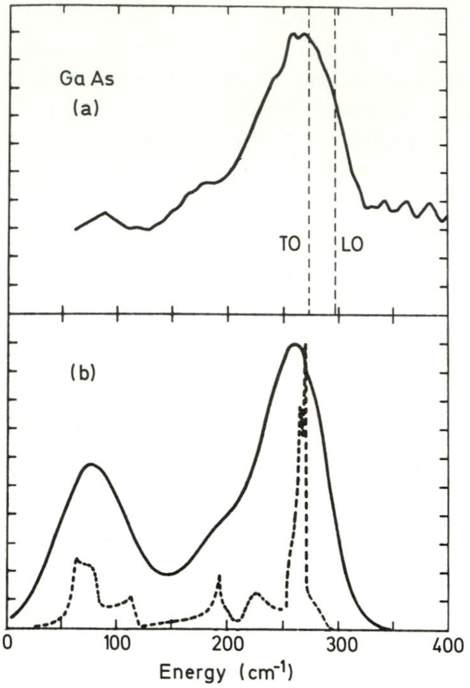

Fig. 5.13. (a) Reduced Raman spectrum of amorphous GaAs from room temperature data [5.23]. The dashed lines indicate the energies of the Raman active crystal GaAs phonons. (b) The density of states of crystalline GaAs (dashed line) from a shell model fit to neutron scattering data [5.24]. The solid line is the crystalline density of states broadened by a convolution with a Gaussian factor of half-width 25 cm^{-1} [5.23]

5.1.6. Amorphous SiC and Group IV Alloys

The Raman spectra of stoichiometric or near-stoichiometric amorphous SiC [5.68], SiGe [5.33, 69], and GeSn [5.31] have been reported. Of these materials only SiC has a stable crystal form as a compound, albeit with many polytypical variations. SiGe as a crystal only occurs as an alloy. There is only trace solubility of Ge in crystalline Sn, or vice versa. It is natural then to think of the "wrong bonds" as a likely occurence in the amorphous group IV alloys as there is no real strong chemical preference for "like bonds" as in the crystal and amorphous III–V's. The Raman spectra of SiC, SiGe, and GeSn each shows evidence of three kinds of bonding. For example, the SiC spectrum [5.68] has peaks characteristic of Si–C bonds and then also Si–Si and C–C bonds. $Si_{1-x}Ge_x$ alloys have been measured over the entire range [5.33] of ($0 \leqq x \leqq 1$) and show the three types of bonds for all x except $x=0$ or 1.

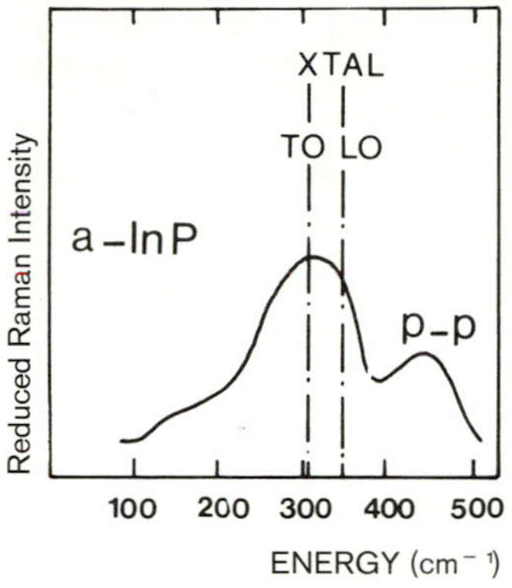

Fig. 5.14. The reduced Raman spectrum of amorphous InP [5.22]. The vertical dashed lines indicate the energies of the Raman active phonons of the crystal. The peak labeled P–P near 440 cm^{-1} is attributed to phosphorus-phosphorus bonds. The spectrum was taken in a back scattering geometry

It is interesting that Raman spectra can give this important information on short range bond variations which is, in general, difficult to deduce from diffraction studies [5.69].

5.2. Amorphous Chalcogens and Chalcogenides

This section is concerned with the description and interpretation of the Raman spectra of the glassy semiconductors containing S, Se, or Te. We first classify the types of materials under consideration and then discuss the general theme of a molecular interpretation of their spectra. We then consider in turn each of three classes of materials: elemental chalcogens, arsenic chalcogenides, and germanium chalcogenides.

5.2.1. Group VI Elements in Amorphous Semiconductors

In contrast to the tetrahedrally bonded semiconductors described above, many of the chalcogens and chalcogenides are glasses in the strictest sense. They can be formed by quenching from the melt, they possess a glass transition temperature at which softening occurs, and there is a continuity of properties from the glassy solid to the true non-viscous liquid. In some cases, (e.g. Te and $Ge_{1-x}Te_x$) the compositional and temperature range of formation can be extended by deposition processes analogous to those listed in Subsection 5.1.1. The motivation for vibrational studies is often structurally oriented here as for the tetrahedrally bonded materials. The general motivation for study of the amorphous chalcogens and their compounds has often been technological. Amorphous Se and $As_{1-x}Se_x$ find use in xerographic processes [5.70]; arsenic chalcogenide glasses are used in infrared optics as lenses, windows, and other components [5.71]; and Ge–Te based chalcogenides are often the basic constituents of the alloys which have been proposed for use as switching and memory elements of electronic circuitry [5.72]. All of these applications are sensitive to structural configurations of the atoms and Raman and infrared spectroscopy have been used extensively to identify various submicroscopic arrangements as well as transformations under external stimuli.

Our prime attention will be on three types of materials. First we shall discuss the elemental amorphous chalcogens, S, Se, and Te [5.1]. Only the liquid form of sulfur is amorphous; the solid has a variety of crystal allotropes. Selenium in its crystalline forms is either monoclinic with eight atom rings stacked in different ways to form the various allotropic modifications or it is trigonal with infinite helical chains closely packed with parallel axes. Crystal tellurium consists of chains and commonly occurs in the same structure as trigonal selenium. Raman studies have been used in an attempt to find remants of various crystal structural elements in the amorphous forms.

Next we shall discuss the arsenic glasses, As_2S_3 and As_2Se_3 [5.1], which have crystalline counterparts that are micaceous, layered structures consisting of strongly bonded AX_3, where $X = S$ or Se, pyramidal units within each plane. The planes are in turn only weakly bonded to each other. Depolarization measurements of their Raman spectra along with four-atom model calculations have indicated that these glasses are nearly molecular even though the reduced Raman spectra look much like a broadened crystalline density of states. Further confirmation of molecular remanence is in the complementarity of infrared and Raman activity.

Thirdly, the germanium chalcogenides, $Ge_{1-x}Y_x$ where x is the fractional composition of S, Se, or Te will be considered. The emphasis here will be somewhat different because there is still considerable controversy of the type of short range order which exists in the amorphous $Ge_{1-x}Y_x$ alloys. The two competing models are GeY_2 molecular units with four-fold coordinated Ge atom and two-fold coordinated chalcogens versus three-fold coordinated Ge and chalcogen atoms. The latter case would be analogous to the crystal arsenic structure and is suggested by the fact that when $x = 0.5$, the $Ge_{1-x}Y_x$ materials have an average number of valence electrons per atom equal to five and thus should be comparable to the group V elements.

5.2.2. Molecular Interpretations of Raman Spectra

We have seen that the Raman spectra of the tetrahedrally bonded amorphous semiconductors can be understood in terms of the entire vibrational density of states of a three-dimensional network that couples the motions of all the atoms to each other. The traditional approach to the study of the vibrational spectra of chalcogens and chalcogenides has emphasized a molecular viewpoint. LUCOVSKY and his colleagues [5.13, 73–75] have discussed at length the relative merits of a molecular versus density of states approach for various amorphous systems. They pointed out that the Raman spectra of molecular solids often have sharper features than those for which the density of states approach is more suitable. BRODSKY et al. [5.47], and SMITH et al. [5.76] indicated that there are limits to the molecular approach because of solid state interactions and that even for presumably molecular solids like amorphous Se and Te a density of states approach is applicable.

The molecular vibrations are more likely to be discernable in the Raman spectra of those amorphous solids which are formed of identifiable, but weakly coupled small molecular units [5.13]. For example, tetrahedrally bonded materials like silica (amorphous SiO_2) is a suitable candidate for a molecular analysis because of the relative softness of the bond bending force constants associated with the oxygen atoms which bridge together the strongly bonded SiO_4 tetrahedra. This is in contrast to non-molecular amorphous silicon where the bridging Si atoms are the same as the center atoms of the $SiSi_4$ tetrahedral units. The fully coupled modes are therefore necessary to understand the a–Si spectrum while some of the high-frequency features of the a–SiO_2 spectrum are discernable as molecular in nature.

Formally, there really is not a great conceptual difference between the two approaches. Figure 5.15 schematically illustrates what the dispersion curves and the density of states might look like in two extreme cases. We have drawn the dispersion curves to represent simple crystals either with (Fig. 5.15b) or without (Fig. 5.15a) molecular units.

We have made $g(E)$ representative of the corresponding amorphous forms by smearing out $g(E)$ to eliminate the sharp critical point singularities which arise only from long range order. The interconnectiveness of the atoms in Fig. 5.15a is reflected in the relative broadness of each band of branches. Neither the acoustic nor the optic branches are narrow in energy space. In Fig. 5.15b the situation is different. Here the separation of groups into molecular units is reflected in some nearly flat optic branches of the dispersion curves. Therefore those branches cover only a narrow range of energy space. In the right hand side of Fig. 5.15, we clearly see, what is now obvious, that broad bands lead to broad peaks

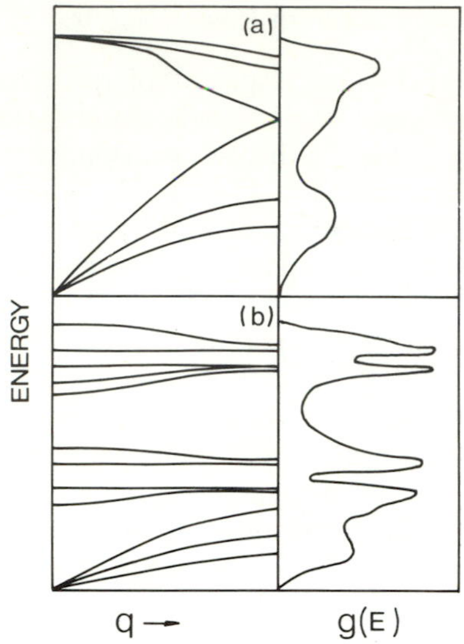

Fig. 5.15a and b. Schematic illustration of the disperison curves (left) and vibrational density of states (right) for solids ameneable to analysis by the density of states approach (upper) or a molecular interpretation (lower)

in $g(E)$ and therefore in the Raman spectrum, while narrow bands lead to sharp molecular type peaks in $g(E)$ and therefore also in the Raman spectrum. The Raman spectra of real molecular systems generally have these relatively sharp and identifiable characteristic frequency peaks which can be used to identify molecular species and their short range ordering. It was then natural for many workers to try analogous approaches for the interpretation of the spectra of amorphous solids. Below we shall discuss some of the successes and failures of the molecular approach.

5.2.3. Raman Scattering from Amorphous S, Se, and Te

Sulfur as a solid occurs in many allotropic crystalline forms but the rhombic form has been the one most studied by Raman spectroscopy [5.77–80]. Rhombic sulfur is composed of puckered 8 atom rings (S_8) which are the molecular units. Raman studies [5.77, 81] have been able to identify the S_8 rings in amorphous (i.e. liquid) sulfur because of the relative strength of the intramolecular covalent forces compared to the much weaker intermolecular interaction in both the crystal and melt. Figure 5.16 shows data of Ward [5.77] for crystal S at 25 °C and liquid S

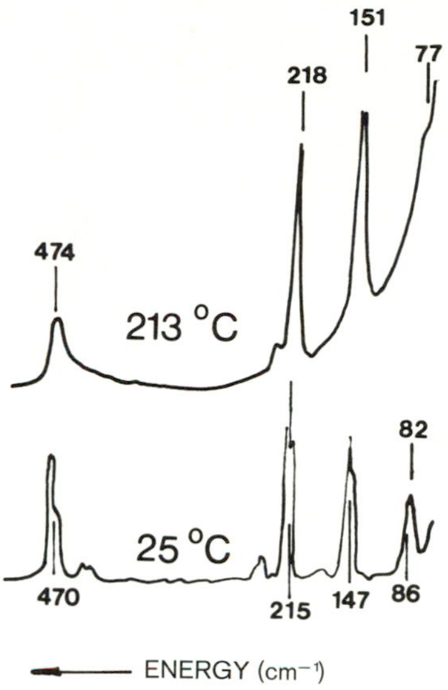

Fig. 5.16. Raman spectra of sulfur at two temperatures [5.77]. At 25 °C sulfur is crystalline at 213 °C it is liquid (amorphous)

at 213 °C. The crystal vibrational modes with frequencies above about 75 cm^{-1} are intramolecular [5.79] and are clearly preserved with some broadening, but with little if any frequency shifts in the liquid. Thus sulfur is an ideal example of the molecular case of Subsection 5.2.2. Details of the mode assignments can be found in [5.77, 78].

The success of the molecular approach for sulfur encouraged many early workers to try an analogous interpretation for Se with the added goal of distinguishing the Se rings from the long helical chains [5.82–85]. Figure 5.17 illustrates the original interpretiation of the Raman spectrum of amorphous Se. Shown in Fig. 5.17 are the liquid He temperature spectra of trigonal (chain), α-monoclinic (ring) and amorphous Se [5.84]. The peak near 235 cm^{-1} in the upper trace is characteristic of the chains in trigonal Se. The structure near 255 cm^{-1} in the lower trace is characteristic of the rings in α-monoclinic Se. The next step in the reasoning was to look at the middle trace and to identify the main peak in amorphous Se as a mixture of the 235 and 255 cm^{-1} peaks and therefore to conclude that amorphous Se had a mixture of rings and chains [5.82, 84]. Further analysis tried to deduce ring concentrations in Se–S, Se–Te, and Se–As amorphous alloys [5.85] from their Raman spectra. Although there are good chemical and physical reasons [5.86, 87] for believing that both rings and chains are present in amorphous Se, we

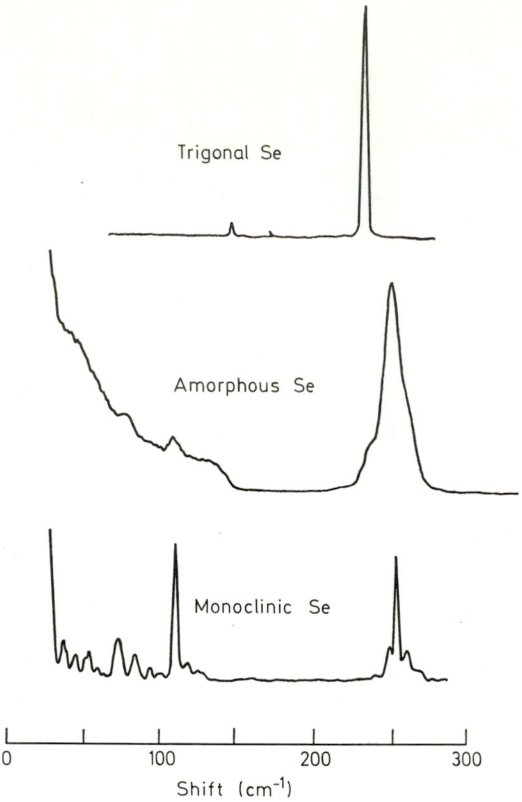

Fig. 5.17. Raman spectra of Selenium in three forms: trigonal crystal, monoclinic crystal, and amorphous [5.84]

shall see from the discussion below that the Raman spectrum can not be so simply interpreted as confirming evidence. Basically the problem is that frequencies which are characteristic of rings or chains in one crystal environment are not necessarily unchanged when the rings and chains are in a new environment such as another crystal polymorph or the amorphous state [5.47, 76, 88].

It is generally believed that the ring to chain ratio of amorphous Se changes upon melting and can also depend on such deposition parameters as quenching rate and the melt temperature when prepared from a melt [5.87] and substrate temperature when prepared by vapor deposition [5.89]. Smith et al. [5.76] observed that the relative strength of the 235 and 255 cm^{-1} contributions to the main peak were unchanged when amorphous Se was heated to above the glass transition temperature. Also the same observation was made on a variety of bulk amorphous

Se samples prepared at different quenching rates and on thin film samples evaporated onto substrates held at temperatures as low as 80 K or as high as 320 K. Under such a variety of conditions, changes in the ring versus chain contributions should have been seen in the Raman spectra if the original interpretation were valid. Indeed it is likely that both rings and chains contribute to the Raman spectrum of amorphous Se, but the different frequencies at which they scatter are not distinguishable. Both rings and chains have all their atoms covalently bonded with two-fold coordination. The bond angle in both cases is about 105°. The predominant frequency observed near 255 cm^{-1} can be well accounted for by a molecular force calculation [5.13, 73, 75, 88] on a three-atom cluster with two bonds separated by 105°. If extra intermolecular forces are included (chain-chain interactions) the frequency shifts in the crystal trigonal Se can be calculated [5.75, 88] and it is

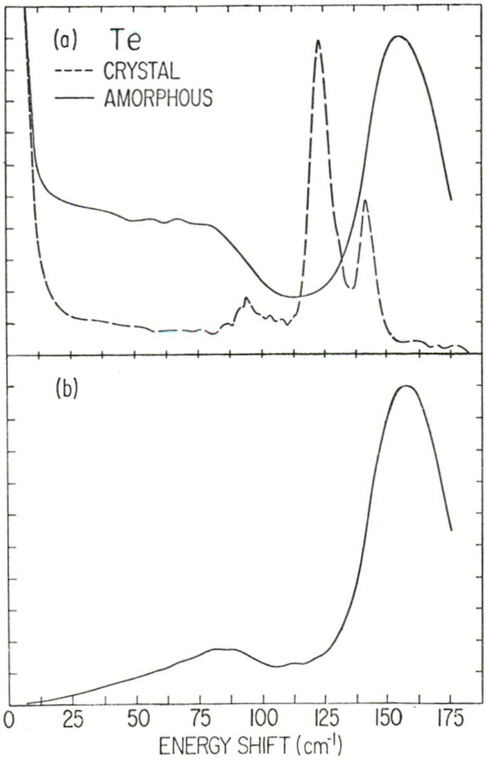

Fig. 5.18. (a) Raman spectra of amorphous (solid line) and polycrystalline (dashed line) tellurium [5.47]. The data were taken in a back scattering geometry. (b) Reduced Raman spectrum of amorphous Te

found to agree with the observations of the upper trace in Fig. 5.17. Both the monoclinic and amorphous forms presumably have very small intermolecular contributions.

In the Te case there is even a bigger difference between the isolated chains in the amorphous form and the closely packed, inter-locked chains of the trigonal crystal. Figure 5.18a shows the measured Raman spectra [5.47] of amorphous and polycrystalline Te at 77 K. Amorphous Te crystallizes near 280 K so it was necessary to prepare and maintain the amorphous sample at low temperatures. Figure 5.18b shows the reduced Raman spectrum (5.5) for amorphous Te. The most significant features of the reduced spectrum are that, like a–Si and a–Ge, the gross features are like those of an entire vibrational density of states spectrum and that, unlike a–Si and a–Ge, the spectrum is not like a broadened version of the corresponding crystalline density of states [5.90]. In fact, the major contribution to the amorphous spectrum occur near the broad peak at 157 cm^{-1} a frequency which is significantly higher than any of the single phonon excitations of crystal Te. Low temperature studies of the crystal phonon spectra show that the density of states peaks near 130 cm^{-1}. The Raman active zone center phonons for the optical branches are the 123 and 143 cm^{-1} peaks seen in the polycrystalline spectrum of Fig. 5.18. Brodsky et al. [5.47] attributed the difference in the vibrational frequencies to inter-chain drag effects which depress the frequencies more in the closely and coherently packed trigonal structure than in the disordered form. Lucovsky and Martin [5.73, 75, 88] have computed the intermolecular and intramolecular contributions for Se and Te in their amorphous and crystalline forms. Their model yields the observed frequency shifts between the essentially untethered vibrations of the chains or rings in amorphous Se and Te and the various depressed frequencies of the same molecular units in the crystals where there are significant intermolecular constraints.

Although our present understanding of the Raman data on a–Se does not allow us to draw firm conclusions on the important question of ring versus chain statistics, there is still the possibility that resonant Raman spectroscopy can be used to obtain such information. Resonant Raman scattering with incident photon energies near resonant electronic absorptions [5.89] characteristic of ring or chain configuration may provide the means.

5.2.4. Arsenic Sulfide and Related Glasses

Arsenic sulphide As_2S_3 and arsenic selenide As_2Se_3 have been studied for many years by Raman and IR spectroscopy both as crystals [5.91, 92] and glasses [5.93–100]. Much of the interest in the crystalline vibrations

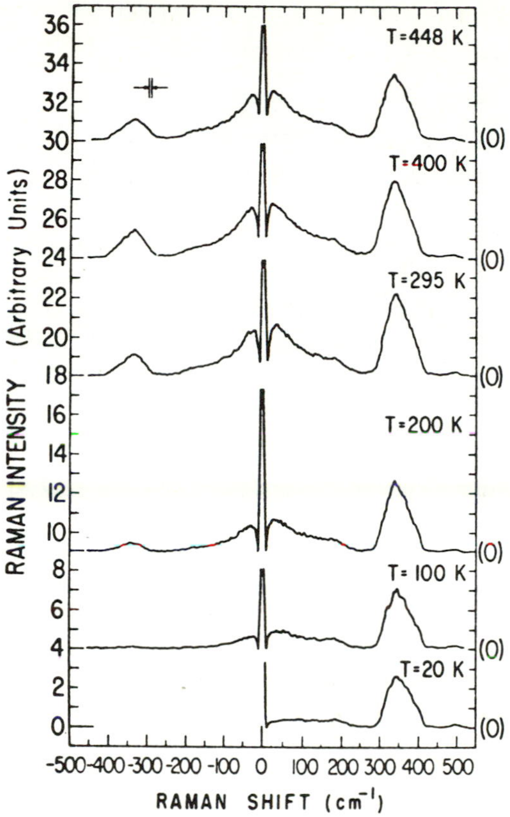

Fig. 5.19. Raman spectra of amorphous As_2S_3 recorded at the temperatures indicated [5.98]. The data was taken in the 90° scattering geometry with a transparent sample. The incident and scattered radiation had polarizations perpendicular to the scattered plane. There is a scale shift in the Raman intensity near the Rayleigh line

exists because of their layer formations with strong intra-layer forces and weak inter-layer forces. Thus, in a way, the arsenic compounds form molecular crystals and are comparable to sulfur [5.80]. Recent measurements by KOBLISKA and SOLIN [5.96–98] and by FINKMAN et al. [5.48, 100] show that the Raman spectra of amorphous As_2S_3 and As_2Se_3 are amenable to analysis both by the reduced Raman density of states approach [5.15] and by the LUCOVSKY and MARTIN's [5.73] molecular model. The reason for the joint applicability is that the main contribution to the density of states comes from modes with a molecular nature, even in the crystalline form.

Figure 5.19 [5.97] shows the temperature dependence of the Stokes and anti-Stokes Raman spectra of As_2S_3 glass from 20 to 448 K. The low

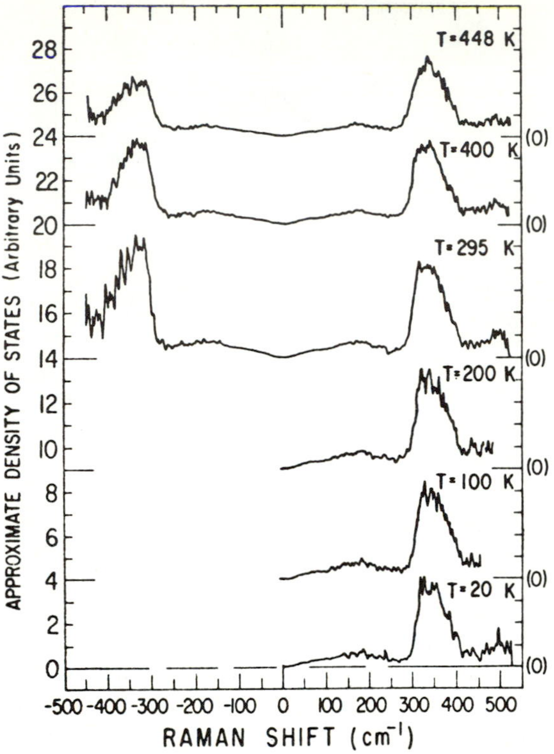

Fig. 5.20. Reduced Raman spectra for the amorphous As_2S_3 data of Fig. 5.19

frequency peak at 30 cm^{-1} is an artifact of the frequency and temperature dependence of the pre-factors in (5.2) and (5.3). There is no low frequency peak in the reduced spectra. The reduced spectra (Fig. 5.20) are all essentially temperature independent and thus verify the Shuker and Gamon [5.15] method of data reduction for glasses. As pointed out above in Subsection 5.1.2, similar artificial low frequency peaks had been observed in oxide glasses and until recently had been incorrectly interpreted as maxima in vibrational spectra. It's particularly important to be careful about low-frequency analyses in As_2S_3 because of the occurrence of real "rigid layer" modes [5.92] in the frequency region below 75 cm^{-1}. Also there is considerable speculation now about the excess specific heat in glasses and much of the speculation hinges on the details of the low-frequency vibrational spectra [5.101]. Finkman et al. [5.100] have extended the measurements of Figs. 5.19 and 5.20 to temperatures up to 1040 K. They found that even in the molten state, As_2S_3

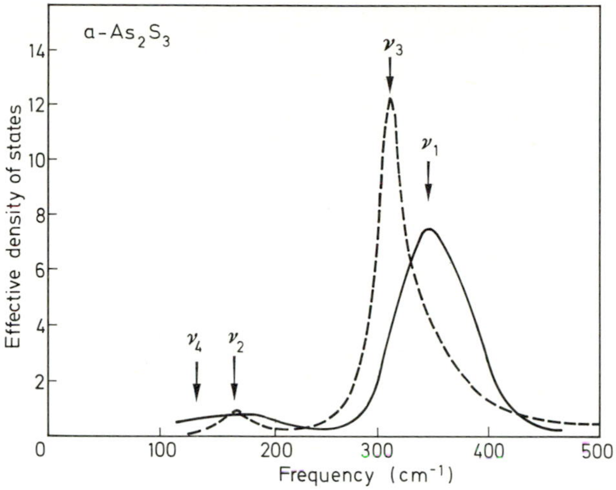

Fig. 5.21. A comparison of the reduced Raman spectrum (solid line) of amorphous As_2S_3 with the infrared absorption spectrum (dashed line), $\omega^2\varepsilon_2$, which also gives an effective density of states [5.95]. The arrows indicate the frequencies of an AsS_3 molecule

has the same Raman frequencies and band widths. They argue that this gives confirming evidence to the retention of a layer structure [5.102] with molecular-like vibrations of pyramidal units of AsS_3 and water-like molecules As_2S within the layers [5.73, 93].

Further confirming evidence for the retention of molecular selection rules in Raman scattering from As_2S_3 has been given by LUCOVSKY [5.13, 95]. As indicated by the arguments presented in Section 5.1, a true test of the density of state arguments is that the reduced Raman spectrum looks like the vibrational density of states except for some envelope of a smoothly varying matrix element factor. It should follow that in such a case the infrared absorption spectrum should also be reducible to a similar effective density of states which should have about the same shape and frequency peaks as the reduced Raman spectrum [5.27]. For a–Si and a–Ge we have shown this to be the case. However, as illustrated in Fig. 5.21, there is a complementarity of the Raman and infrared deduced effective densities of states. In the high frequency region where we expect the modes to be more nearly localized [5.13], the molecular model applies and gives two bond stretching modes, one which is a anti-symmetric stretch and more strongly IR active and the other which is a breathing mode and more strongly Raman active [5.95]. Another test of the retention of the molecular character of the local vibrations is by the depolarization Raman spectrum. KOBLISKA and SOLIN [5.98] have

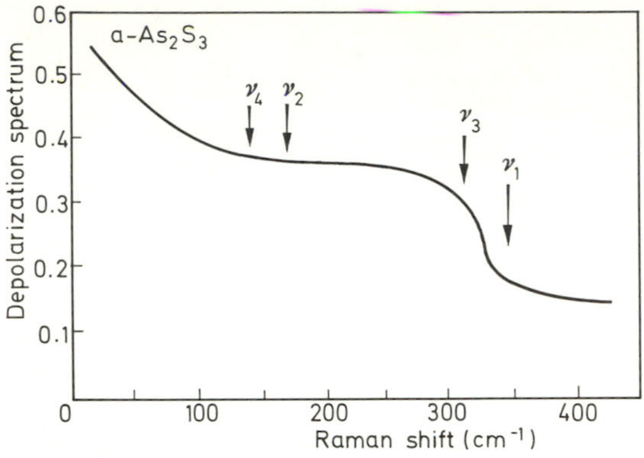

Fig. 5.22. The depolarization spectrum of amorphous As_2S_3 [5.98]. The arrows indicate the optic mode frequencies of an AsS_3 molecule [5.95]

measured the ratio of HV to VV scattering for amorphous As_2S_3 and the result is shown in Fig. 5.22. As is to be expected, the higher frequency symmetric vibrations give more nearly polarized scattering as indicated by the drop in the depolarization ratio at the peak of the scattering intensity. (Compare Figs. 5.21 and 5.22.)

Arsenic sulfide is the only amorphous semiconductor to date in which resonant Raman scattering has been reported (see, however [Ref. 8.1, Fig. 2.18]). Kobliska and Solin [5.96] used a fixed laser frequency and temperature tuned the absorption edge of their sample. Although the value of a characteristic energy in amorphous As_2S_3 is somewhat arbitrary, Kobliska and Solin chose a band gap of

$$E_g(T) = [2.32 + 6.70 \times 10^{-4}(300 - T)] \quad [\text{eV}], \tag{5.17}$$

where T is the temperature of measurement. The room temperature gap $E_g(300) = 2.32$ eV is not a well defined quantity in terms of the usual crystal concepts of a demarcation between allow and non-allowed states but for the sake of displaying the data it serves as a useful reference point. Figure 5.23 shows the normalized Raman scattering efficiency plotted against the incident laser photon energy as measured from E_g of (5.17). The scattering efficiencies are shown for three different regions of the Raman spectra of Fig. 5.19, namely 62, 161, and 342 cm^{-1}, for both Stokes and anti-Stokes scattering. It is clear that there is a an enhancement effect in the region of E_g as well as weaker effects of unknown origin about 0.4 to 0.5 eV below. No theory is yet available to analyze resonant Raman scattering in amorphous semiconductors.

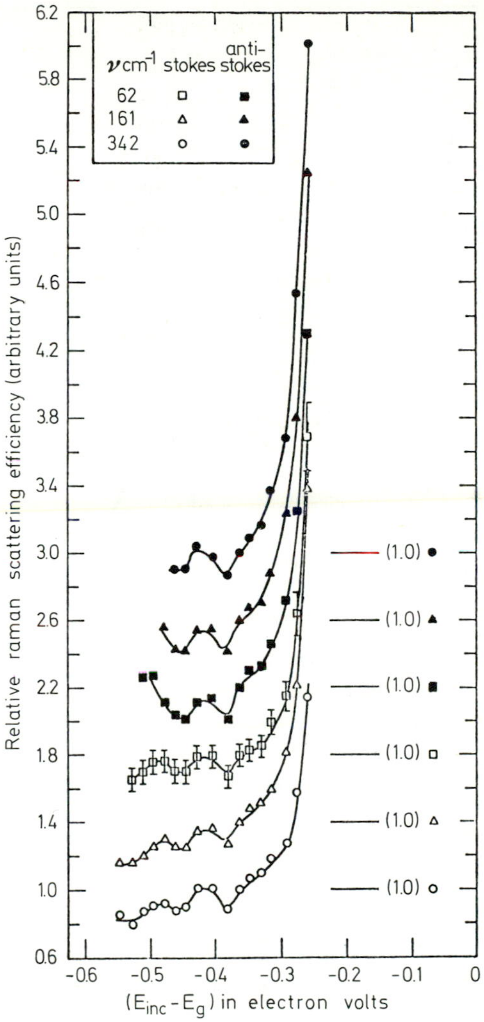

Fig. 5.23. The resonant Raman spectra of amorphous As_2S_3 recorded in the 90° scattering geometry. Resonance was obtained by temperature tuning the optical absorption spectrum. The absicca is measured relative to E_g, as defined by (5.17). The scattered light was measured at several energy shifts as indicated in the insert [5.96]

The crystallization of As_2Se_3 has been studied in real time by Raman scattering [5.48]. Crystallization in the temperature region above the glass transition T_g and below the melting point T_m occured only when the glass was illuminated. As_2Se_3 is an extremely stable glass and does not normally crystallize in the dark in this temperature range. The illumina-

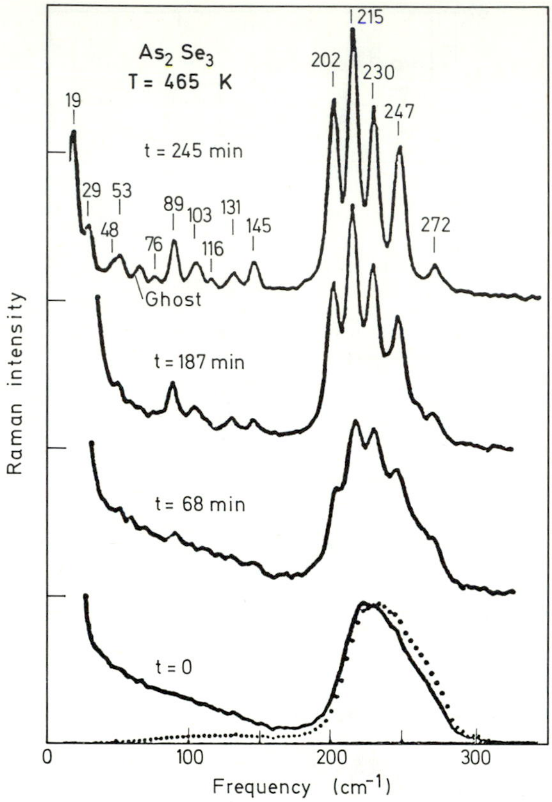

Fig. 5.24. Changes in the Raman spectrum of As_2Se_3 during crystallization. The dotted line shows the reduced spectrum of amorphous As_2Se_3 [5.48]

tion was provided by the same incident laser beam (5145 Å argon line) used for the scattering measurements. Figure 5.24 shows that the Raman spectrum gradually converts in time from that of amorphous As_2Se_3 to that of crystal As_2Se_3. By examination of the shift in the weight of the spectrum during crystallization, Finkman et al. [5.48] deduced that the phonon dispersion curves of crystal As_2Se_3 are more closely grouped together at the Brillouin zone edge than at the zone center. Figure 5.25 shows a schematic representation of their deductions. They were able to reason out these results because the crystal spectrum contains information only about zone center phonons, while the amorphous spectrum is a density of states measurement and weights the remnants of the zone edge region because of the extra phase space with increasing wavevector.

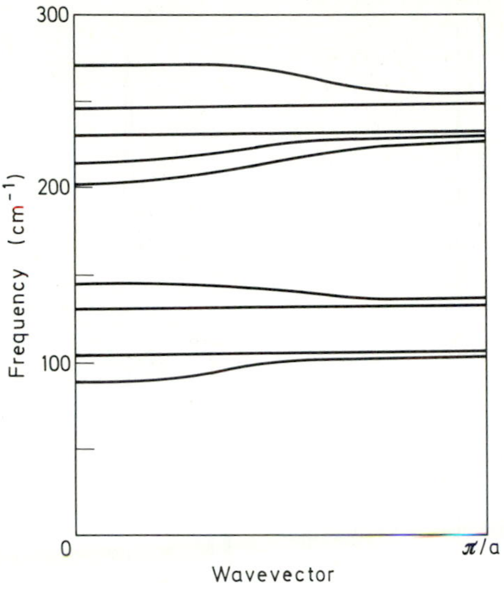

Fig. 5.25. Schematic representation of the dispersion curves of crystal As_2Se_3, as deduced from Fig. 5.24. The bands which change little with wavevector are drawn as horizontal lines [5.48]

The technique of using real time Raman scattering should find extensive application in chalcogenide glasses where photo-crystallization phenomena [5.103] are of interest. Here, as also for the tetrahedral semiconductors, Raman scattering is capable of detecting traces of crystallization. In addition for compound glasses, phase separation is also detectable in some cases. For example, As_2Te_3 is more difficult to prepare as a glass than As_2S_3 or As_2Se_3. Raman scattering has been used to test for traces of amorphous Te or crystalline Te segregation [5.47].

5.2.5. The Structure of Germanium Chalcogenides

Raman studies have been reported on the three germanium-chalcogenide alloy glass systems $Ge_{1-x}Y_x$, where Y = S, Se, or Te. Most of the results are on chalcogenide rich alloys, $x > 0.5$, with emphasis on the most stable glass forming region near $x = 2/3$. For GeY_2, the structure is well established, by Raman scattering [5.104–107] and other means [5.108, 109], to consist of four fold coordinated Ge atoms and two fold coordinated chalcogens arranged in GeY_4 tetrahedra with bridging Y atoms. The analogy is to the amorphous quartz structure [5.110].

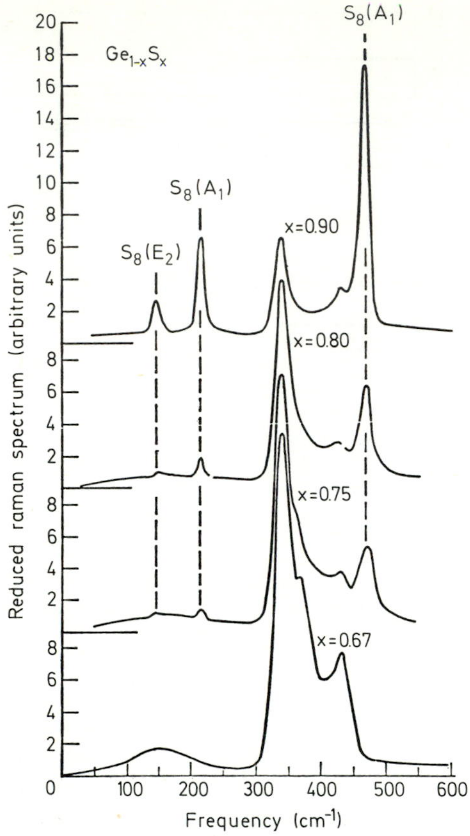

Fig. 5.26. The Raman spectra of amorphous $Ge_{1-x}S_x$ for sulfur rich compositions conpared with GeS_2 [5.105]. The peaks identified with S_8 molecules are indicated

For alloy compositions more germanium rich that the $x=\frac{2}{3}$ case, there still is some speculation that as x approaches $\frac{1}{2}$, then perhaps there can be more than just an average of three-fold coordination. Perhaps there really are a significant numbers of three-fold coordinated Ge and chalcogen atoms.

LUCOVSKY et al. [5.104, 105] have studied the infrared and Raman vibrational spectra of $Ge_{1-x}S_x$ over the range $0.55 < x < 0.90$. They used the fact (see Subsection 5.2.4) that for molecular vibrations there can be a complementarity between IR and Raman activity to discriminate between various structural alternatives. In this case, depolarization Raman spectra also helped, but the depolarization results were not fully trustworthy because of optical inhomogeneities in some of the samples. First LUCOVSKY et al. [5.105] pointed out (see Figs. 5.26 or 5.27)

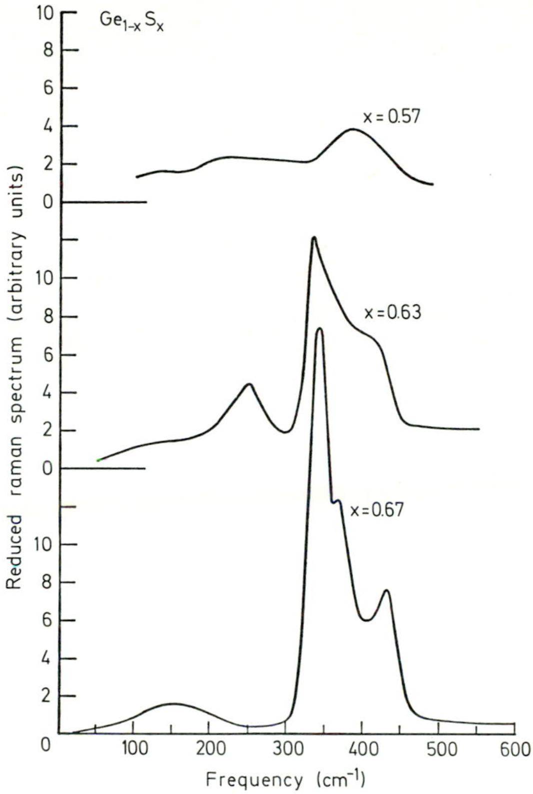

Fig. 5.27. The Raman spectra of amorphous $Ge_{1-x}S_x$ for germanium rich compositions compared with GeS_2 [5.105]

that the GeS_2 ($x=0.67$) glass has a strong polarized 342 cm^{-1} Raman peak which along with a 367 cm^{-1} IR absorption peak is characteristic of the GeS_4 tetrahedral unit. The 342 cm^{-1} peak can that be tracked in the Raman spectra for other compositions. For example, the 342 cm^{-1} peak (Fig. 5.26) is present for all the sulfur rich ($x>0.67$) glasses. In addition, there are the frequencies characteristic of S_8 rings in the sulfur rich spectra. The IR data were used to establish the presence of rings over the mere formation of some S–S bonds between Ge atoms. Figure 5.27 shows some spectra for Ge rich (relative to GeS_2) compositions. There is a gradual loss of the characteristic GeS_4 peak with increasing Ge content. The results are interpreted in terms of a restricted random network model [5.111] with GeS_4, GeS_3Ge, GeS_2Ge_2, $GeSGe_3$, and $GeGe_4$ tetrahedra which allows for a gradual increase in the number of Ge–Ge bonds. The composition range studied did not allow the

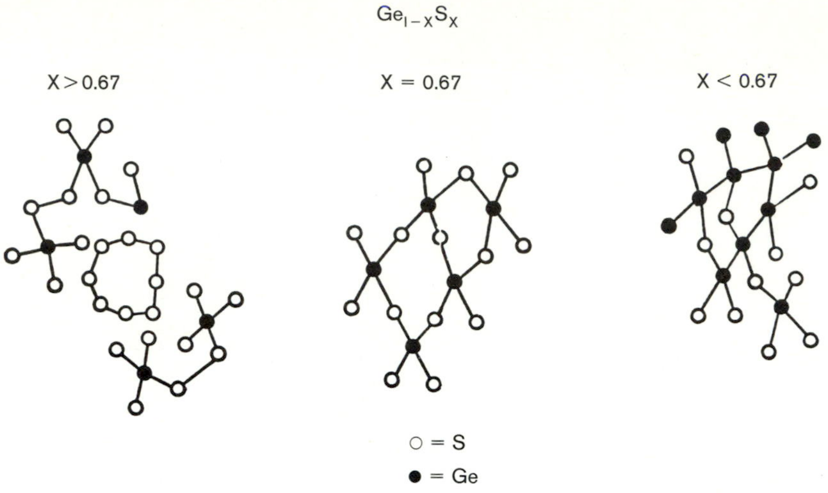

Fig. 5.28. Schematic representation of the model of $Ge_{1-x}S_x$ amorphous alloys for Ge rich ($x<0.67$), GeS_2 ($x=0.67$), and S rich ($x>0.67$) compositions [5.105]

authors to make a definitive statement about the possibility of three-fold coordination in the vicinity of a compound corresponding to GeS ($x=0.5$). Figure 5.28 summarizes in schematic form the conclusions about the structure of $Ge_{1-x}S_x$ [5.105].

Tronc et al. [5.106] have studied the Se rich ($x>0.67$) region of $Ge_{1-x}Se_x$ glasses by Raman spectroscopy along with measurements of the fundamental optical-absorption edge. They conclude that for this system there are also no Ge–Ge bonds and that the excess Se is accomodated by the removal of Ge–Se–Ge sequences. At present they have not yet been able to extend to measurements to $x<0.6$ to test the three-fold coordination hypothesis.

The Raman and IR spectra of $Ge_{1-x}Te_x$ have been obtained by Fisher et al. [5.107] over a composition range of $0.33<x<0.8$ and they were able to discuss the question of the coordination in amorphous GeTe ($x=0.5$). Figure 5.29 shows their data for the extreme values of x and for $x=0.5$. They could interpret the data in term of contributions for Ge–Te ($\approx 230\,cm^{-1}$), Te–Te (≈ 85 and $165\,cm^{-1}$), and Ge–Ge ($\approx 275\,cm^{-1}$) bonds. Because they were able to interpret all their results in terms of two-fold coordinated Te and four-fold coordinated Ge they excluded the possibility of three-fold coordination. The fact that over a large composition range, the spectra are smoothly related one to another, also supports their contention. However, for the following reasons we

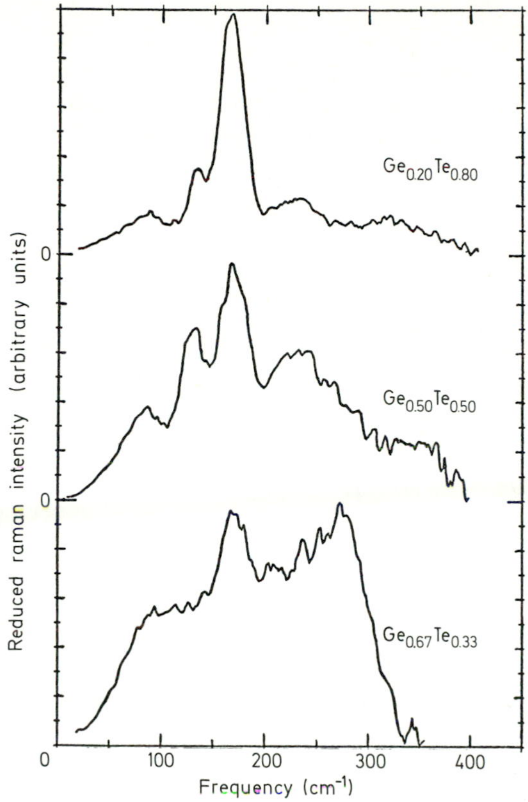

Fig. 5.29. Reduced Raman spectra of $Ge_{1-x}Te_x$ amorphous alloys for three values of x [5.107]

feel that a case really has yet to be established against or for three-fold coordination near $x=0.5$.

It is important to have more controllable experiment parameters than the composition if one is to use Raman spectra to reach truly unambiguous conclusions about structure. While the compositional dependence of various Raman active modes are important steps to building structural models for a–$Ge_{1-x}X_{2x}$ and a–$As_{2(1-x)}X_{3x}$ alloys it is also important to check model self-consistency by, for example, temperature and pressure dependence. In the case of a–Se it was the immutability of the Raman spectra with the temperature of sample preparation which lead BRODSKY et al. [5.47] to suspect the ring and chain mixture model since the ring to chain ratio was supposed to depend on the temperature of preparation [5.87].

Acknowledgements

This chapter incorporates the ideas developed over several years in many discussions and interactions with my colleagues. I am particularly grateful to R. ALBEN, E. BURSTEIN, G. LUCOVSKY, J. E. SMITH, JR., and D. WEAIRE for their interest, collaboration, and initiation of much that is described here.

References

5.1. N. F. MOTT, E. A. DAVIS: *Electronic Processes in Non-Crystalline Materials* (Oxford University Press, London, 1971).
5.2. J. STUKE, W. BRENNIG (Eds.): *Proc. 5th Intern. Conf. Liquid and Amorphous Semiconductors*, Garmisch, 1973 (Taylor and Francis, London, 1974).
5.3. J. M. ZIMAN: J. Phys. C **2**, 1704 (1969).
5.4. R. J. BELL: Rep. Progr. Phys. **35**, 1315 (1972).
5.5. P. DEAN: Rev. Mod. Phys. **44**, 127 (1972).
5.6. J. WONG, C. A. ANGELL: Appl. Spectrosc. Rev. **4**, 155 (1971) and references therein.
5.7. M. HASS: J. Phys. Chem. Solids **31**, 415 (1970).
5.8. W. B. GRANT, H. SCHULZ, S. HÜFNER, J. PELZL: Phys. Stat. Sol. (b), **60**, 331 (1973) and references therein.
5.9. M. BALKANSKI (Ed.): *Proc. 2nd Intern. Conf. Light Scattering in Solids*, Paris, 1971 (Flammarion, Paris, 1971), Chapter IV.
5.10. P. CHAUDARI, J. CUOMO, R. J. GAMBINO: Appl. Phys. Letters **22**, 337 (1973).
5.11. H. O. HOOPER, A. M. DE GRAAF (Eds.): *Amorphous Magnetism* (Plenum, New York, 1973).
5.12. A. L. ROBINSON: Science **182**, 908 (1973).
5.13. G. LUCOVSKY: In *Proc. 5th Intern. Conf. Liquid and Amorphous Semiconductors*, Garmisch, 1973 (Taylor and Francis, London, 1974), p. 1099.
5.14. H. BÖTTGER: Phys. Stat. Sol. (b) **62**, 9 (1974).
5.15. R. SHUKER, R. GAMON: Phys. Rev. Letters **25**, 222 (1970).
5.16. R. SHUKER, R. W. GAMON: In *Proc. 2nd Intern. Conf. Light Scattering in Solids*, Paris, 1971 (Flammarion Sciences, Paris, 1971), p. 334.
5.17. M. H. BRODSKY, S. KIRKPATRICK, D. WEAIRE (Eds.): *Proc. Intern. Tetrahedrally Bonded Amorphous Semiconductors* (American Institute of Physics, New York, 1974).
5.18. M. H. BRODSKY: J. Vac. Sci. Technol. **8**, 125 (1971).
5.19. S. C. MOSS, D. ADLER: Comments in Solid State Phys. **5**, 47 (1973).
5.20. J. E. SMITH, JR., M. H. BRODSKY, B. L. CROWDER, M. I. NATHAN, A. PINCZUK: Phys. Rev. Letters **26**, 642 (1971).
5.21. R. ALBEN, J. E. SMITH, JR., M. H. BRODSKY, D. WEAIRE: Phys. Rev. Letters **30**, 1141 (1973).
5.22. M. WIHL, M. CARDONA, J. TAUC: J. Non-Cryst. Solids **8—10**, 172 (1972).
5.23. J. E. SMITH, JR., M. H. BRODSKY, B. L. CROWDER, M. I. NATHAN: In *Proc. 2nd. Intern. Conf. Light Scattering in Solids*, Paris, 1971 (Flammarion, Paris, 1971), p. 330.
5.24. G. DOLLING, R. A. COWLEY: Proc. Phys. Soc., London **88**, 463 (1966).
5.25. R. W. STIMETS, J. WALDMAN, J. LIN, T. S. CHANG, R. J. TEMKIN, G. A. N. CONNELL: Sol. Stat. Comm. **13**, 1485 (1973).
5.26. W. PRETTL, N. J. SHEVCHIK, M. CARDONA: Phys. Stat. Sol. (b) **59**, 241 (1973).

5.27. M. H. Brodsky, A. Lurio: Phys. Rev. B**9**. 1646 (1974).
5.28. F. R. Ladan, A. Zylbersztejn: Phys. Rev. Letters **28**, 1198 (1972).
5.29. B. Schröder: In *Proc. Intern. Conf. Tetrahedrally Bonded Amorphous Semiconductors*, Yorktown, 1974 (American Institute of Physics, New York, 1974), p. 114.
5.30. J. D. Axe, D. T. Keating, G. S. Cargill, III, R. Alben: In *Proc. Intern. Conf. Tetrahedrally Bonded Amorphous Semiconductors*, Yorktown, 1974 (American Institute of Physics, New York, 1974), p. 279.
5.31. J. S. Lannin: Sol. Stat. Comm. **11**, 1523 (1972).
5.32. J. S. Lannin: Sol. Stat. Comm. **12**, 947 (1973).
5.33. J. S. Lannin: In *Proc. 5th Intern. Conf. Amorphous and Liquid Semiconductors*, Garmisch, 1973 (Taylor and Francis, London, 1974), p. 1245.
5.34. A. J. Martin, W. Brenig: Phys. Stat. Sol. (b) **64**, 163 (1974).
5.35. L. von Heimendahl: In *Proc. Intern. Conf. Tetrahedrally Bonded Amorphous Semiconductors*, Yorktown, 1974 (American Institute of Physics, New York, 1974), p. 274.
5.36. G. A. N. Connell: Phys. Stat. Solidi (b) **69**, 9 (1975).
5.37. C. N. King, W. A. Phillips, J. P. de Neufville: Phys. Rev. Letters **32**, 538 (1974).
5.38. P. Flubacher, A. J. Leadbetter, J. A. Morrison, B. P. Stoicheff: J. Phys. Chem. Solids **12**, 53 (1959).
5.39. A. S. Barker: In *Far Infrared Properties of Solids*, ed. by S. S. Mitra and S. Nudelman (Plenum Press, New York, 1970), p. 247.
5.40. R. Shuker, R. Gamon: Phys. Lett. **33**A, 96 (1970).
5.41. B. L. Crowder, J. E. Smith, Jr., M. H. Brodsky, M. I. Nathan: In *Proc. Second Intern. Conf. Ion Implantation*, Garmisch (1971), p. 255.
5.42. J. E. Smith, Jr., M. H. Brodsky, B. L. Crowder, M. I. Nathan: J. Non-Cryst. Solids **8—10**, 179 (1972).
5.43. J. C. Bourgoin, J. F. Morhange, R. Beserman: Rad. Effects **22**, 205 (1974).
5.44. M. H. Brodsky, R. S. Title: Phys. Rev. Letters **23**, 581 (1969).
5.45. R. S. Title, M. H. Brodsky, B. L. Crowder: In *Proc. 10th Intern. Conf. Physics of Semiconductors*, Cambridge, Mass., 1970 (U.S. Atomic Energy Comm., Oak Ridge, 1970). p. 794.
5.46. W. Beyer, J. Stuke: In *Proc. 5th Intern. Conf. Amorphous and Liquid Semiconductors*, Garmisch, 1973 (Taylor and Francis, London, 1974), p. 251.
5.47. M. H. Brodsky, R. J. Gambino, J. E. Smith, Jr., Y. Yacoby: Phys. Stat. Sol. (b) **52**, 609 (1972).
5.48. E. Finkman, A. P. Defonzo, J. Tauc: In *Proc. 12th Intern. Conf. on the Physics of Semiconductors*, Stuttgart (1974), p. 1022.
5.49. A. Solin, R. J. Kobliska: In *Proc. 15th Inter. Conf. Liquid and Amorphous Semiconductors*, Garmisch, 1973 (Taylor and Francis, London, 1974), p. 1251.
5.50. J. F. Morhange, R. Beserman, J. C. Bourgoin, P. R. Brosius, Y. H. Lee, L. J. Cheng, J. W. Corbett: Proc. Intern. Conf. Ion Implantation, Kyoto (1974)
5.51. M. I. Nathan, J. E. Smith, Jr., K. N. Tu: J. Appl. Phys. **45**, 2370 (1974).
5.52. R. Alben, D. Weaire, J. E. Smith, Jr., M. H. Brodsky: In *Proc. 5th Intern. Conf. Amorphous and Liquid Semiconductors*, Garmisch, 1973 (Taylor and Francis, London, 1974), p. 1231.
5.53. R. Alben, D. Weaire, J. E. Smith, Jr., M. H. Brodsky: Phys. Rev. B**11**, 2271 (1975).
5.54. P. N. Keating: Phys. Rev. **145**, 637 (1966).
5.55. R. H. Wentorf, Jr., J. S. Kasper: Science **139**, 338 (1963).
5.56. M. F. Thorpe: Phys. Rev. B**8**, 5352 (1973).
5.57. R. Zallen: Phys. Rev. **113**, 824 (1968).

5.58. M. F. Thorpe: In *Proc. Intern. Conf. Tetrahedrally Bonded Amorphous Semiconductors*, Yorktown, 1974 (American Institute of Physics, New York, 1974), p. 267.
5.59. M. Born: Ann. Phys. (Leipzig) **44**, 605 (1914).
5.60. S. S. Mitra, D. K. Paul, Y. F. Tsay, B. Bendow: In *Proc. Intern. Conf. Tetrahedrally Bonded Amorphous Semiconductors*, Yorktown, 1974 (American Institute of Physics, New York, 1974), p. 284.
5.61. D. Weaire, R. Alben: Phys. Rev. Lett. **29**, 1505 (1972).
5.62. D. Weaire, M. F. Thorpe: Phys. Rev. B**4**, 2508 (1971).
5.63. M. F. Thorpe, D. Weaire, R. Alben: Phys. Rev. B**7**, 3777 (1972).
5.64. R. Alben, D. Weaire, P. Steinhardt: J. Phys. C**6**, L384 (1973).
5.65. R. Alben: In *Proc. Intern. Conf. Tetrahedrally Bonded Amorphous Semiconductors*, Yorktown, 1974 (American Institute of Physics, New York, 1974), p. 249.
5.66. G. A. N. Connell: In *Proc. 12th Intern. Conf. Physics of Semiconductors*, Stuttgart (1974), p. 1003.
5.67. J. S. Lannin: In *Proc. Intern. Conf. Tetrahedrally Bonded Amorphous Semiconductors*, Yorktown, 1974 (American Institute of Physics, New York, 1974), p. 260.
5.68. M. Gorman, S. A. Solin: Solid State Comm. **15**, 761 (1974).
5.69. N. J. Shevchik, J. S. Lannin, J. Tejeda: Phys. Rev. B**7**, 3987 (1973).
5.70. J. Mort: In *Proc. 5th Intern. Conf. Amorphous and Liquid Semiconductors*, Garmisch, 1973 (Taylor and Francis, London, 1974), p. 1361.
5.71. A. R. Hilton: Appl. Optics **5**, 1877 (1966).
5.72. S. R. Ovshinsky: J. Non-Cryst. Solids **2**, 99 (1970).
5.73. G. Lucovsky, R. M. Martin: J. Non-Cryst. Solids **8—10**, 185 (1972).
5.74. G. Lucovsky, R. M. White: Phys. Rev. B**8**, 660 (1973).
5.75. G. Lucovsky, R. M. Martin: To be published.
5.76. J. E. Smith, Jr., M. H. Brodsky, R. J. Gambino: Bull. Am. Phys. Soc. II **17**, 336 (1972).
5.77. A. T. Ward: J. Phys. Chem. **72**, 4133 (1968).
5.78. I. Srb, A. Vasko: Czech. J. Phys. B**13**, 827 (1963).
5.79. A. Anderson, Y. T. Loh: Canad. J. Chem. **44**, 879 (1969).
5.80. R. Zallen: Phys. Rev. B**9**, 4485 (1974).
5.81. H. Gerding, R. Westrik: Recueil Trav. Chim. Pays-Bas **62**, 68 (1943).
5.82. G. Lucovsky, A. Mooradian, W. Taylor, G. B. Wright, R. C. Keezer: Solid State Comm. **5**, 113 (1967).
5.83. G. Lucovsky: In *Physics of Selenium and Tellurium*, ed. by W. C. Cooper (Pergamon Press, Oxford, 1964), p. 255.
5.84. A. Mooradian, G. B. Wright: In *Physics of Selenium and Tellurium*, ed. by W. C. Cooper (Pergamon Press, Oxford, 1969), p. 169.
5.85. J. Schottmiller, M. Tabak, G. Lucovsky, A. Ward: J. Non-Cryst. Solids **4**, 80 (1970).
5.86. G. Brielieb: Z. Phys. Chem. A**144**, 321 (1929).
5.87. A. Eisenberg, A. V. Tobalsky: J. Polym. Sci. **46**, 19 (1960).
5.88. R. M. Martin, G. Lucovsky: In *Proc. 12th Intern. Conf. Physics of Semiconductors*, Stuttgart (1974).
5.89. N. B. Zakharova, Yu. A. Chekasov: Sov. Phys.-Solid State **12**, 1572 (1971).
5.90. A. S. Pine, G. Dresselhaus: Phys. Rev. B**4**, 356 (1971).
5.91. R. Zallen, M. L. Slade, A. T. Ward: Phys. Rev. B**3**, 4257 (1971).
5.92. R. Zallen, M. Slade: Phys. Rev. B**9**, 1629 (1974).
5.93. I. G. Austin, E. S. Garbett: Phil. Mag. **23**, 17 (1971).
5.94. A. T. Ward, M. B. Myers: J. Phys. Chem. **73**, 1374 (1969).
5.95. G. Lucovsky: Phys. Rev. B**6**, 1480 (1972).
5.96. R. J. Kobliska, S. A. Solin: Solid State Somm. **10**, 231 (1972).

5.97. R. J. Kobliska, S. A. Solin: J. Non-Cryst. Solids **8—10**, 91 (1972).
5.98. R. J. Kobliska, S. A. Solin: Phys. Rev. B**8**, 756 (1973).
5.99. Yu. F. Markov, N. B. Reshetnyak: Fiz. Tverd. Tela **14**, 1242 (1972), translated Sov. Phys.-Solid State **14**, 1063 (1972).
5.100. E. Finkman, A. Defonzo, J. Tauc: In *Proc. 5th Intern. Conf. Amorphous and Liquid Semiconductors*, Garmisch, 1973 (Taylor and Francis, London, 1974), p. 1275.
5.101. R. O. Pohl, W. F. Love, R. B. Stephens: In *Proc. 5th Intern. Conf. Amorphous and Liquid Semiconductors*, Garmisch, 1973 (Taylor and Francis, London, 1974), p. 1121.
5.102. D. C. Taylor, S. G. Bishop, D. L. Mitchell: Phys. Rev. Letters **27**, 414 (1973).
5.103. J. P. de Neufville: In *Proc. 5th Intern. Conf. Amorphous and Liquid Semiconductors*, Garmisch, 1973 (Taylor and Francis, London, 1974), p. 1351.
5.104. G. Lucovsky, J. P. de Neufville, F. L. Galeener: Phys. Rev. B**9**, 1591 (1974).
5.105. G. Lucovsky, F. L. Galeener, R. C. Keezer, R. H. Geils, H. A. Six: Phys. Rev. B**10**, 5134 (1974).
5.106. P. Tronc, M. Bensoussan, A. Brenac, C. Sebenne: Phys. Rev. B**8**, 5947 (1973).
5.107. G. B. Fisher, J. Tauc, Y. Verhelle: In *Proc. 5th Intern. Conf. Liquid and Amorphous Semiconductors*, Garmisch, 1973 (Taylor and Francis, London, 1974), p. 1259.
5.108. S. C. Rowland, S. Narasimhan, A. Bienenstock: J. Appl. Phys. **43**, 2741 (1972).
5.109. L. Cervinka, A. Hruby: In *Proc. 5th Intern. Conf. Liquid and Amorphous Semiconductors*, Garmisch, 1973 (Taylor and Francis, London, 1974), p. 431.
5.110. R. L. Mozzi, B. E. Warren: J. Appl. Cryst. **2**, 164 (1969).
5.111. H. Phillip: J. Non-Cryst. Solids **8—10**, 627 (1972).
5.112. D. Bermejo, M. Cardona: J. Non-Cryst. Solids **32**, 421 (1979).

6. Brillouin Scattering in Semiconductors*

A. S. PINE

With 8 Figures

Brillouin scattering is known to be a valuable probe of acoustic phonons in gases, liquids and solids. The effects peculiar to Brillouin scattering in semiconductors will be the subject of this chapter. The velocity of sound is determined directly from the Brillouin shift; so, generally, elastic constants and velocity anisotropy, relaxation processes, phase transitions and a variety of acoustic interactions with other low-frequency excitations may be accurately measured. The spectral width of the scattered light, examined with high resolution, yields information about acoustic attenuation arising from anharmonicity, carrier damping, structural relaxations or other possible mechanisms. The light scattering intensity and selection rules give insight into the fundamental nature of the coupling of the phonons to the electronic states responsible for the optical properties and can be related to the characteristics of a variety of piezo-optical and acousto-optical devices [Ref. 8.2, Chaps. 6 and 7].

6.1. Background on Brillouin Scattering

Typically Brillouin scattering is complementary to conventional ultrasonics techniques for the study of acoustical properties. Usually higher frequency phonons with higher attenuations are accessible by Brillouin scattering, smaller volumes may be probed, and often it is more convenient experimentally to contact the sample optically than piezoelectrically. Though historically Brillouin scattering has been confined to transparent media, the effect has now been observed in semiconductors above the bandgap and there is expectation that techniques will be improved for the study of metals.

The properties of semiconductors to be discussed in this chapter will be the determination of acoustic velocities and anharmonicities, the interaction of acoustic waves with free carriers, the effect of opacity on the scattered light spectral distribution, the resonance effects with near-bandgap light, and stimulated Brillouin scattering. Light scattering

* This work was sponsored by the Department of the Air Force.

from thermally excited phonons will be emphasized, though some pertinent studies with piezoelectrically and acoustoelectrically transduced ultrasonic waves will be included. To start, the scattering kinematics, the photoelastic coupling rules and some typical research apparatus will be reviewed briefly.

6.1.1. Kinematics, Sound Velocity, Phonon Lifetime

The kinematics of the scattering process follow directly from the conservation of frequency (energy) and wavevector (momentum)

$$\omega_s = \omega_i \pm \omega_q , \tag{6.1}$$

$$\boldsymbol{k}_s = \boldsymbol{k}_i \pm \boldsymbol{q} , \tag{6.2}$$

between the incident (i) and scattered (s) photons and the phonon (q). The + sign indicates that the phonon is absorbed (anti-Stokes shift); the − sign that the phonon is emitted (Stokes shift). Since $k_{i,s} = n_{i,s}\omega_{i,s}/c$, where $n_{i,s}$ is the refractive index for the frequency and polarization of the respective light waves, and $q = \omega_q/v_q$ where v_q is the acoustic phase velocity, the conservation laws may be solved for the Brillouin shift

$$\omega_q \cong \omega_i (v_q/c) \, [(n_i - n_s)^2 + 4 n_i n_s \sin^2(\theta/2)]^{\frac{1}{2}} . \tag{6.3}$$

Here θ is the scattering angle, $\cos^{-1}(\hat{k}_i \cdot \hat{k}_s)$. Classically these scattering relations are identical to Bragg reflection from a grating of spacing $2\pi/q$ moving with velocity v_q.

Equation (6.3) neglects corrections to higher order in (v_q/c). For example, the exact expression for backscattering, $\theta = \pi$, from (6.1) and (6.2) is

$$\omega_{q\pm} = \omega_i (v_q/c) \, (n_i + n_s) \, [1 \mp (n_s v_q/c)]^{-1} \tag{6.4}$$

with the anti-Stokes shift, ω_{q+}, slightly greater than the Stokes shift, ω_{q-}. This asymmetry in the frequency shifts has been predicted by several authors but it has yet to be experimentally resolved [Ref. 8.2, Chap. 7]. Another asymmetry may occur because of the dispersion of n_s between the Stokes and anti-Stokes frequencies. Semiconductors are good candidates for observing these asymmetries because of their high refractive indices and their large dispersion, particularly near resonance.

The spectral shape of the Brillouin scattered light with monochromatic light incident on a transparent crystal contains contributions from the instrumental resolution function, the scattering geometry, the velocity anisotropy, and the phonon lifetime. The influence of the geometry on lineshape arises from the finite solid angle, $\delta\theta^2$, collecting a range of frequencies, $\delta\omega_q^\theta$, as seen from the explicit angular dependence in (6.3) and in the implicit angular variation due to velocity anisotropy. To second order in $\delta\theta$ this frequency spread is given by

$$\delta\omega_q^\theta/\omega_q = \left(\frac{A}{2} + \frac{1}{2}\cot\frac{\theta}{2}\right)\delta\theta + \left(\frac{B}{4} - \frac{1}{8}\right)\delta\theta^2 . \tag{6.5}$$

Here it has been taken that $n_i = n_s$ in (6.3) and the anisotropic velocity surface has been expressed as $\delta v_q/v_q = A\delta\theta' + B\delta\theta'^2$ where the range of phonon directions, $\delta\theta'$, is half that of the scattered light. $\delta\theta$. The frequency spread from the term linear in $\delta\theta$ usually dominates the natural linewidth due to phonon damping except for backscattering ($\theta = \pi$) at a velocity surface extremum ($A = 0$).

The normalized spectral lineshape contributed by the phonon lifetime τ_q is a Lorentzian

$$S(\omega_s, q) = \Gamma_q/\pi\left[(\omega_s - \omega_i \mp \omega_q)^2 + \Gamma_q^2\right], \tag{6.6}$$

where the exponential damping constant $\Gamma_q = \tau_q^{-1} = \alpha_q v_q$, and α_q is the sound attenuation coefficient. Various damping mechanisms may be studied by measuring this natural linewidth using high-resolution techniques.

6.1.2. Photoelastic Constants [Ref. 8.1, Sects. 2.2.12 and 2.3.4]

The coupling between light and sound is phenomenologically described by the photoelastic or Pockels tensor [6.1], which gives the overall intensity and polarization properties of the scattering. Benedek and Fritsch [6.2], and Born and Huang [6.3] dealt thoroughly with the scattering cross-sections in cubic crystals, and Landau and Lifshitz [6.4] treated the case of amorphous solids.

Generally the power P_s of light scattered into a solid angle $\partial\Omega$ in the frequency interval $\partial\omega_s$ is related to the Brillouin scattering coefficient σ_B and the incident power P_i according to $\partial^2 P_s/\partial\Omega\,\partial\omega_s = \sigma_B P_i L S(\omega_s, q)/2$, where L is the length of the scattering region. The Brillouin scattering coefficient is given by

$$\sigma_B = \frac{\pi^2}{\lambda_s^4}\frac{n_s}{n_i}(n_q + \tfrac{1}{0})\frac{\hbar\omega_q}{\varrho v_q^2}|\boldsymbol{m}|^2 \sin^2\varphi . \tag{6.7}$$

Here $\lambda_s = 2\pi/k_s$; n_q is the phonon mode occupation number, $n_q + 1$ for anti-Stokes and n_q for Stokes; ϱ is the density and $\boldsymbol{m}$ is a vector in the direction of the dipole moment induced by the interacting sound wave and incident light field $\hat{\boldsymbol{E}}_i$

$$\boldsymbol{m} = \overleftrightarrow{\varepsilon} \cdot \overleftrightarrow{p} \cdot \overleftrightarrow{\varepsilon} \,\vdots\, \hat{E}_i \hat{q} \hat{u}_q \; ; \tag{6.8}$$

where $\boldsymbol{u}$ is the displacement vector of the acoustic wave, and $\overleftrightarrow{\varepsilon}$ and $\overleftrightarrow{p}$ are the optical dielectric and Pockels tensors of the medium. The contracted product $\overleftrightarrow{\varepsilon} \cdot \overleftrightarrow{p} \cdot \overleftrightarrow{\varepsilon}$, a fourth-rank tensor, is contracted with the three unit vectors $\hat{E}_i, \hat{q}, \hat{u}_q$ (symbol $\vdots$) so as to obtain the vector $\boldsymbol{m}$. The angle φ is $\arccos(\hat{m} \cdot \hat{k}_s)$. For thermal scattering $n_q = [\exp(\hbar\omega_q/kT) - 1]^{-1}$. Usually $kT \gg \hbar\omega_q$, so $(n_q + 1) \cong (n_q) = kT/\hbar\omega_q$ and the Stokes and anti-Stokes intensities are the same. For piezoelectrically or acoustoelectrically transduced sound, n_q can be orders of magnitude higher than the thermal value; and either Stokes or anti-Stokes shifts can be selected by choice of a traveling, rather than a standing, wave geometry.

The symmetrized Pockels tensors given in standard texts such as NYE [6.1] are based on strain-optical coupling. NELSON and LAZAY [6.5] have shown that the rotational components of transverse acoustic waves yield an additional antisymmetrical tensor coupling in birefringent crystals. NELSON and LAX [6.6] gave the phenomenological theory of this effect along with the form of the antisymmetrical Pockels tensors in the various crystal classes and estimates of their magnitudes. Microscopic model calculations of the Pockels coefficients for the Brillouin scattering coefficient illustrate the intermediary role played by the electronic states; these will be discussed in the Section 6.4 in connection with resonance scattering.

6.1.3. Apparatus [Ref. 8.2, Sect. 6.2]

The experimental apparatus for Brillouin scattering from thermal phonons will be discussed. Scattering from acoustoelectrically or piezoelectrically driven ultrasonic waves is easily observed with lasers or conventional light sources, and the angular dependence establishes either the sound velocity or frequency if the other is known. A particularly simple, but versatile, apparatus of this kind was used by GARROD and and BRAY [6.7] to study resonant Brillouin scattering in GaAs from acoustoelectric domains. Thermal Brillouin scattering is much weaker, so laser sources are generally used with interferometric spectral analysis. High-resolution low-light-level Brillouin spectrometers have been well developed for thermal scattering from transparent media [6.8, 9]. In semiconductors modifications are necessary to observe the Brillouin-

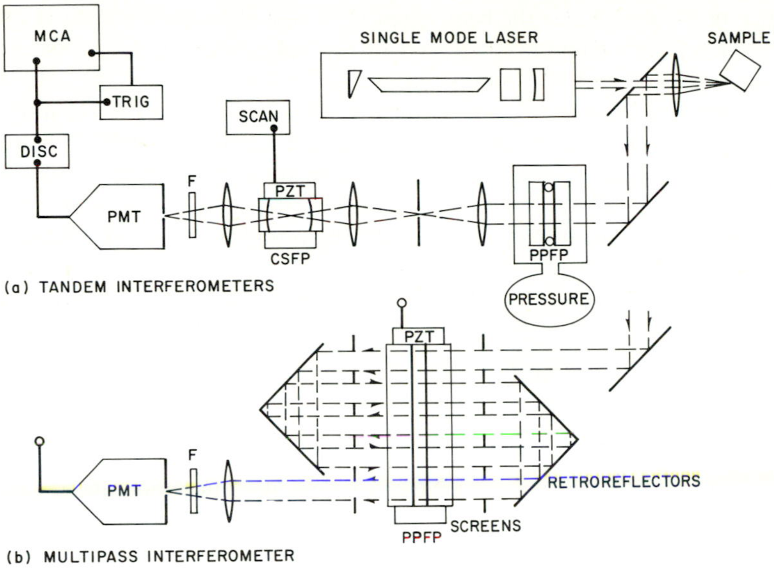

Fig. 6.1a and b. High contrast thermal Brillouin spectrometers

shifted light in the stronger unshifted scattered background due to crystal surfaces and defects. Two high contrast spectrometers developed for this purpose are illustrated in Fig. 6.1. The tandem interferometer system consisting of a high-resolution confocal spherical Fabry-Perot (CSFP) and a plane parallel Fabry-Perot (PPFP) prefilter was used by PINE [6.10] to analyze below-bandgap light in CdS. The free spectral range of the prefilter is adjusted for overlap between the Stokes and anti-Stokes shifts of adjacent interorders, and the filter is pressure-tuned to this overlap. Instrumental and geometry-induced line-broadening is minimized by back-scattering with a single mode laser. Multiscanned collection of data in a multichannel analyzer (MCA) compensates for laser or interferometer drift as discussed by DURAND and PINE [6.9].

The multipassed interferometer system in Fig. 6.1 was devised, and described in detail, by SANDERCOCK [6.11] to observe scattering of above-bandgap light in Si and Ge. For a single pass finesse F_1 and p passes, he obtained the overall finesse $F_p = F_1/\sqrt{2^{1/p}-1}$, and the contrast $C_p = (2F_1/\pi)^{2p}$. F_1 is slightly degraded from the reflectivity finesse $F_R = \pi\sqrt{R}/(1-R)$ because of lack of surface flatness or parallelism. The acceptance angle α_{FP} for the central fringe of the multipassed Fabry-Perot is $(2\lambda/lF_p)^{\frac{1}{2}}$, where λ is the incident wavelength and l the mirror spacing. The étendue, or light collection efficiency, is proportional to

$\alpha_{FP}^2 \times$ beam area. Since the beam area is at least a factor of p less than the mirror area, the étendue is reduced by the same factor from a single pass interferometer with the same plates coated for a finesse equal to F_p. Because of this reduced light gathering efficiency for a given resolution, the multipass system is better suited for lower resolution studies, whereas the single pass or tandem system is more suitable for high resolution. Of course, where the extreme high contrast is necessary, such as for opaque, translucent or small samples, the multipassed interferometer is unsurpassed.

6.2. Acoustoelectric Effects

6.2.1. Domain Probe

The interaction of phonons with free carriers in semiconductors has been the subject of many Brillouin scattering studies. In particular the technique has been used to probe the dynamical characteristics of acoustoelectric domains. In piezoelectric semiconductors, where the interaction is particularly strong, Hutson and White [6.12] demonstrated that acoustic waves could be amplified by carriers with drift velocity v_d exceeding the sound velocity. When high drift fields are applied, intense packets of acoustic waves are observed to travel along the sample at the sound velocity. These acoustoelectric domains theoretically [6.12] consist of amplified thermal phonons in a narrow band of frequencies around $\omega_{MAX} = (\omega_c \omega_D)^{\frac{1}{2}}$ where $\omega_c = \sigma/\varepsilon$ is the dielectric relaxation frequency, $\omega_D = v_q^2/D$ is the diffusion frequency, σ is the conductivity, and D the diffusion constant of the semiconductor. Zucker and Zemon [6.13] were first to examine the spectrum of these domains in CdS by Brillouin scattering, and they found frequencies an order-of-magnitude lower than ω_{MAX}. They postulated that the initial phonon flux was generated piezoelectrically from the Fourier components of the current pulse, rather than from thermal phonons, and were amplified first to saturation. Many subsequent light scattering studies of acoustoelectric domains showed that the phonon frequency distribution initially peaks near ω_{MAX} and downshifts as the packet propagates through the crystal-probably due to anharmonic parametric conversion. These experiments up to 1970 were reviewed comprehensively by Meyer and Jørgensen [6.14], so they need not be included here. Thereafter, Spears [6.15], in a thorough Brillouin scattering study of acoustoelectric domains in GaAs, established the validity of a generalized Hutson-White theory due to Jacoboni and Prohofsky [6.16] for the $ql_e \sim 1$ regime where l_e is the electron mean-free-path. He also observed the ensuing nonlinear parametric downconversion in the strong flux limit. MANY and GELBART [6.17] later showed that the thermal flux could be uniformly amplified with the Hutson-White spectrum by carefully tailoring the current pulse to avoid domain-inducing shocks. Several recent higher sensitivity acoustoelectric studies of the effects of carriers and fields on the thermal phonon Brillouin shift, linewidth and intensity will be discussed in the next section. Also some experiments on the near-bandgap wavelength dependence of the scattering from acoustoelectric domains will be covered in the section on resonance effects.

6.2.2. Effect on Thermal Phonons

A critical test of the small-signal theories of the acoustoelectric interaction comes in the study of the carrier and field effects on the thermal phonons. To this end thermal Brillouin scattering has been employed by several workers. PINE [6.10] measured the velocity and attenuation of 35 GHz, c-axis, longitudinal acoustic waves in CdS as a function of temperature in high and low conductivity samples using the tandem spectrometer illustrated in Fig. 6.1. A spectrum from the high σ sample at 95 K is shown in Fig. 6.2. The acoustoelectric and the anharmonic contributions to the Brillouin linewidth—hence the acoustic attenuation—are distinguished by their dependence on temperature and electron concentration, as shown in Fig. 6.3. The anharmonic damping exhibited by the low σ sample increases monotonically with temperature similar to that observed previously in α-quartz. Also the velocity, which is proportional

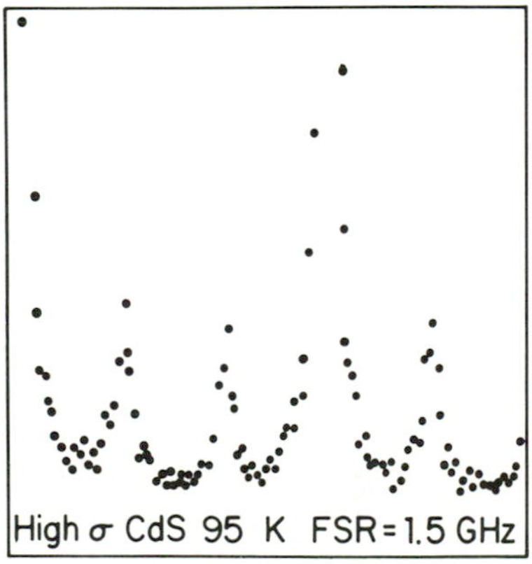

Fig. 6.2. High resolution Brillouin data; backscattering from c-axis LA phonons in CdS ($\sigma_0 = 0.07$ mho/cm) with 6328 Å light. The free spectral range of the interferometer is 1.5 GHz

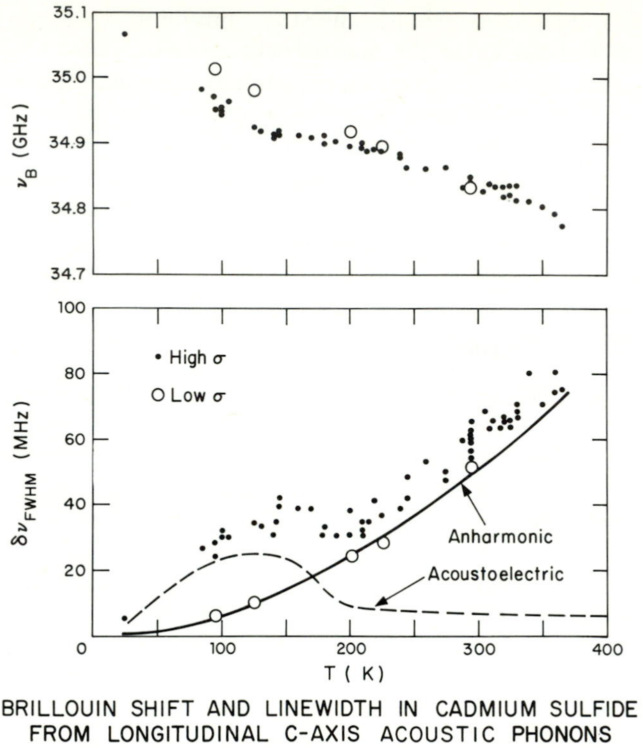

Fig. 6.3. Temperature dependence of Brillouin shift and linewidth in high and low conductivity CdS

to the Brillouin shift, is measurably lower in the temperature range where the electrons can screen the piezoelectric stiffening of the elastic constants.

The acoustoelectric theory of Hutson and White [6.12] for the velocity relaxation and phonon damping is implicitly restricted to $q l_e \ll 1$ since they assumed a local relation between currents and fields. Several authors [6.16, 18, 19] have used a Boltzmann-equation formulation to generalize the theory for all $q l_e$. The resultant theory with no external fields is

$$(v_q - v_0)/v_0 = -K_{dp}^2 + (K_{pe}^2 + K_{dp}^2)\,\mathrm{Re}\{H\}\,, \tag{6.9}$$

$$\Gamma_q/\omega_q = \alpha_q v_q/\omega_q = -(K_{pe}^2 + K_{dp}^2)\,\mathrm{Im}\{H\}\,, \tag{6.10}$$

$$H = [1 + i f\,\omega_q/\omega_D]/[1 + i(\omega_c/\omega_q + f\,\omega_q/\omega_D)]\,, \tag{6.11}$$

were v_0 is the low frequency ($\omega_q \ll \omega_c$) sound velocity with the piezoelectric stiffening totally screened by the free carriers, and f denotes the untrapped fraction of the space charge. The electromechanical coupling constants for piezoelectric and deformation potential interaction are given by

$$K_{pe}^2 = d_{33}^2/2\varepsilon_3 c_{33}, \tag{6.12}$$

$$K_{dp}^2 = \varepsilon_3 q^2 \chi^2/2e^2 c_{33}. \tag{6.13}$$

The piezoelectric constant d_{33}, the *dc* dielectric constant ε_3, and the elastic constant c_{33} are pertinent to the *c*-axis LA phonon in this case. For CdS $K_{pe}^2 = 0.012$; so for a reasonable deformation potential $\chi \sim 10$ eV, $K_{dp}^2 \sim 10^{-2} K_{pe}^2$ and piezoelectric coupling dominates.

The dielectric relaxation frequency ω_c and the diffusion frequency ω_D are given in terms of the frequency- and wavevector-dependent conductivity by

$$\omega_c = \sigma(\omega_q, q)/\varepsilon_3, \tag{6.14}$$

$$\omega_D = v_q^2/D(\omega_q, q), \tag{6.15}$$

with

$$D(\omega_q, q) = kT\sigma(\omega_q, q)/[n_e e^2(1 - i\omega_q \tau_e)]. \tag{6.16}$$

Here n_e is the free carrier concentration and τ_e is the electron lifetime obtained from the measured crystal mobility $\mu = e\tau_e/m^*$, where m^* is the effective mass. The electron mean-free-path for nondegenerate statistics is given by

$$l_e = \tau_e(2kT/m^*)^{\frac{1}{2}}. \tag{6.17}$$

Defining a dimensionless parameter $x = (1 - i\omega_q \tau_e)/q l_e$, the generalized conductivity may be written

$$\sigma(\omega_q, q) = \sigma_0(2x/q l_e)[1 - \pi^{\frac{1}{2}} \times F(x)], \tag{6.18}$$

where $F(x)$ is given by the plasma-dispersion function [6.20]

$$F(x) = 2\pi^{-\frac{1}{2}} \exp(x^2) \int_x^{\infty} \exp(-t^2)\, dt\,. \tag{6.19}$$

The Hutson-White limit is reached by setting $ql_e \to 0$ above, with the result [6.20] that $\sigma(\omega_q, q) \to \sigma_0 = n_e e \mu$ is the dc conductivity.

It was found [6.10] that the acoustoelectric contributions to the phonon damping in Fig. 6.3 could not be explained by the above theory with the expected $f=1$ corresponding to no trapping. A temperature-dependent trapping model was required to fit the data. Since the trapping at these high frequencies is not directly measurable, but is expected to be negligible, the discrepancy is significant and may be due to the single-relaxation time assumption for the Boltzmann collision integral used in the theory. JACOBONI and PROHOFSKY [6.21] have extended the theory of acoustoelectric damping by using energy-dependent relaxation times. However, only the average mobility lifetime has been measured, so it is difficult to confirm any distributed τ_e model for application to this experiment.

WAKITA et al. [6.22] have also studied thermal Brillouin scattering in CdS. They determined three elastic constants, c_{11}, c_{44}, and c_{12}, by single-pass interferometer analysis of 90° scattered 6328 Å light. Their resolution was insufficient to measure velocity relaxation or acousto-electric damping. However, WAKITA et al. [6.23] also observed the acoustoelectric build-up of the phonon energy density from the thermal flux when high fields were applied. The acoustoelectric flux could then be determined relative to the thermal flux since the light scattering efficiency was obtained from thermal phonons in the same geometry. SMITH [6.24] accomplished the same experiment somewhat earlier and was able to follow the build-up of flux from the thermal level for at least six decades.

As an extension of his previous work, SMITH [6.25] painstakingly showed, with low drift fields, that not only does the thermal flux increase for $\boldsymbol{q} \parallel \boldsymbol{v}_d$, but it decreases for $\boldsymbol{q}$ anti-$\parallel \boldsymbol{v}_d$. The latter arises since the electronic component of the acoustic attenuation is increased for carriers drifting opposite to the phonon propagation. This is a difficult experiment because of extraneous asymmetries introduced by interferometer mirror misalignment and curvature.

Before closing this section on acoustoelectric effects, it should be noted that free carriers also scatter light—either as single particles or as collective plasmons [when $\omega_p \tau_e > 1$ where the plasma frequency $\omega_p = (4\pi n_e e^2/m^* \varepsilon)^{\frac{1}{2}}$] [6.26, 27]. Scattering from the bunched carriers in acoustoelectric domains may be distinguished from the phonons by polarization selection rules or wavelength-dependence and may provide an additional tool for the study of electron-phonon interaction in semiconductor.

6.3. Effect of Opacity on Lineshape

In opaque crystals, a broadening of the Brillouin linewidth may occur in excess of the instrumental, geometry and lifetime effects discussed earlier. This broadening is due to the distribution of wavevectors associated with the Fourier transform of the exponentially absorbed incident and scattered light. For backscattering at the boundary of a semi-infinite medium having complex refractive indices $n_{i,s} = \eta_{i,s} + i\varkappa_{i,s}$ the distribution of wavevectors transferred to the phonons is [6.27a–d] (see, however, [Ref. 8.2, Sect. 6.3.1])

$$S(q) = 4qq_0\alpha_0/\pi[(q^2 - q_0^2 - \alpha_0^2)^2 + (2q\alpha_0)^2]\,. \tag{6.20}$$

Here the optical propagation vector and absorption constant are given by $q_0 = 2\pi(\eta_i/\lambda_i + \eta_s/\lambda_s)$ and $\alpha_0 = 2\pi(\varkappa_i/\lambda_i + \varkappa_s/\lambda_s)$. This wavevector distribution is directly related to the frequency spread by the velocity of sound. For the low absorption case $\varkappa_{i,s} \ll \eta_{i,s}$ and for $q \sim q_0$, (6.20) reduces to the simple Lorentzian $S(q) \sim \alpha_0/\pi[(q-q_0)^2 + \alpha_0^2]$ so that the excess halfwidth is $\delta\omega_0 \cong \alpha_0 v_q$.

The clearest demonstration of this opacity broadening has been given by SANDERCOCK [6.28] for Brillouin scattering well above the bandgaps of Si and Ge where $\varkappa \leqq \eta$. His spectra, taken with the multipass interferometer, are shown in Fig. 6.4. SANDERCOCK obtained accurate values for the complex refractive indices in Si and Ge from these spectra since the sound velocities are well known. These values are in general agreement with more conventional reflectivity and transmission measurements. Note that the asymmetric lineshape given by (6.20) fits the data of SANDERCOCK better than the symmetric Lorentzian when $\varkappa \sim \eta$.

PINE [6.29] also observed optical absorption broadening in CdS with light slightly below the bandgap. Spectra using 5145 Å light at temperatures of 295 K and 203 K, where the crystal is, respectively, absorbing and transparent, are shown in Fig. 6.5. The broadening, evident at higher temperature with this resolution, yields an absorption constant twice as high as the direct transmission measurements on thin platelets by DUTTON [6.30]. This agreement is reasonable allowing for sample and surface preparation variations and for possible laser-induced local heating and carrier generation.

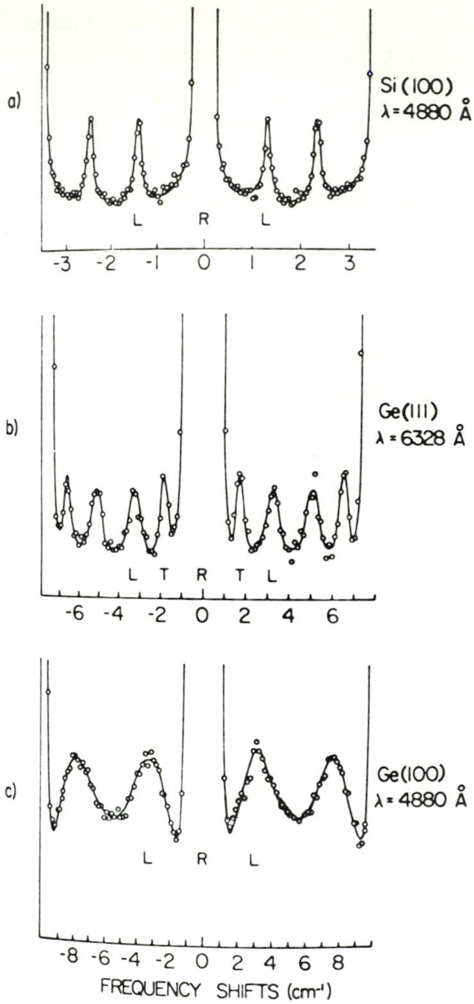

Fig. 6.4a—c. Opacity-broadened Brillouin peaks in Si and Ge

6.4. Resonance Scattering Effects

Brillouin scattering is expected to be resonantly enhanced when the incident or scattered light approaches the electronic transitions in the medium. This enhancement has been observed for light below the band-gap of several semiconductors. The theories of this effect, analogous to those developed for resonant Raman scattering in earlier chapters, will

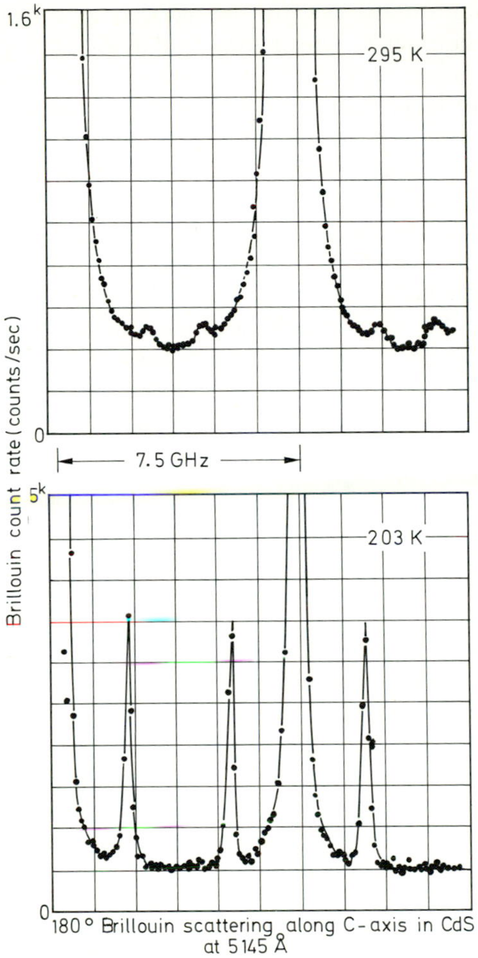

Fig. 6.5. Near-resonance Brillouin backscattering in CdS with 5145 Å light

be summarized in this section. The experimental results in the next section will be shown to be in qualitative agreement with theory though the enhancement seems to be less than predicted and certainly is weaker than observed for resonance Raman scattering.

6.4.1. Theoretical Scattering Coefficients

Several models of varying sophistication exist to explain the frequency dependence of Brillouin scattering. The simplest of these assumes that the light couples only to the density fluctuations accompanying the LA phonon. Then the Pockels tensor may be calculated from the Lorentz-Lorenz law as given by $|\overleftrightarrow{\varepsilon}\cdot\overleftrightarrow{\boldsymbol{p}}\cdot\overleftrightarrow{\varepsilon}| \sim (\varepsilon-1)(\varepsilon+2)/3$. This approximation is appropriate for incident and scattered light of frequencies well below the optical absorption edges in liquids and solids. However, it underestimates the resonance behavior near the bandgap since the $\mathrm{Re}\{\varepsilon\}$ is dominated by higher energy transitions whereas the scattering is influenced strongly by the effect of strains on the nearby bandgap. Naturally the Lorentz-Lorenz derivative with respect to density also fails to predict piezobirefringence effects or scattering from shear waves.

A more detailed microscopic formulation of the light-scattering problem in crystals has been given by LOUDON [6.31]. He obtained

$$\sigma_{\mathrm{B}} = \left(\frac{e}{\hbar m c}\right)^4 \frac{kT}{2\varrho v_{\mathrm{q}}^2} \frac{\omega_{\mathrm{s}}}{\omega_{\mathrm{i}}} |R_{\mathrm{is}}^{jq}|^2 , \tag{6.21}$$

$$R_{\mathrm{is}}^{jq} = \frac{1}{V} \sum_{\mu\mu' k} \frac{\langle 0|p_{\mathrm{i}}|\mu k\rangle \, \chi_{\mu\mu'}^{jq} \langle \mu' k|p_{\mathrm{s}}|0\rangle}{(\omega_{\mu k}-\omega_{\mathrm{i}})(\omega_{\mu' k}-\omega_{\mathrm{s}})} . \tag{6.22}$$

This is a perturbation-theory description of a process whereby a photon (ω_{i}, $\boldsymbol{k}_{\mathrm{i}}$) incident on a crystal in ground state $|0\rangle$ creates a virtual electron-hole pair in band μ with wavevector $\boldsymbol{k}$. The electron or hole then interacts with the phonon (ω_{q}, $\boldsymbol{q}$, branch j) via its deformation potential $\chi_{\mu\mu'}^{jq}$, possibly changing its state to μ', and subsequently recombines emitting a shifted scattered photon (ω_{s}, $\boldsymbol{k}_{\mathrm{s}}$). The photon and phonon wavevectors are neglected in comparison with those of the electron-hole pair (vertical transitions). Here V is the crystal volume, $\langle|p|\rangle$ is the momentum matrix element, and $\hbar\omega_{\mu k}$ is the energy of the pair state above $|0\rangle$. This definition of the deformation potential is related to LOUDON's by $\chi = qa\Xi$. The resonance effects arise from the denominator of (6.22) as the optical frequencies approach the excited electronic states. Terms with denominators of the form, $(\omega_{\mu k}+\omega_{\mathrm{i,s}})$, have been omitted.

Taken by itself, LOUDON's theory contains enough flexibility in the band parameters of (6.22) to explain the resonance Brillouin data to be given later. However, many of these parameters are specified by the optical absorption and it is felt that the details of the theory are less important than the relationship between the scattering cross-section and the absorption. The absorption as given by SEITZ [6.32] is

$$\alpha_{\mathrm{i}}(\omega_{\mathrm{i}}) = \frac{2\pi e^2 \omega_{\mathrm{i}}}{\hbar m^2 c \eta_{\mathrm{i}}(\omega_{\mathrm{i}})} \frac{1}{v} \sum_{\mu k} \frac{|\langle 0|p|\mu k\rangle|^2 (\gamma_{\mu k}/\omega_{\mu k}^2)}{(\omega_{\mu k}-\omega_{\mathrm{i}})^2+\gamma_{\mu k}^2} . \tag{6.23}$$

Here $\gamma_{\mu k}$ is the electronic damping and $\eta_i(\omega_i)$ varies slowly over the frequency range of interest. Again the major frequency dependence comes from the resonance denominator. It is clear from (6.22) and (6.23) that if $(\omega_{\mu k}-\omega_i)$ is much greater than ω_q and $\gamma_{\mu k}$ then $\sigma_B \propto \alpha_i^2$ if $\chi_{\mu\mu}^{jq} \neq 0$ (intraband electron-phonon scattering allowed), or $\sigma_B \propto \alpha_i$ if $\chi_{\mu\mu}^{jq}=0$ (interband electron-phonon contribution). These are the two and three band terms discussed in Chapter 2. Experimentally, however, the Brillouin scattering coefficient can increase sublinearly with the absorption. This is readily evident in Fig. 6.5 where the net scattering intensity decreases as the bandgap is temperature-tuned close to the exciting light. The apparent divergence of σ_B as $\omega_i \to \omega_{\mu k}$ does not occur, because of the vanishing density of states, as shown by LOUDON [6.33] for a spherical band model. PINE [6.29] showed that the spherical band model preserves the relation between σ_B and α_i for the intraband scattering as indicated above.

More recently, BURSTEIN et al. [6.34] have extended LOUDON's theory of resonance Brillouin scattering to excitonic insulators. They treated the strongly coupled light and the quasilocalized electron-hole pair as a polariton and considered the scattering of polaritons by phonons. Their expression for the scattering efficiency may be written

$$\sigma_B = \left(\frac{\omega_s}{2\pi\hbar}\right)^2 \frac{kT}{2\varrho v_q^2} \frac{|M_{is}^{jq}|^2}{v_p^2(\omega_s)\, v_g(\omega_s)\, v_g(\omega_i)}, \tag{6.24}$$

$$M_{is}^{jq} = \frac{1}{2} \sum_{\mu\mu'} \left(1 + \frac{\omega_\mu \omega_{\mu'}}{\omega_s \omega_i}\right) S_\mu^{\frac{1}{2}}(\omega_i)\, \chi_{\mu\mu'}^{jq}\, S_{\mu'}^{\frac{1}{2}}(\omega_s), \tag{6.25}$$

where $v_p(\omega)$ and $v_g(\omega)$ are the phase and group velocities of the polariton at frequency ω, and $S_\mu(\omega)$ is the exciton strength of the polariton. These are given by

$$\frac{v_p(\omega)\, v_g(\omega)}{c^2} = \left[1 + \sum_{\mu'} \frac{4\pi\beta_{\mu'}\omega_{\mu'}^4}{(\omega_{\mu'}^2-\omega^2)^2}\right]^{-1}, \tag{6.26}$$

$$S_\mu(\omega) = \frac{4\pi\beta_\mu \omega_\mu^3 \omega}{(\omega_\mu^2-\omega^2)^2} \left[\frac{v_p(\omega)\, v_g(\omega)}{c^2}\right], \tag{6.27}$$

where the oscillator strength β_μ may be written in terms of the momentum matrix elements as

$$\beta_\mu = \left(\frac{2e^2 N}{m^2 \hbar V}\right) \frac{|p_{0\mu}|^2}{\omega_\mu^3}. \tag{6.28}$$

LOUDON's perturbation theory [6.31] in the limit of narrow dispersionless bands (such that $\Sigma_k \to N$) reduces to this polariton theory, apart from a factor $v_p(\omega_i)/v_p(\omega_s)$ which is always close to unity. Therefore there is no distinct difference in resonance behavior between the two theoretical approaches. On the other hand, if one considers the electron-phonon coupling via the electric field accompanying a polar or piezoelectric phonon, then it is useful to distinguish between exciton and band states. This gives rise to the electro-optical contribution to the light scattering which is usually dominated by excitonic intermediaries. The relative resonant Raman scattering of longitudinal-optic (LO) and transverse-optic (TO) phonons in CdS attests to this [6.35]. However, BURSTEIN et al. [6.34] showed that the piezoelectric field strength of the LA phonon in CdS is only a few percent that of the field of the LO phonon, so the electro-optic Brillouin scattering terms are not expected to be large. Experimentally [6.29] one observes the same resonance-scattering effects from LA phonons along either the piezo inactive *a*-axis or the piezo active *c*-axis in CdS, affirming this expectation.

6.4.2. Experimental Scattering Coefficients

There have been three types of Brillouin scattering experiments exhibiting resonant dispersion of the scattering coefficient near the bandgap of semiconductors. The first measurements were by TELL et al. [6.36] using small angle diffraction by transducer driven 50 MHz LA phonons along the *a*-axis of CdS and ZnO. There, incident light from a monochromator filtered xenon arc lamp was tuned over several thousand Å to the transparency limit of the crystals. Over this range they obtained almost a decade of dispersion in the p_{31} and p_{21} tensor configurations and less in the p_{11}.

The second measurement was by thermal Brillouin backscattering from LA phonons along the *c*-axis in CdS by PINE [6.29]. Here the bandgap was thermally tuned well into the absorption tail for 5145 Å argon ion laser light, and a large enhancement of the scattering coefficient was observed. The scattering coefficient was obtained from the scattered intensity data by correcting for the absorption of the optical path,

$$I_s/I_i = \sigma_B[1 - \exp(-2\alpha_i L)]/2\alpha_i\,. \tag{6.29}$$

For transparent samples then $I_s/I_i \sim \sigma_B L$ whereas for opaque samples $I_s/I_i \sim \sigma_B/2\alpha_i$. In the latter case if $\sigma_B \propto \alpha_i$ (or α_i^2) as in LOUDON's theory for interband (or intraband) electron-phonon coupling, then the measured intensity would be independent of (or proportional to) the absorption. Instead it was found that σ_B was roughly proportional to $\alpha_i^{\frac{1}{2}}$, as shown

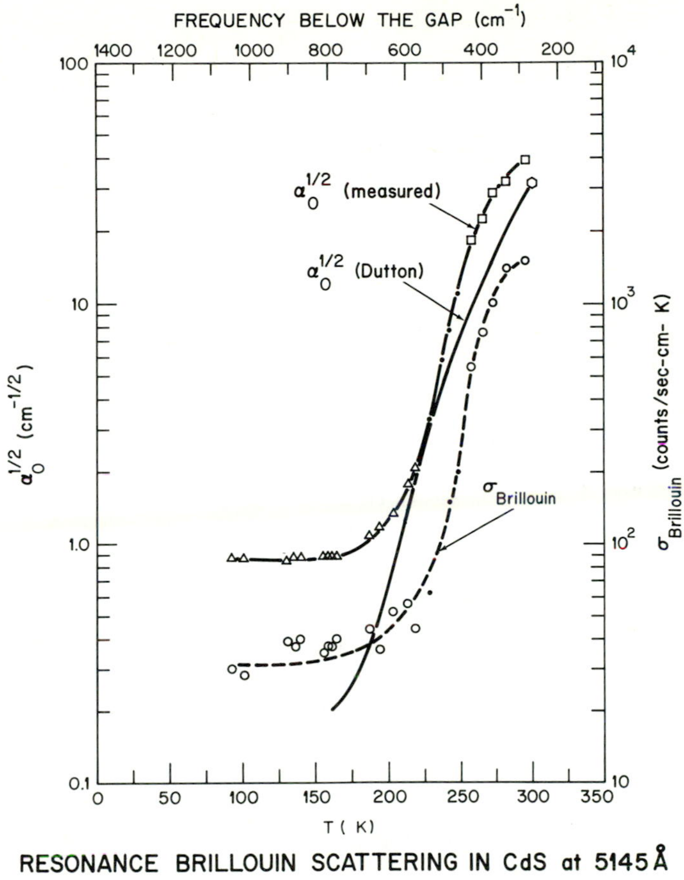

Fig. 6.6 Optical absorption and resonance Brillouin scattering coefficient in CdS

in Fig. 6.6. The scattering coefficient plotted there is also linearly corrected for temperature according to (6.7), (6.21), or (6.24) since the bandgap is thermally tuned. The scale marked "frequency below the gap" is obtained from $E_g(T) = (20840 - 3.82\,T)\,\text{cm}^{-1}$, as measured by DUTTON [6.30].

The third set of resonant Brillouin experiments involved scattering of monochromator filtered arc lamp light from transverse acoustic phonons in acoustoelectric domains. GARROD and BRAY [6.7] found a sharp dip in the scattering dispersion about 30 meV below the gap in GaAs, as shown in Fig. 6.7. This minimum was attributed to a cancellation between contributions to (6.22) from the resonant band and non-

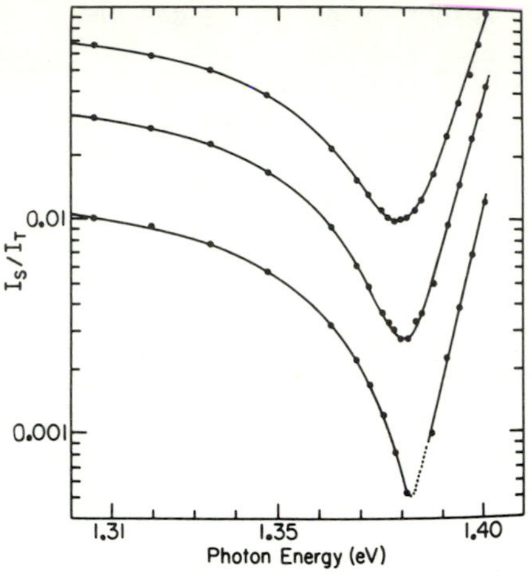

Fig. 6.7. Brillouin scattering dispersion curves from acoustoelectrically generated shear waves in GaAs. Higher curves for higher acoustic flux

resonant states with opposite sign. However, the residual scattering at the minimum was not zero and in fact depended on the acoustic intensity. It is believed that spatial inhomogeneities in the bandgap energy caused by non-uniform acoustic flux washed out the expected null. The dispersion of the Brillouin scattering coefficient closely followed piezobirefringence data on p_{44} available for energies below 1.38 eV. Note that the data in Fig. 6.7 is normalized to the light transmitted by the sample. This automatically compensates for the absorption along the optical path. However, I_s/I_t neves exceeds unity and eventually both disappear as the bandgap is approached. This implies that the resonant portion of the Brillouin scattering coefficient increases less slowly than the absorption just as for $|p_{33}|^2$ in CdS, YAMADA et al. [6.37] observed a similar resonant cancellation in the Brillouin dispersion for $|p_{44}|^2$ in CdS using acoustoelectrically generated shear waves.

6.4.3. Predictions of New Modes

The exciton-polariton description of Brillouin scattering [6.34] has lead BRENIG et al. [6.38] to predict some possible new scattering modes above a free exciton resonance. A free exciton has a finite effective mass; so the coupled photon-exciton polariton exhibits spatial dispersion, as illustrated schematically in Fig. 6.8. In the absence of coupling, the exciton frequency at zero wavevector is ω_{1s}. Below this resonance ordinary Brillouin scattering occurs designated by the lower $K_2 \rightarrow K_2'$

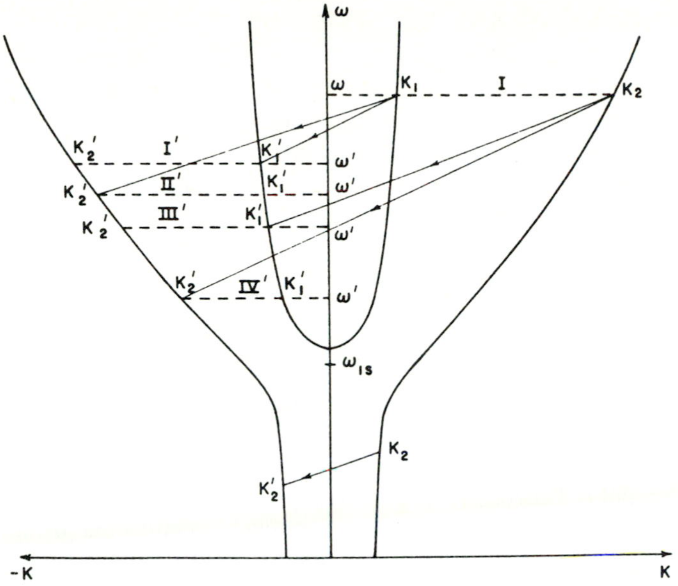

Fig. 6.8. Dispersion curves for photon-free exciton polaritons illustrating Brillouin backscattering

line. Above the resonance an incident photon at frequency ω can scatter into four ω' channels according to the wavevector pairs $K_1 \rightarrow K_1'$, $K_1 \rightarrow K_2'$, $K_2 \rightarrow K_2'$, and $K_2 \rightarrow K_1'$; the last three combinations are the new modes involving the exciton-like branch. Only the Stokes shift for backscattering from a single phonon branch is shown; the slopes of the arrowed lines indicate the phonon phase velocities. Observation of these new modes could provide a means of measuring the acoustic or the exciton dispersion at large wavevector assuming that one of the other is known from independent measurements.

The scattering cross section for each of these channels involves the relative admixture of the photon and exciton strengths at the initial and final polariton states such as given in (6.27, 26) for a bound exciton. Brenig et al. [6.38] calculated that the cross section for the new modes should be observable very near a sharp free-exciton resonance.

It should be noted that the dispersion of Fig. 6.8 kinematically allows forward scattering between polariton branches or on a single branch near $v_g(\omega) = v_q$. Such scattering could probably not be observed except for very thin crystals because of optical absorption. Similarly the backscattering processes may not be observable if the cross section increases sublinearly with absorption as seen in previous resonant Brillouin experiments. These modes have recently been observed [Ref. 8.2, Chap. 7].

6.5. Stimulated Brillouin Scattering

Under intense optical excitation the Brillouin scattered Stokes light and the acoustic wave may be exponentially amplified by a parametric process. This effect, known as stimulated Brillouin scattering, was first observed in α-quartz and sapphire by CHIAO et al. [6.39]. The exponential gain factor given by their steady state theory [6.39] for backscattering is

$$g_{\mathrm{B}} = 2\pi^2 |\overleftrightarrow{\varepsilon} \cdot \overleftrightarrow{p} \cdot \overleftrightarrow{\varepsilon}|^2 / n_{\mathrm{i}} \lambda_{\mathrm{s}}^2 \varrho v_{\mathrm{q}} \Gamma_{\mathrm{q}} \,, \tag{6.30}$$

so it is seen to relate to the Brillouin scattering coefficient σ_{B} and the phonon damping. Usually longitudinal acoustic waves have the highest gain and are most strongly excited. Under transient pulse excitation, where the pulse time is less than Γ_{q}^{-1}, the gain is derated.

Though the effect has now been reported in many solids and liquids, semiconductors, which have large Pockels coefficients usually show parasitic heating effects which limit or prevent stimulated Brillouin scattering. However, ASAM et al. [6.40] did observe intensity limiting and stimulated Brillouin backscattering in Ge using Q-switched CO_2 laser radiation. Their measured gain agreed with (6.30) using the Lorentz approximation for the Pockels coefficient and they observed up to 20% conversion of incident light to Stokes radiation. The stimulated acoustic flux was estimated to be $\sim 500\ \mathrm{W/cm^2}$ though it was not directly measured.

KRESSEL and MIEROP [6.41] have suggested that catastrophic degradation of GaAs injection lasers may be due to stimulated Brillouin scattering. Pitting of the mirror surface was attributed to the strong acoustic flux. The threshold for damage increased with temperature consistent with a higher Γ_{q}, and it increased for shorter pulses indicating transient gain reduction. However, direct observation of the Brillouin shifted light was not made; so the mechanism for damage was not conclusively established.

References

6.1. J. F. NYE: *Physical Properties of Crystals* (Clarendon Press, Oxford, 1964), Chapter 13.
6.2. G. B. BENEDEK, K. FRITSCH: Phys. Rev. **149**, 647 (1966).
6.3. M. BORN, K. HUANG: *Dynamical Theory of Crystal Lattices* (Clarendon Press, Oxford, 1954), Section 50.
6.4. L. D. LANDAU, E. M. LIFSHITZ: *Electrodynamics of Continuous Media* (Pergamon Press, New York, 1958), Chapter 14.

6.5. D. F. NELSON, P. D. LAZAY: Phys. Rev. Letters **25**, 1187 (1970).
6.6. D. F. NELSON, M. LAX: Phys. Rev. B **3**, 2778 (1971).
6.7. D. K. GARROD, R. BRAY: Phys. Rev. B **6**, 1314 (1972).
6.8. S. M. SHAPIRO, R. W. GAMMON, H. Z. CUMMINS: Appl. Phys. Letters **9**, 157 (1966).
6.9. G. E. DURAND, A. S. PINE: IEEE J. Quant. Electron. QE **4**, 523 (1968).
6.10. A. S. PINE: Phys. Rev. B **5**, 2997 (1972).
6.11. J. R. SANDERCOCK: In *2nd Intern. Conf. on Light Scattering in Solids*, ed. by M. BALKANSKI (Flammarion Paris, 1971), p. 9.
6.12. A. R. HUTSON, D. L. WHITE: J. Appl. Phys. **33**, 40 (1962).
6.13. J. ZUCKER, S. ZEMON: Appl. Phys. Letters **9**, 398 (1966).
6.14. N. I. MEYER, M. H. JØRGENSEN: *Advances in Solid State Physics* (Pergamon Press, Vieweg, 1970), Chapter 2.
6.15. D. L. SPEARS: Phys. Rev. B **2**, 1931 (1970).
6.16. C. JACOBONI, E. W. PROHOFSKY: J. Appl. Phys. **40**, 454 (1969).
6.17. A. MANY, U. GELBART: Appl. Phys. Letters **19**, 192 (1971).
6.18. H. N. SPECTOR: In *Solid State Physics*, ed. by F. SEITZ and D. TURNBULL (Academic Press, New York, 1966), vol. 19, p. 291.
6.19. K. W. NILL: Ph. D. dissertation (MIT Electr. Engrng. Dept., 1966), unpublished.
6.20. B. D. FRIED, S. D. CONTE: *The Plasma Dispersion Function* (Academic Press, New York, 1961).
6.21. C. JACOBONI, E. W. PROHOFSKY: Phys. Rev. B **1**, 697 (1970).
6.22. K. WAKITA, M. UMENO, K. TAKAGI, S. MIKI: J. Phys. Soc. Japan **35**, 149 (1973).
6.23. K. WAKITA, M. UMENO, S. HAMADA, S. MIKI: Jap. J. Appl. Phys. **12**, 706 (1973).
6.24. R. W. SMITH: In *1st Intern. Conf. on Light Scattering in Solids*, ed. by G. B. WRIGHT (Springer, New York, Heidelberg, Berlin, 1969), p. 611.
6.25. R. W. SMITH: J. Acoust. Soc. Am. **49**, 1033 (1970).
6.26. A. MOORADIAN, G. B. WRIGHT: Phys. Rev. Letters **16**, 999 (1966).
6.27. J. F. SCOTT, T. C. DAMEN, R. C. C. LEITE, J. SHAH: Phys. Rev. B **1**, 4330 (1970).
6.27a. G. DRESSELHAUS, A. S. PINE: Solid State Commun. **16**, 1001 (1975).
6.27b. A. S. PINE, G. DRESSELHAUS: In *3rd Intern. Conf. on Light Scattering in Solids*, ed. by M. BALKANSKI (Flammarion, Paris, 1975), p. 138.
6.27c. B. I. BENNETT, A. A. MARADUDIN, L. R. SWANSON: Ann. Phys. (N.Y.) **71**, 357 (1972).
6.27d. A. DERVISH, R. LONDON: J. Phys. C **9**, L 669 (1976).
6.28. J. R. SANDERCOCK: Phys. Rev. Letters **28**, 237 (1972).
6.29. A. S. PINE: Phys. Rev. B **5**, 3003 (1972).
6.30. D. DUTTON: Phys. Rev. **112**, 785 (1958).
6.31. R. LOUDON: Proc. Roy. Soc. (London) A **275**, 218 (1963).
6.32. F. SEITZ: *Modern Theory of Solids* (McGraw-Hill, New York, 1940), Chapter 17.
6.33. R. LOUDON: J. Phys. Radium **26**, 677 (1965).
6.34. E. BURSTEIN, R. ITO, A. PINCZUK, M. SHAND: J. Acoust. Soc. Am. **49**, 1013 (1971).
6.35. R. C. C. LEITE, T. C. DAMEN, J. F. SCOTT: In *1st Intern. Conf. on Light Scattering in Solids*, ed. by G. B. WRIGHT (Springer, New York, Heidelberg, Berlin, 1969), p. 359.
6.36. B. TELL, J. M. WORLOCK, R. J. MARTIN: Appl. Phys. Letters **6**, 123 (1965).
6.37. M. YAMADA, K. ANDO, C. HAMAGUCHI, J. NAKAI: J. Phys. Soc. Japan **34**, 1696 (1973).
6.38. W. BRENIG, R. ZEYHER, J. L. BIRMAN: Phys. Rev. B **6**, 4617 (1972).
6.39. R. Y. CHIAO, C. H. TOWNES, B. P. STOICHEFF: Phys. Rev. Letters **12**, 592 (1964).
6.40. P. ASAM, P. DEUFLHARD, W. KAISER: Phys. Letters **27** A, 78 (1968).
6.41. H. KRESSEL, H. MIEROP: J. Appl. Phys. **38**, 5419 (1967).

7. Stimulated Raman Scattering

Y.-R. SHEN

With 24 Figures

Stimulated Raman scattering was accidentally discovered by WOODBURY and NG in 1962 [7.1]. In studying Q-switching of a ruby laser by a nitrobenzene Kerr cell, they detected intense infrared radiation emitted from the Kerr cell, whose origin was not immediately identified. ECKHARDT [7.2] first proposed the correct interpretation as stimulated Raman emissions from nitrobenzene; this was soon thereafter verified experimentally by ECKHARDT et al. [7.3].

Subsequently, a similar effect was observed in many other liquids by ECKHARDT et al. [7.3], GELLER et al. [7.4], and STOICHEFF [7.5], in several solids by ECKHARDT et al. [7.6], and in hydrogen gas by MINCK et al. [7.7]. An early theoretical treatment of stimulated Raman scattering was given by HELLWARTH [7.8].

In Table 7.1 we list a number of Raman lines for some of the materials in which both spontaneous and stimulated Raman scattering have been measured. As seen from this table, one would need a laser beam of 1 GW/cm^2 propagating over a 15-cm nitrobenzene cell in order to generate e^{30} Raman photons from a noise photon. These numbers are

Table 7.1. Frequency shift, linewidth, and scattering cross-section of spontaneous Raman scattering for a number of substances and the corresponding stimulated Raman gain

Substance	Raman shift [cm^{-1}]	Linewidth 2Γ [cm^{-1}]	Cross-section $d\sigma/d\Omega \times 10^8$ [cm^{-1} – ster^{-1}]	Raman gain $G_R \times 10^3$ [cm/MW]
Gas H_2 [a]	4155	0.2		1.5 (300 K, 10 atm)
Liquid O_2	1522	0.177	0.48 ± 0.14	14.5 ± 4
Liquid N_2	2326.5	0.067	0.29 ± 0.09	17 ± 5
Benzene	992	2.15	3.06	2.8
CS_2	655.6	0.50	7.55	24
Nitrobenzene	1345	6.6	6.4	2.1
$LiNbO_3$	258	7	262	28.7
InSb [b]	0—300	0.3	10	1.7×10^4

[a] E. E. HAGENLOCKER, R. W. MINCK, W. G. RADO: Phys. Rev. **154**, 226 (1967).
[b] For a carrier concentration $n_e \simeq 10^{16}$ cm^{-3}.

roughly the same for many other liquids from which stimulated Raman emission has been observed. However, in the earlier experiments on stimulated Raman scattering, laser beams of less than 100 MW/cm^2 were used, yet more than e^{30} Raman photons per sec were recorded. This anomaly, together with a number of other observed anomalies [7.5] such as the extremely sharp stimulated Raman threshold, the asymmetry of forward-backward Raman intensity, the appreciable spectral broadening of the Raman radiation, etc., has baffled research workers in the field for quite a few years. As we shall see in a later sections, these anomalies are now understood as due to self-focusing of the incident laser beam in the medium.

Both Stokes and anti-Stokes radiation are normally observed in stimulated Raman scattering. The observation of stimulated anti-Stokes scattering [7.9] was quite surprising since from the simple theory of two-photo transitions, the anti-Stokes Raman gain is negative at thermal equilibrium, as we shall see later. The anti-Stokes radiation is actually generated through parametric coupling with the laser and Stokes radiation. This also explains why the anti-Stokes generated in liquids and solids always has intense components radiated in the off-axis directions.

Higher-order Stokes and anti-Stokes radiation were also frequently observed [7.9, 10]. They were presumably generated by stepwise processes. Because of high laser intensity in the medium (usually in local regions through either external focusing or self-focusing), intense first-order Stokes and anti-Stokes radiation are first generated. They are then intense enough to generate second-order Raman radiation which in turn may become intense enough to generate higher-order Raman radiation. The theoretical description of this stepwise Raman production is, however, very difficult, as we shall see.

Early interest in stimulated Raman scattering arose because it can provide intense coherent radiation at new frequencies and because it is a possible loss mechanism in transmitting high-power laser beams through a medium, for example, the atmosphere. Later, it was demonstrated that stimulated Raman scattering via phonons which are both Raman and infrared active can have its Raman frequency tuned continuously over a certain range by varying the directions of the beam propagation in a crystal [7.11]. This is known as stimulated polariton scattering. Tunable far-infrared radiation is generated simultaneously with the Raman radiation in stimulated polariton scattering [7.12]. Then, it was discovered that stimulated Raman scattering can also occur via two-photon spin-flip transitions in semiconductors (first observed in InSb) [7.13]. The Raman frequency is again continuously tunable by adjusting the Zeeman splitting with an applied magnetic field. Since the spin-flip transitions can also be induced directly by far-infrared radiation, the problem is rather similar to the problem of stimulated polariton scattering. We can in fact use the theory of stimulated polariton scattering to correctly describe both stimulated Raman emission and far-infrared generation via spin-flip transitions in InSb [7.14].

The relaxation times of the Raman effect in liquids and solids are usually very short, in the picosecond range. Therefore, with nanosecond incident laser pulses, the Raman generation (in the absence of self-focusing) can certainly be regarded as quasi-steady-state. But when picosecond laser pulses are used, the transient effect in the Raman generation may become important [7.15]. The theory of transient stimulated Raman scattering is now well understood [7.16]. Experimentally, the transient Raman effect has been used to excite coherent molecular or phonon vibration in a medium. By monitoring the decay of such a forced vibration, one can then measure the corresponding vibrational relaxation times [7.17, 18]. So far, this has been the only existing method for measuring vibrational relaxation times of liquids and solids directly.

In Section 7.2, we shall discuss the classical theory of stimulated Raman scattering. We shall use the coupled-wave approach [7.19] to describe the generation of first- and higher-order Stokes and anti-Stokes radiation [7.20]. In Section 7.3, we shall review the experimental results on stimulated Raman effect and show how we can explain the various anomalous effects observed in stimulated Raman scattering. In Sections 7.4 and 7.5, we shall discuss two special cases of stimulated Raman scattering, namely, stimulated polariton scattering and stimulated spin-flip Raman emission. In Section 7.6 we shall consider the transient behavior of stimulated Raman emission when the pulsewidth of the pump field is either narrower than, or comparable with the relaxation times of the Raman excitation. In Section 7.7, we shall discuss a number of possible applications of stimulated Raman scattering. These include measurements of phonon or vibrational relaxation times, measurements of third-order nonlinear refractive indices, detection of substances of low concentration, spectroscopic studies of low-energy excitations, heating of plasmas, and transmission of high-power laser beams in a medium. Finally, in Section 7.8, we shall give brief concluding remarks on the anticipated future progress in this field.

There already exist in the literature several review articles on the stimulated Raman effect [7.21, 22]. In this paper, we shall put more emphasis on the basic understanding of the effect and on the more recent progress in the field. The references quoted here are admittedly far from complete. For a more complete reference list, the readers should consult [7.21, 22], also [Ref. 8.1, Chap. 4].

7.1. Basic Principles

It is well known that Raman scattering is a direct two-photon process. Simultaneously in the process, one photon at $\omega_1(\boldsymbol{k}_1)$ is absorbed and one photon at $\omega_2(\boldsymbol{k}_2)$ is emitted while the material makes a transition from the initial state $|i\rangle$ to the final state $|f\rangle$ (see Fig. 7.1). Energy conservation requires $\hbar(\omega_1-\omega_2)$ to be equal to the energy difference of the two states $E_f-E_i=\hbar\omega_{fi}$ within the uncertainty limit of the linewidth. We can have either $\omega_1>\omega_2$ or $\omega_1<\omega_2$. The former is known as Stokes scattering and the latter anti-Stokes scattering.

A straightforward second-order perturbation calculation leads to the following Raman transition probability per unit time per unit volume per unit energy interval [7.23]

$$\frac{dW_{fi}}{d(\hbar\omega)}=(8\pi^3N\omega_1\omega_2/\varepsilon_1\varepsilon_2)|\langle f|M|i\rangle|^2\langle\alpha_f|a_2^+a_1|\alpha_i\rangle|^2g(\Delta\omega)$$

$$M=\frac{e^2}{m^2\omega_1\omega_2}\sum_s\left[\frac{e^{-i\boldsymbol{k}_2\cdot\boldsymbol{r}}(\boldsymbol{p}\cdot\hat{e}_2)|s\rangle\langle s|(\boldsymbol{p}\cdot\hat{e}_1)e^{i\boldsymbol{k}_1\cdot\boldsymbol{r}}}{\hbar(\omega_1-\omega_{si})}-\frac{(\boldsymbol{p}\cdot\hat{e}_1)e^{i\boldsymbol{k}_1\cdot\boldsymbol{r}}|s\rangle\langle s|e^{-i\boldsymbol{k}_2\cdot\boldsymbol{r}}(\boldsymbol{p}\cdot\hat{e}_2)}{\hbar(\omega_2+\omega_{si})}\right]. \tag{7.1}$$

Here, N is the density of molecules or unit cells in the medium, ε is the dielectric constant, $\boldsymbol{p}$ is the momentum operator, $\hat{e}$ denotes the field

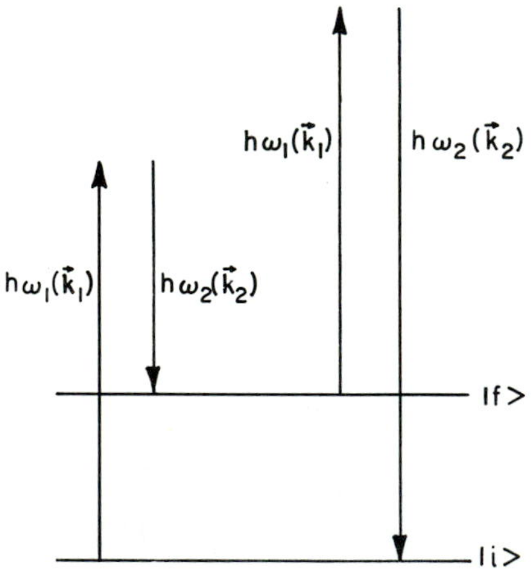

Fig. 7.1. Schematic drawing showing the Stokes $(\omega_1>\omega_2)$ and anti-Stokes $(\omega_1<\omega_2)$ Raman transition from the initial state $|i\rangle$ to a final excited state $|f\rangle$

polarization, $|s\rangle$ is the intermediate state of the material system, $|\alpha\rangle$ denotes the state of the radiation field, a^+ and a are the photon creation and annihilation operators, respectively, and finally $g(\Delta\omega=\omega_1-\omega_2-\omega_{fi})$ is the joint density of states of the transition. If the Raman transition has a Lorentzian lineshape, then $g(\Delta\omega)=\Gamma/\pi[(\Delta\omega)^2+\Gamma^2]$, where Γ is the halfwidth of the line.

The transition probability W_{fi} in (7.1) is proportional to $|\langle\alpha_f|a_2^+a_1|\alpha_i\rangle|^2$. If the Raman process begins with an initial state which contains practically no photon at ω_2, it is known as spontaneous Raman emission. Then, W_{fi} is simply proportional to $|\langle\alpha_f|a_1|\alpha_i\rangle|^2$. Otherwise, it is known as stimulated Raman emission. In the particular case where the states contain integer numbers of photons at ω_1 and ω_2 such that $|\alpha_i\rangle=|n_1,n_2\rangle$ and $|\alpha_f\rangle=|n_1-1,n_2+1\rangle$, we have $W_{fi}\propto n_1(n_2+1)$; spontaneous Raman scattering corresponds to $n_2=0$ in this case. In general, the radiation states are more complex [7.24], and no such simple relation can be obtained. However, if the average numbers of photons $\bar{n}_1$ and $\bar{n}_2$ at ω_1 and ω_2 are much larger than 1, then the approximation $|\langle\alpha_f|a_2^+a_1|\alpha_i\rangle|^2\cong\bar{n}_1\bar{n}_2$ is excellent [7.24].

Thus we expect that the spontaneous Raman scattering cross-section should be linearly proportional to the stimulated Raman gain coefficient. By definition, the differential Raman cross-section $d^2\sigma/d(\hbar\omega_2)d\Omega$ is the probability of a unit volume of material scattering an incident photon at ω_1 per unit area into a Raman photon of one polarization at ω_2 in a unit solid angle around Ω and a unit energy interval around $\hbar\omega_2$. Since the density of radiation modes per unit solid angle is $g_{\mathrm{E}}d\omega_2=k_2^2dk_2/(2\pi)^3$, we have

$$\begin{aligned} d^2\sigma/d(\hbar\omega_2)d\Omega &= \varrho_i g_{\mathrm{E}}[dW_{fi}/d(\hbar\omega_2)]/|\langle\alpha_f|a_2^+a_1|\alpha_i\rangle|^2 c \\ &= N\frac{\omega_1\omega_2^3\varepsilon_2^{\frac{1}{2}}}{c^4\varepsilon_1}|M_{fi}|^2 g(\Delta\omega)\varrho_i\,, \end{aligned} \tag{7.2}$$

where $M_{fi}=\langle f|M|i\rangle$ and ϱ_i is the population of $|i\rangle$. In the stimulated Raman amplification, the change of the number of Raman photons in one mode per unit length of propagation is given by [7.8]

$$\frac{d\bar{n}_2}{dz}=\left(\frac{dW_{fi}}{d\omega_2}\varrho_i-\frac{dW_{if}}{d\omega_2}\varrho_f\right)\varepsilon_2^{\frac{1}{2}}/c-\alpha_2\bar{n}_2 \tag{7.3a}$$

$$\cong(G_{\mathrm{R}}-\alpha_2)\bar{n}_2 \quad \text{if} \quad \bar{n}_1,\bar{n}_2\gg 1\,, \tag{7.3b}$$

$$\begin{aligned} G_{\mathrm{R}} &= 8\pi^3 N\hbar(\omega_1\omega_2/\varepsilon_1\varepsilon_2)|M_{fi}|^2 g(\Delta\omega)(\varrho_i-\varrho_f)\bar{n}_1(\varepsilon_2^{\frac{1}{2}}/c) \\ &= \frac{4\pi^2c^3\varepsilon_1}{\omega_1\omega_2^2\varepsilon_2\varrho_i}(\varrho_i-\varrho_f)\left(\frac{d^2\sigma}{d(\hbar\omega_2)d\Omega}\right)|E_1|^2\,, \end{aligned}$$

where α_2 is the absorption coefficient at ω_2, and $|E_1|^2 \varepsilon_{1,r}/2\pi = \bar{n}_1 \hbar\omega_1$ is the field energy per unit volume and $\varepsilon_{1,r}$ is the real part of the dielectric constant. Equation (7.3) shows that G_R is the Raman gain and is proportional to $d^2\sigma/d(\hbar\omega_2)d\Omega$. In the next section, we shall show that we can obtain the same expression for G_R from a third-order nonlinear optical susceptibility, known as the Raman susceptibility.

Note in Table 7.1 that the material which has the strongest Raman gain is InSb. From (7.3), we have $n_2(l) = n_2(0)\exp[(G_R - \alpha_2)l]$. Then, even in InSb, in order to generate e^{30} Raman photons from one noise photon in a 1-cm crystal, we need an incident CO_2 laser intensity of $2\ MW/cm^2$. For this reason, stimulated Raman scattering was only observed after the high-power laser was invented and developed.

7.2. Theory of Stimulated Raman Scattering

We shall use only semiclassical descriptions in the following theoretical treatment of stimulated Raman scattering (SRS), i.e., we shall avoid quantization of electromagnetic fields. This is of course not valid when the number of phonons in the Raman mode is small, for example, when SRS is first initiated by the spontaneously scattered photons. Therefore, the descriptions are valid only for stimulated Raman amplification where the input Raman radiation is sufficiently intense. For Raman oscillation which builds up from spontaneous scattering, we must use the full quantum description, for example (7.3a), with $W_{fi} \propto n_1(n_2+1)$ and $W_{if} \propto (n_1+1)n_2$ [7.25]. There are numerous papers in the literature on the theory of SRS, see, for example [7.8; 26–30]. In this section, we shall discuss only the coupled-wave approach for stimulated Raman amplification [7.19, 20].

7.2.1. Coupling of Pump and Stokes Waves

Consider the problem of SRS in a medium with energy levels shown in Fig. 7.1. Let us first assume that only two frequency components, ω_1 and ω_2 with $\omega_1 > \omega_2$, are present (i.e., we consider only first-order Stokes scattering). In the semiclassical description, these two components of the fields are represented by the waves

$$\begin{aligned} E_1 &= \mathscr{E}_1 \exp(\mathrm{i}\boldsymbol{k}_1 \cdot \boldsymbol{r} - \mathrm{i}\omega_1 t) \\ E_2 &= \mathscr{E}_2 \exp(\mathrm{i}\boldsymbol{k}_2 \cdot \boldsymbol{r} - \mathrm{i}\omega_2 t)\,. \end{aligned} \tag{7.4}$$

In the steady-state case, they obey the wave equations

$$\nabla \times (\nabla \times \boldsymbol{E}_1) - \frac{\omega_1^2}{c^2} \overleftrightarrow{\varepsilon}_1 \cdot \boldsymbol{E}_1 = \frac{4\pi\omega_1^2}{c^2} \boldsymbol{P}^{(3)}(\omega_1)$$
$$\nabla \times (\nabla \times \boldsymbol{E}_2) - \frac{\omega_2^2}{c^2} \overleftrightarrow{\varepsilon}_2 \cdot \boldsymbol{E}_2 = \frac{4\pi\omega_2^2}{c^2} \boldsymbol{P}^{(3)}(\omega_2), \quad (7.5)$$

where the nonlinear polarizations $\boldsymbol{P}^{\mathrm{NL}}$ in a medium with inversion symmetry are given by [7.31]

$$\boldsymbol{P}^{(3)}(\omega_1) \cong (\chi_1^{(3)}|E_1|^2 + \chi_{\mathrm{R}1}^{(3)}|E_2|^2)\boldsymbol{E}_1$$
$$\boldsymbol{P}^{(3)}(\omega_2) \cong (\chi_{\mathrm{R}2}^{(3)}|E_1|^2 + \chi_2^{(3)}|E_2|^2)\boldsymbol{E}_2, \quad (7.6)$$

where the $\chi^{(3)}$'s are third-order nonlinear susceptibilities. For simplicity, we assume $\chi^{(3)}$ to be scalar.

We then see clearly that the two wave equations in (7.5) are actually coupled with each other through the $\chi_{\mathrm{R}}^{(3)}$ terms in $\boldsymbol{P}^{(3)}$. This coupling between $\boldsymbol{E}_1$ and $\boldsymbol{E}_2$ causes effective energy transfer between the two waves. In this respect, the $\chi_{\mathrm{R}}^{(3)}$'s are known as Raman susceptibilites. The $\chi_1^{(3)}$ and $\chi_2^{(3)}$ terms in $\boldsymbol{P}^{(3)}$ simply modify the dielectric constants $\varepsilon(\omega_1)$ and $\varepsilon(\omega_2)$ in (7.5). They are responsible for self-focusing of high-intensity beams with finite cross-sections. In the following, we shall, however, assume infinite plane wave propagation in the medium and hence we shall neglect $\chi_1^{(3)}$ and $\chi_2^{(3)}$.

Consider the waves propagating along $\hat{z}$. For cubic or isotropic media, $\nabla \times (\nabla \times \boldsymbol{E}) = -\partial^2 \boldsymbol{E}/\partial z^2$. If the energy transfer between $\boldsymbol{E}_1$ and $\boldsymbol{E}_2$ is not so rapid, $\mathscr{E}_1$ and $\mathscr{E}_2$ can be considered as slowly varying functions ($|\partial^2\mathscr{E}/\partial z^2| \ll k|\partial\mathscr{E}/\partial z|$). Then (7.5) reduces to

$$\partial\mathscr{E}_1/\partial z = \mathrm{i}(2\pi\omega_1^2/c^2 k_1)\chi_{\mathrm{R}1}^{(3)}|E_2|^2\mathscr{E}_1$$
$$\partial\mathscr{E}_2/\partial z = \mathrm{i}(2\pi\omega_2^2/c^2 k_2)\chi_{\mathrm{R}2}^{(3)}|E_1|^2\mathscr{E}_2, \quad (7.7)$$

where $k = \omega\varepsilon^{\frac{1}{2}}/c$. From (7.7), we obtain

$$\partial|\mathscr{E}_1|^2/\partial z = -(4\pi\omega_1^2/c^2 k_1)(\mathrm{Im}\,\chi_{\mathrm{R}1}^{(3)})|\mathscr{E}_1|^2|\mathscr{E}_2|^2$$
$$\partial|\mathscr{E}_2|^2/\partial z = -(4\pi\omega_2^2/c^2 k_2)(\mathrm{Im}\,\chi_{\mathrm{R}2}^{(3)})|\mathscr{E}_1|^2|\mathscr{E}_2|^2. \quad (7.8)$$

We can now compare the second equation in (7.8) directly with (7.3), knowing $|\mathscr{E}_2|^2 \varepsilon(\omega_1)/2\pi = \bar{n}_2 \hbar\omega_2$. We then immediately find

$$
\begin{aligned}
G_{\mathrm{R}} &= -(4\pi\omega_2^2/c^2 k_2)(\mathrm{Im}\,\chi_{\mathrm{R}2}^{(3)})|\mathscr{E}_1|^2 \\
\chi_{\mathrm{R}2}^{(3)} &= \frac{-c^4\varepsilon_1}{\omega_1\omega_2^3\varepsilon_2^{\frac{1}{2}}\varrho_i}(\varrho_i-\varrho_f)\frac{1}{\hbar[(\omega_1-\omega_2-\omega_{fi})-\mathrm{i}\Gamma]}\frac{d\sigma}{d\Omega}+(\chi_{\mathrm{R}2}^{(3)})_{\mathrm{NR}}\,.
\end{aligned}
\tag{7.9}
$$

We assume here $g(\Delta\omega)$ is a Lorentzian and $(\chi_{\mathrm{R}2}^{(3)})_{\mathrm{NR}}$ is the non-resonant term due to non-resonant virtual transition. This yields a microscopic expression for $\chi_{\mathrm{R}2}^{(3)}$

$$
\chi_{\mathrm{R}2}^{(3)} = -N|M_{fi}|^2(\varrho_i-\varrho_f)/\hbar[(\omega_1-\omega_2-\omega_{fi})-\mathrm{i}\Gamma]+(\chi_{\mathrm{R}2}^{(3)})_{\mathrm{NR}}\,. \tag{7.10}
$$

If the field at ω_1 is not at resonance with some direct transitions in the medium, then for each photon absorbed at ω_1, there is a photon emitted at ω_2. Applying this conservation law to (7.8), we readily find

$$
\mathrm{Im}\{\chi_{\mathrm{R}1}^{(3)}\} = -\mathrm{Im}\{\chi_{\mathrm{R}2}^{(3)}\}
$$

and with the help of the Kramers-Kronig relation, we obtain

$$
\chi_{\mathrm{R}1}^{(3)} = (\chi_{\mathrm{R}2}^{(3)})^* \tag{7.11}
$$

which is the well-known symmetry relation for Raman susceptibilities [7.31].

With the conservation of number of photons, $[\varepsilon^{\frac{1}{2}}(\omega_1)|\mathscr{E}_1|^2/\omega_1 + \varepsilon^{\frac{1}{2}}(\omega_2)|\mathscr{E}_2|^2/\omega_2] = K$, the solution of (7.8) is

$$
\begin{aligned}
&\frac{|\mathscr{E}_1(z)|^2}{|\mathscr{E}_1(z)|^2-\omega_1 K/\varepsilon^{\frac{1}{2}}(\omega_1)} \\
&\quad = \frac{|\mathscr{E}_1(0)|^2}{|\mathscr{E}_1(0)|^2-\omega_1 K/\varepsilon^{\frac{1}{2}}(\omega_1)}|\mathscr{E}_1|^2\exp[-\omega_1 K G_{\mathrm{R}} Z/\varepsilon^{1/2}(\omega_1)] \\
&\frac{|\mathscr{E}_2(z)|^2}{|\mathscr{E}_2(z)|^2-\omega_2 K/\varepsilon^{\frac{1}{2}}(\omega_2)} \\
&\quad = \frac{|\mathscr{E}_2(0)|^2}{|\mathscr{E}_2(0)|^2-\omega_2 K/\varepsilon^{\frac{1}{2}}(\omega_2)}|\mathscr{E}_1|^2\exp[+\omega_2 K G_{\mathrm{R}} Z/\varepsilon^{1/2}(\omega_2)]\,.
\end{aligned}
\tag{7.12}
$$

If $|\mathscr{E}_1|^2 \gg |\mathscr{E}_2|^2$, we have the familiar result

$$|\mathscr{E}_2(z)|^2 = |\mathscr{E}_2(0)|^2 \exp(G_R z) \tag{7.13}$$

which shows explicitly the exponential growth of the Stokes field.

In the case where the pump and Stokes waves propagate into the medium with a plane boundary at $z=0$ along directions other than $\hat{z}$, the wave vectors k_1 and k_2 in (7.8) should be replaced by the projections of $\boldsymbol{k}_1$ and $\boldsymbol{k}_2$ along $\hat{z}$. This is obvious since $\mathscr{E}_1$ and $\mathscr{E}_2$ are only functions of z.

7.2.2 Raman Susceptibilities

The microscopic expression for $\chi_R^{(3)}$ can of course be obtained directly from quantum mechanical perturbation calculation. The derivation is straightforward but tedious [7.20, 31, 32]. We can, however, derive it very simply by realizing that so far as the response of the material to the field is concerned, a two-photon process can be considered as an equivalent one-photon process.

Consider the two-photon transition probability given by (7.1). It can be obtained directly from first-order perturbation if we regard the transition as a direct transition with an effective interaction Hamiltonian

$$\mathscr{H}'_{\text{eff}} = -ME_1E_2^* + \text{complex conjugate (c}\cdot\text{c)}. \tag{7.14}$$

This can be proved readily if we remember $E=(2\pi\hbar c^2/\omega)^{\frac{1}{2}}a$ and $E^*=(2\pi\hbar c^2/\omega)^{\frac{1}{2}}a^+$, and use the golden rule to derive $dW_{fi}/d(\hbar\omega)$.

Similarly, we can also derive $\chi_R^{(3)}$ by using $\mathscr{H}'_{\text{eff}}$ in the equation for linear polarization [7.33]. We then have for the third-order nonlinear polarization $\boldsymbol{P}^{(3)}$,

$$\hat{e}_1\cdot\boldsymbol{P}^{(3)}_{(\omega_1)} = \hat{e}_1\cdot\overleftrightarrow{\boldsymbol{\chi}}^{(3)}_{R1}\cdot\boldsymbol{E}_1\boldsymbol{E}_2\boldsymbol{E}_2^* = -N\frac{|\langle f|ME_2^*|i\rangle|^2(\varrho_i-\varrho_f)}{\hbar(\omega_1-\omega_2-\omega_{fi}+i\Gamma)}E_1\,, \tag{7.15}$$

where, for simplicity, we have neglected the non-resonant term. This equation together with $\chi_{R2}^{(3)}=\chi_{R1}^{(3)*}\equiv\hat{e}_1\cdot\overleftrightarrow{\boldsymbol{\chi}}^{(3)*}_{R1}:\hat{e}_1\hat{e}_2\hat{e}_2$ leads immediately to the same microscopic equation for χ_{R2} given in (7.10) with $(\chi_{R2}^{(3)})_{NR}$ neglected.

More generally, (7.10) can also be derived by considering the states $|i\rangle$ and $|f\rangle$ being coherently admixed by the interaction Hamiltonian $\mathscr{H}'_{\text{eff}}$. Let us denote the perturbed states as $|i'\rangle$ and $|f'\rangle$. They should obey the Schrödinger equation

$$\begin{aligned} -\mathrm{i}\hbar\partial\langle f'|/\partial t &= \langle f'|(\mathscr{H}_0+\mathscr{H}'_{\text{eff}}) \\ \mathrm{i}\hbar\partial|i'\rangle/\partial t &= (\mathscr{H}_0+\mathscr{H}'_{\text{eff}})|i'\rangle\,, \end{aligned} \tag{7.16}$$

where $\mathscr{H}_0$ is the unperturbed Hamiltonian for the material system. If we use the interaction representation, we then have in the first-order approximation

$$\begin{aligned} -\hbar\left(\mathrm{i}\frac{\partial}{\partial t}-\omega_{fi}+\mathrm{i}\Gamma\right)\langle f'|i\rangle &= \hbar\left(\mathrm{i}\frac{\partial}{\partial t}-\omega_{fi}+\mathrm{i}\Gamma\right)\langle f|i'\rangle \\ &= \langle f|\mathscr{H}'_{\text{eff}}|i\rangle\,. \end{aligned} \tag{7.17}$$

Here, we have inserted a phenomenological damping constant Γ for ω_{fi}. With $\mathscr{H}'_{\text{eff}}$ in (7.14), the solution of (7.17) at frequency $\omega_1-\omega_2$ is

$$-\langle f'|i\rangle = \langle f|i'\rangle = -M_{fi}E_1E_2^*/\hbar(\omega_1-\omega_2-\omega_{fi}+\mathrm{i}\Gamma)\,. \tag{7.18}$$

The nonlinear polarization induced by this perturbation is given by

$$\begin{aligned} P^{(3)}(\omega_2) &= N\langle -\partial\mathscr{H}'_{\text{eff}}/\partial E_2^*\rangle \\ &\cong (\langle f|i'\rangle^*\varrho_i+\langle f'|i\rangle^*\varrho_f)NM_{fi}E_1 \\ &= \langle f'|i'\rangle^*(\varrho_i-\varrho_f)NM_{fi}E_1\,. \end{aligned} \tag{7.19}$$

Here again, we have considered only the resonant term. Using the expression for $\langle f|i'\rangle$ in (7.18), we obtain again the same microscopic expression for $\chi^{(3)}_{\mathrm{R}2}$.

A matter of interest here is that we can consider $\psi_i=\langle f|i'\rangle$ and $\psi_f=\langle f'|i\rangle$ $(\psi\equiv\psi_i=-\psi_f)$ physically as excitational waves (at frequency $\omega_1-\omega_2$) in the medium [7.16]. They are coupled with E_1 and E_2 via the coupling energy

$$\begin{aligned} \langle\mathscr{H}'_{\text{eff}}\rangle &= \langle i'|\mathscr{H}_{\text{eff}}|i'\rangle\varrho_i+\langle f'|\mathscr{H}_{\text{eff}}|f'\rangle\varrho_f \\ &= -M_{fi}(\varrho_i-\varrho_f)E_1E_2^*\psi^*+\mathrm{c\cdot c}\,. \end{aligned} \tag{7.20}$$

Following (7.17), these two waves obey the driven wave equations

$$\begin{aligned} \hbar\left(\mathrm{i}\frac{\partial}{\partial t}-\omega_{fi}+\mathrm{i}\Gamma\right)\psi_i &= +\frac{\partial\langle\mathscr{H}'_{\text{eff}}\rangle}{\partial(\varrho_i\psi_i^*)} = -M_{fi}E_1E_2^* \\ \hbar\left(\mathrm{i}\frac{\partial}{\partial t}-\omega_{fi}+\mathrm{i}\Gamma\right)\psi_f &= -\frac{\partial\langle\mathscr{H}'_{\text{eff}}\rangle}{\partial(\varrho_f\psi_f^*)} = -M_{fi}E_1E_2^*\,. \end{aligned} \tag{7.21}$$

Through the coupling with E_1 and E_2, these excitational waves induce the nonlinear polarizations

$$\begin{aligned} P^{(3)}(\omega_1) &= -N\partial\langle\mathscr{H}'_{\text{eff}}\rangle/\partial E_1^* = N M_{fi}^* E_2 \psi(\varrho_i - \varrho_f) \\ P^{(3)}(\omega_2) &= -N\partial\langle\mathscr{H}'_{\text{eff}}\rangle/\partial E_2^* = N M_{fi} E_1 \psi^*(\varrho_i - \varrho_f) \end{aligned} \quad (7.22)$$

which then in turn act as the sources in the driven wave equations for E_1 and E_2 in (7.5). Stimulated Raman scattering (SRS) is now simply the result of nonlinear coupling of the three waves $\psi_i = -\psi_f$, E_1, and E_2, and can be obtained from the solution of the coupled Equations (7.5) and (7.21). Note that except the population factors ϱ_i and ϱ_f which appear to have quantum origins, we can now treat both the material excitation and the radiation as classical waves and SRS as the result of nonlinear coupling of these classical waves.

We have so far assumed ϱ_i and ϱ_f constant. However, when the field intensities are high so that the Raman transition probability W_{if} is large, the populations ϱ_i and ϱ_f can be changed appreciably during the Raman process. Directly from physical consideration, we can write the rate equations for ϱ_i and ϱ_f as [7.32, 34]

$$\begin{aligned} \frac{\partial \varrho_i}{\partial t} &= -(W_{fi}\varrho_i - W_{if}\varrho_f) + \left(\frac{\partial \varrho_i}{\partial t}\right)_{\text{damping}} \\ \frac{\partial \varrho_f}{\partial t} &= (W_{fi}\varrho_i - W_{if}\varrho_f) + \left(\frac{\partial \varrho_f}{\partial t}\right)_{\text{damping}} \\ \hbar\omega_{fi}(W_{fi}\varrho_i - W_{if}\varrho_f) &= \frac{1}{2}\left[\frac{\partial P^{(3)}(\omega_1)^*}{\partial t} E_1 + \frac{\partial P^{(3)}(\omega_2)^*}{\partial t} E_2\right] + \text{c}\cdot\text{c} \\ &= \tfrac{1}{2}\mathrm{i}\omega_{fi} N[M_{fi} E_1 E_2^* \psi^* \\ &\quad - M_{fi}^* E_1^* E_2 \psi](\varrho_i - \varrho_f)\,. \end{aligned} \quad (7.23)$$

In the simple case where $(\partial\varrho/\partial t)_{\text{damping}}$ is dominated by random relaxation between $|i\rangle$ and $|f\rangle$, we have

$$\begin{aligned} \Delta\varrho &= \varrho_f - \varrho_f^0 = -(\varrho_i - \varrho_i^0) \\ \frac{\partial \Delta\varrho}{\partial t} + \frac{\Delta\varrho}{T_1} &= (W_{fi}\varrho_i - W_{if}\varrho_f)\,, \end{aligned} \quad (7.24)$$

where ϱ_i^0 and ϱ_f^0 are the populations at thermal equilibrium, and T_1 is the well-known longitudinal relaxation time as distinguished from the transverse relaxation time $T_2 \equiv 1/\Gamma$. Raman transitions and Raman susceptibilities in this saturation limit are the subject of discussion in [7.32].

The above discussion deals with localized electronic excitations but the general formalism is, of course, valid for any material excitation, e.g., molecular vibration, phonon [7.20], magnon [7.35], exciton, plasmon [7.36], polariton [7.37], etc. The wave equations are generally different for different types of excitations; Eq. (7.21) should therefore be replaced by wave equations appropriate for the excitation involved in the problem. The coupling constant M_{fi} in (7.21) and (7.22) should also be changed accordingly. There is the difference between boson-type excitations and localized excitations between two levels. For localized excitations, we use (7.24), but for boson-type excitations, we have $W_{fi}\varrho_i \propto 1+n_\psi$, $W_{if}\varrho_f \propto n_\psi$, and $(\partial/\partial t+1/T_1)(n_\psi-\bar{n}_\psi)=W_{fi}\varrho_i-W_{if}\varrho_f$, where n_ψ is the average number of bosons at thermal equilibrium. The general formalism is also valid for any two-photon transition process. For example, it can be applied to the problem of two-photon absorption and second-harmonic generation near an excitonic resonance, e.g., in CuCl [7.38–40].

7.2.3 Parametric Coupling Between Photons and Phonons

In this section, we consider the special case where molecular vibrations or phonons in the medium are being excited in the Raman process. This is the most important case since at least 90% of the published Raman work deals with Raman scattering by molecular vibrations or phonons. We can, of course, use quantum mechanics to describe molecular vibration or phonons [7.8, 25]. However, as shown in the previous section, we can describe stimulated Raman amplification classically as the result of parametric coupling between electromagnetic waves and material excitational waves. We shall present only the classical description here [7.20].

Let us consider first the coupling between photon and phonon waves. It is governed by the interaction energy of (7.20) with, (in order to follow the usual convention, [7.21]) ψ replaced by $(2\omega_3/\hbar)^{\frac{1}{2}}Q$, where Q is the phonon displacement wave at frequency $\omega_3\equiv\omega_1-\omega_2$. Comparing with the usual expression of $\langle\mathscr{H}'_{\text{eff}}\rangle=-(\partial\alpha/\partial Q^*)E_1E_2^*Q^*+\text{c}\cdot\text{c}.$, we have $M_{fi}=(\partial\alpha/\partial Q)/(2\omega_3/\hbar)^{\frac{1}{2}}$ where α is the polarizability [7.20]. The coupling constant M_{fi} can also be obtained from the general expression of M in (7.1). As shown in (7.9) and (7.10), the spontaneous Raman scattering cross-section $d\sigma/d\Omega$ is directly proportional to $|M_{fi}|^2$. We shall not go further into the detailed microscopic theory of $d\sigma/d\Omega$ for this case since it has already been dealt with in the other chapters of this book. For our purposes, we only need to remember that the coupling constant M_{fi} can be easily deduced from the spontaneous Raman cross-sections $d\sigma/d\Omega$.

The wave equation for phonons is given by [7.20]

$$\left[\beta\nabla^2+\frac{\partial^2}{\partial t^2}+2\Gamma\frac{\partial}{\partial t}+\omega_0^2\right]Q=-\partial\langle\mathscr{H}'_{\text{eff}}\rangle/\partial Q^*(\varrho_i-\varrho_f)$$
$$=\left(\frac{2\omega_3}{\hbar}\right)^{\frac{1}{2}}M_{fi}E_1E_2^*\,, \tag{7.25}$$

where β is a constant which characterizes the phonon dispersion near $k=0$, ω_0 is the phonon frequency at $k=0$, and Γ is again the damping constant. SRS by phonons is then fully described by the solution of the three coupled wave equations (7.25), and (7.5), where $P^{(3)}(\omega_1)$ and $P^{(3)}(\omega_2)$ are given by (7.22) with $\psi_i=-\psi_f$ replaced by $(2\omega_3/\hbar)^{\frac{1}{2}}Q$.

Several different cases arise depending on the values of β and ω_0. If $\beta<0$, and $\omega_0=0$, then (7.25) is the equation for an acoustic wave. Stimulated light scattering by acoustic phonons is known as stimulated Brillouin scattering [7.41]. From the above discussion we see clearly that stimulated Brillouin scattering is simply a special class of SRS. We shall not discuss stimulated Brillouin scattering any further in this chapter but refer the reader to Chapter 6. For optical phonon waves, we have $\omega_0\neq 0$ and $\beta\neq 0$ (usually $\beta>0$). In the limit where interaction between molecules is negligible, the optical phonons become dispersionless and reduce essentially to molecular vibrations. It is SRS by optical phonons or molecular vibrations which one normally encounters in the studies of SRS.

The wave vector of photons and phonons involved in SRS is of the order of $10^5\,\text{cm}^{-1}$ or less. Hence, the $\beta\nabla^2 Q$ term in (7.25) for optical phonons is often negligible. We then have

$$Q=-(2\omega_3/\hbar)^{\frac{1}{2}}M_{fi}E_1E_2^*/(\omega_3^2-\omega_{fi}^2+\mathrm{i}2\omega_3\Gamma)\,, \tag{7.26}$$

where $\omega_3=\omega_1-\omega_2$. From (7.22) with the non-resonant term included, we find for $\omega_3\sim\omega_0$,

$$\chi_{\mathrm{R}2}^{(3)}=-N|M_{fi}|^2(\varrho_i-\varrho_f)/\hbar(\omega_3-\omega_{fi}-\mathrm{i}\Gamma)+(\chi_{\mathrm{R}2}^{(3)})_{\mathrm{NR}} \tag{7.27}$$

which is identical to (7.10). Consequently, the expression for the Raman gain G_{R} in (7.9) and the solution of the coupled wave equations in (7.12) (assuming ϱ_i and ϱ_f constant) are still valid for the present case.

7.2.4 Stokes—Anti-Stokes Coupling

We have assumed so far that in SRS only electromagnetic waves at ω_1 and ω_2 are present in the medium. We shall now see that in general Stokes and anti-Stokes waves $(\omega_1\pm\omega_3)$ are in fact simultaneously generated in the stimulated Raman process, even at 0 K. This is very different from the spontaneous scattering case where no anti-Stokes scattering occurs at 0 K.

We can see most clearly the simultaneous Stokes—anti-Stokes generation from the coupled wave approach. The incoming wave at ω_1 first beats with the waves at $\omega_1 \pm \omega_3$ to drive the material excitational wave at ω_3. The material excitation in turn beats with the incoming wave at ω_1 to create nonlinear polarizations at $\omega_1 \pm \omega_3$ [7.26, 27]. These nonlinear polarizations then act as the sources to amplify Stokes $(\omega_1 - \omega_3)$ and anti-Stokes $(\omega_1 + \omega_3)$ waves in the medium.

Let us use the scripts s and a to denote Stokes and anti-Stokes, respectively. Then, we have $\omega_1 - \omega_s = \omega_a - \omega_1 = \omega_3$. The interaction energy for coupling between E_1, E_s, E_a, and ψ is

$$\langle \mathscr{H}'_{\text{eff}} \rangle = -M^{s}_{fi}(\varrho_i - \varrho_f) E_1 E^*_s \psi^* - M^{a}_{fi}(\varrho_i - \varrho_f) E_a E^*_1 \psi^* + \text{c} \cdot \text{c}\,, \tag{7.28}$$

where M^s_{fi} and M^a_{fi} are identical to M_{fi}, except that ω_s replaces ω_2 in M^s_{fi} and ω_a and ω_1 replace ω_1 and ω_2 respectively, in M^a_{fi}. If dispersion in M_{fi} is negligible, then $M^s_{fi} \cong M^a_{fi}$. The nonlinear polarizations at ω_s, ω_a, and ω_1 are now given by

$$\begin{aligned} \hat{e}_s \cdot \boldsymbol{P}^{(3)}(\omega_s) &= -N\partial\langle \mathscr{H}'_{\text{eff}} \rangle/\partial E^*_s = N M^s_{fi}(\varrho_i - \varrho_f) E_1 \psi^* \\ \hat{e}_a \cdot \boldsymbol{P}^{(3)}(\omega_a) &= -N\partial\langle \mathscr{H}'_{\text{eff}} \rangle/\partial E^*_a = N(M^a_{fi})^*(\varrho_i - \varrho_f) E_1 \psi \\ \hat{e}_1 \cdot \boldsymbol{P}^{(3)}(\omega_1) &= -N\partial\langle \mathscr{H}'_{\text{eff}} \rangle/\partial E^*_1 \\ &= [(M^s_{fi})^* E_s \psi + M^a_{fi} E_a \psi^*]\,(\varrho_i - \varrho_f)\,, \end{aligned} \tag{7.29}$$

where we have not included the non-resonant terms. The driven wave equation for ψ is

$$\hbar\left(\mathrm{i}\frac{\partial}{\partial t} - \omega_{fi} + \mathrm{i}\Gamma\right)\psi = -M^s_{fi} E_1 E^*_s - M^a_{fi} E_a E^*_1\,. \tag{7.30}$$

Stokes and anti-Stokes generation in SRS is now governed by the solution of the four coupled wave equations, i.e., (7.30) and the wave equations for E_1, E_s, and E_a.

We can first obtain ψ from (7.30)

$$\psi = -(M^s_{fi} E_1 E^*_s + M^a_{fi} E_a E^*_1)/\hbar(\omega_3 - \omega_{fi} + \mathrm{i}\Gamma)\,. \tag{7.31}$$

Then, with the above expression for ψ inserted into (7.29), the wave equations for E_1, E_s, E_a in the steady state becomes [7.20]

$$\hat{e}_1 \cdot \left[\nabla \times (\nabla \times \boldsymbol{E}_1) - \frac{\omega_1^2}{c^2} \boldsymbol{\varepsilon}_1 \cdot \boldsymbol{E}_1 \right]$$

$$= \frac{4\pi\omega_1^2}{c^2} [\chi_{ss}^{(3)} |E_s|^2 E_1 + (\chi_{sa}^{(3)} + \chi_{sa}^{(3)*}) E_s E_a E_1^* + \chi_{aa}^{(3)} |E_a|^2 E_1]$$

$$\hat{e}_s \cdot \left[\nabla \times (\nabla \times \boldsymbol{E}_s) - \frac{\omega_s^2}{c^2} \boldsymbol{\varepsilon}_s \cdot \boldsymbol{E}_s \right] = \frac{4\pi\omega_s^2}{c^2} [\chi_{ss}^{(3)} |E_1|^2 E_s + \chi_{sa}^{(3)} E_1^2 E_a^*] \tag{7.32}$$

$$\hat{e}_a \cdot \left[\nabla \times (\nabla \times \boldsymbol{E}_a) - \frac{\omega_a^2}{c^2} \boldsymbol{\varepsilon}_a \cdot \boldsymbol{E}_a \right] = \frac{4\pi\omega_a^2}{c^2} [\chi_{sa}^{(3)*} E_1^2 E_s^* + \chi_{aa}^{(3)} |E_1|^2 E_a]$$

where

$$\begin{aligned} \chi_{ss}^{(3)} &= (\chi_{ss}^{(3)})_{NR} - N |M_{fi}^{s}|^2 (\varrho_i - \varrho_f)/\hbar(\omega_3 - \omega_{fi} - i\Gamma) \\ \chi_{sa}^{(3)} &= (\chi_{sa}^{(3)})_{NR} - N M_{fi}^{s} M_{fi}^{a*} (\varrho_i - \varrho_f)/\hbar(\omega_3 - \omega_{fi} - i\Gamma) \\ \chi_{aa}^{(3)} &= (\chi_{aa}^{(3)})_{NR} - N |M_{fi}^{a}|^2 (\varrho_i - \varrho_f)/\hbar(\omega_3 - \omega_{fi} - i\Gamma) . \end{aligned} \tag{7.33}$$

The solution of (7.32) is greatly simplified if we can neglect the depletion of the incoming wave E_1 in the SRS process, since then we have just a set of two linearly coupled equations for E_s and E_a. Assuming an isotropic medium with a plane boundary at $z=0$ and slowly varying amplitudes for E_s and E_a, we readily find [7.20] (see Fig. 7.2)

$$\begin{aligned} E_s &= [\mathscr{E}_{s+} \exp(i\Delta K_+ z) + \mathscr{E}_{s-} \exp(i\Delta K_- z)] \exp(i\boldsymbol{k}_s \cdot \boldsymbol{r} - \alpha_{sz} z) \\ E_a^* &= [\mathscr{E}_{a+}^* \exp(i\Delta K_+ z) + \mathscr{E}_{a-}^* \exp(i\Delta K_- z)] \\ &\quad \cdot \exp[-i\boldsymbol{k}_a \cdot \boldsymbol{r} - (i\Delta k + \alpha_{az}) z] \end{aligned} \tag{7.34}$$

where

$$\begin{aligned} k^2 &= \omega^2 \varepsilon'/c^2 \\ 0 &= 2k_{1x,y} - k_{sx,y} - k_{ax,y} \\ \Delta k &= 2k_{1z} - k_{sz} - k_{az} , \qquad k_z \equiv \boldsymbol{k} \cdot \hat{z} \\ \Delta K_\pm &= \Delta k/2 \pm [(\Delta k/2)^2 - (\Delta k)\lambda]^{\frac{1}{2}} \\ \lambda &= (2\pi\omega_s^2/c^2 k_{sz}) \chi_{ss}^{(3)} |E_1|^2 , \qquad \alpha_z = \omega^2 \varepsilon''/c^2 k_z = \alpha(k/k_z) . \end{aligned}$$

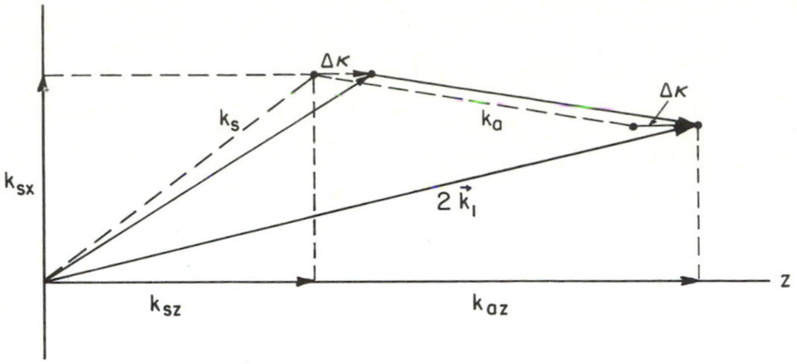

Fig. 7.2. General relationship between the wave vectors of Stokes, anti-Stokes, and laser waves, as stated in (7.34). (After [7.22])

For simplicity, we have neglected here the dispersions of the absorption coefficients α_z and $(2\pi\omega^2/c^2\,\mathrm{Re}\{\boldsymbol{k}\cdot\hat{z}\})\chi^{(3)}$. The more general solution with the dispersion included can of course also be written down quite straightforwardly. If $\mathscr{E}_{s0}$ and $\mathscr{E}_{a0}$ are the boundary values of $\mathscr{E}_s$ and $\mathscr{E}_a$ at $z=0$, then we have

$$\begin{aligned}\mathscr{E}_{a\pm}/\mathscr{E}_{s\pm} &= (\Delta K_\pm - \lambda)/\lambda \\ \mathscr{E}_{s\pm} &= [(-\Delta K_\mp + \lambda)\mathscr{E}_{s0} + \lambda\mathscr{E}^*_{a0}]/(\Delta K_\pm - \Delta K_\mp)\,.\end{aligned} \tag{7.35}$$

A number of physical results follow immediately from the solution in (7.34) and (7.35). a) If the phase mismatch Δk is sufficiently large or $\chi^{(3)}_{ss}|E_1|^2$ sufficiently small so that $|\Delta k| \gg |\lambda|$, then the Stokes and anti-Stokes fields are effectively decoupled. The two parts of the solution reduce to

$$\begin{cases}\Delta K_- = \lambda \\ |\mathscr{E}^*_a/\mathscr{E}_s| = |\lambda/\Delta k| \ll 1\end{cases} \quad \text{and} \quad \begin{cases}\Delta K_+ = -\lambda + \Delta k \\ |\mathscr{E}^*_a/\mathscr{E}_s| = |\Delta k/\lambda| \gg 1.\end{cases}$$

(Note that $-\mathrm{Im}\{\lambda\} = G_R/2$.) The first part corresponds to an almost pure Stokes wave with an exponential gain G_R and the second part corresponds to an almost pure anti-Stokes wave with a gain $-G_R$. These results are what we should expect when there is little Stokes—anti-Stokes coupling as discussed in previous sections. We notice that at thermal equilibrium, the almost pure anti-Stokes wave has a negative gain since its energy is now used in SRS to amplify the pump wave. b) If the linear phase matching condition $\Delta k = 0$ is satisfied, we find $\Delta K_\pm = 0$ and $|\mathscr{E}^*_{a\pm}/\mathscr{E}_{s\pm}| = 1$. There is no exponential gain for both Stokes and anti-Stokes fields. Although the Stokes—anti-Stokes coupling is maximum in this case, the

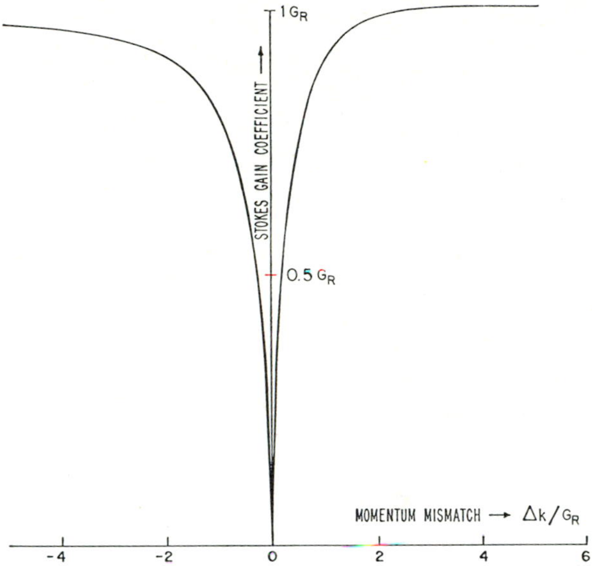

Fig. 7.3. The Stokes power gain as a function of the normalized linear momentum mismatch $\Delta k/G_R$ in the z direction. The asymmetry is due to the nonresonant part $\chi_{NR}^{(3)} = 0.1\,|\mathrm{Im}\,\chi_R^{(3)}|_{max}$. (After [7.22])

positive work done on the Stokes field is exactly compensated by the negative work done on the anti-Stokes field. This is well known in parametric amplifier theory where no gain can be obtained at $\omega_s = \omega_1 - \omega_3$ if the other side band at $\omega_a = \omega_1 + \omega_3$ is not suppressed. c) As Δk gradually deviates from zero, the positive exponential gain increases rapidly towards the value G_R, as shown in Fig. 7.3, while the corresponding $|\mathscr{E}_a^*/\mathscr{E}_s|$ decreases from 1 towards 0. Consequently, at some value of $|\Delta k|$, the anti-Stokes power generated in this SRS process goes through a maximum. This is shown in Fig. 7.4 for two different values of $G_R z$. We therefore expect that the anti-Stokes radiation appears in the form of double cones in k-space. For more details on the calculation, the readers should consult [7.20]. We shall discuss the experimental results in terms of these theoretical predictions in a later section.

7.2.5 Higher-Order Raman Effects

Intense higher-order Stokes and anti-Stokes fields can also be generated in SRS [7.3–9]. They are generated by the successively induced third-order nonlinear polarizations $P^{(3)}$ at appropriate frequencies. For ex-

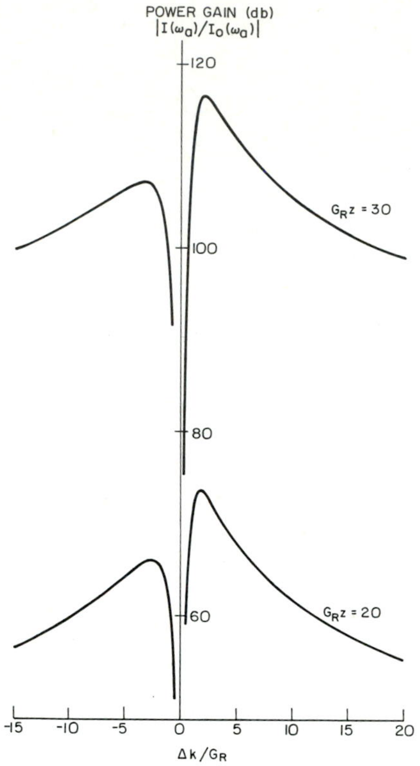

Fig. 7.4. Anti-Stokes intensity versus the linear momentum mismath Δk (normalized by the Stokes power gain G_R). The asymmetry is due to $\chi_{NR} = 0.1\,|\mathrm{Im}\,\chi_R^{(3)}|_{max}$. (After [7.22])

ample, the second Stokes field E_{s2} can be generated by

$$P^{(3)}(\omega_{s2}) = \chi_{s\alpha}^{(3)} E_s^2 E_1^* + \chi_{s\beta}^{(3)} E_1 E_s E_a^* \,.$$

They should appear most strongly around the phase-matching directions given by $\boldsymbol{k}_{s2} = \boldsymbol{k}_s + \boldsymbol{k}'_s - \boldsymbol{k}_1$ and $\boldsymbol{k}_{s2} = \boldsymbol{k}_1 + \boldsymbol{k}_s - \boldsymbol{k}_a$. The generation of second Stokes is clearly a higher-order effect since $P^{(3)}(\omega_{s2})$ here is linearly proportional to the pump field $|E_1|$ while $P^{(3)}(\omega_s)$ and $P^{(3)}(\omega_a)$ are proportional to $|E_1|^2$. When the first Stokes $E(\omega_s)$ becomes intense enough, E_{s2} can also be generated through $P^{(3)}(\omega_{s2}) = \chi_{s\gamma}^{(3)} |E_s|^2 E_{s2} + \chi_{s\delta}^{(3)} E_s^2 E_{a2}^*$, where E_{a2} is the second anti-Stokes. Similarly, the nonlinear polarizations responsible for the generation of other higher-order Stokes and anti-Stokes fields can be written down easily. The complete description of these higher-order stimulated Raman effects should then be

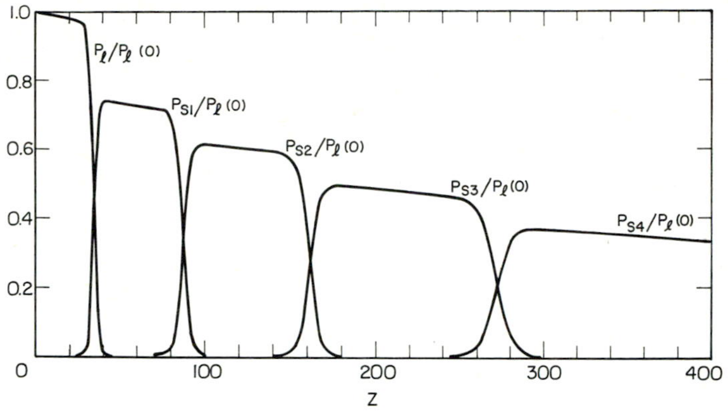

Fig. 7.5. The saturation effect of a single-mode laser beam of infinite extent. The intensities of the various orders of Stokes waves are normalized by the incoming laser intensity. The distance is also normalized to a dimensionless one $Z=(16\pi^3\omega_1^2\,\mathrm{Im}\{\chi_R\}/C^3k_1)P_1(0)z$. (After [7.22])

obtained from the solution of the many wave equations which are nonlinearly coupled through the nonlinear polarizations. This is of course a formidable task in general.

In the special case where we can assume that only Stokes waves along $+\hat{z}$ are generated, the set of coupled wave equations is [7.20]

$$[\partial^2/\partial z^2+(\omega_1\varepsilon_1/c^2)]E_1=-(4\pi\omega_1^2/c^2)\chi_{s1}^{(3)*}|E_s|^2E_1$$

$$[\partial^2/\partial z^2+(\omega_s\varepsilon_s/c^2)]E_s=-(4\pi\omega_s^2/c^2)[\chi_{s1}^{(3)}|E_1|^2E_s+\chi_{s2}^{(3)*}|E_{s2}|^2E_s]$$

$$[\partial^2/\partial z^2+(\omega_{s2}\varepsilon_{s2}/c^2)]E_{s2}=-(4\pi\omega_{s2}^2/c^2)$$

$$\cdot[\chi_{s2}^{(3)}|E_{s2}|^2E_{s2}+\chi_{s3}^{(3)*}|E_{s3}|^2E_{s2}]\,. \tag{7.36}$$

The solution of (7.36) obtained from numerical calculation for infinite plane waves is shown in Fig. 7.5. It is seen that the first Stokes power first increases gradually and then suddenly builds up to a maximum value while the pump power gets almost completely depleted. As z increases further, the first Stokes power remains nearly constant for a while, and then again suddenly gets depleted into the second Stokes, and so on. This was actually demonstrated in a properly designed experiment by VON DER LINDE et al. [7.42], as we shall see later.

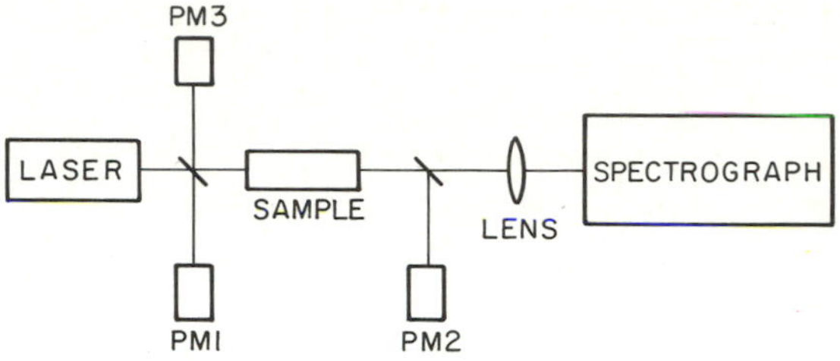

Fig. 7.6. A typical experimental set-up for investigation of stimulated Raman scattering. PM1, PM2, and PM3 are photo-detectors measuring the laser, the forward Raman, and the backward Raman radiation, respectively

7.3. Experimental Observations

After the accidental discovery of SRS in nitrobenzene [7.1] in 1962, the subject of SRS soon attracted much attention. The earlier detailed experimental studies of SRS were mostly on liquids with large Kerr constants. A typical set-up is shown in Fig. 7.6. It was then realized that the observed Raman output was much too intense to be accounted for by the theory. For example, from the spontaneous Raman data, one can calculate, by using (7.3), the maximum Raman gain for nitrobenzene to be 2.8×10^{-3} cm/MW (see Table 7.1), i.e., in order to amplify the Raman radiation from the noise level by e^{28} times in a 10-cm cell, one would need an input laser intensity of 1000 MW/cm^2. The Q-switched laser pulse intensity used in the experiments was, however, always around 100 MW/cm^2 or less, which should not be sufficient to generate any detectable Raman radiation. This observed gain anomaly [7.43] together with many other related anomalous effects such as forward-backward asymmetry [7.44], spectral broadening [7.44], anomalous anti-Stokes rings [7.44, 45], etc., had stimulated a lot of research acitivites on the subject. Only several years later, people began to understand that most of these anomalous effects were actually induced by self-focusing of the input beam [7.46]. Therefore, before we compare the experimental results with the theory, we should discuss briefly how self-focusing affects SRS.

7.3.1 Anomalous Effects Due to Self-Focusing

The gain anomaly is illustrated in Fig. 7.7. The curve of first Stokes intensity versus input power shows a sharp threshold. The slope at the threshold yields an anomalously large Raman gain which cannot be understood from the theory of SRS. Now, this is known to be due to self-focusing. Self-focusing arises because of the existence of a positive field-induced refractive index in the medium. Consider a beam with finite cross-section. The central portion, which is more intense, sees a larger refractive index and therefore propagates more slowly than the edge. Consequently, the wavefront is distorted. Since the rays should always propagate perpendicular to the wavefront, they then appear to bend towards the axis and self-focus to a point. The focus is at the position [7.47]

$$z_{\mathrm{f}}(t) \cong K/[\sqrt{P(t)} - \sqrt{P_{\mathrm{cr}}}]\,,$$

where K and P_{cr} are constants depending on the beam characteristics and material properties. As the laser power P increases, z_{f} decreases. When z_{f} first appears in the medium, the intense focus generates the Raman radiation readily. This then explains the observed sharp Raman threshold. It has now been clearly established that SRS in self-focusing liquids is in fact always initiated from the focal region [7.48, 49].

Self-focusing also explains the forward-backward asymmetry in SRS. For an input beam with a finite cross-section, the Raman intensities in different directions are expected to be different because of the different

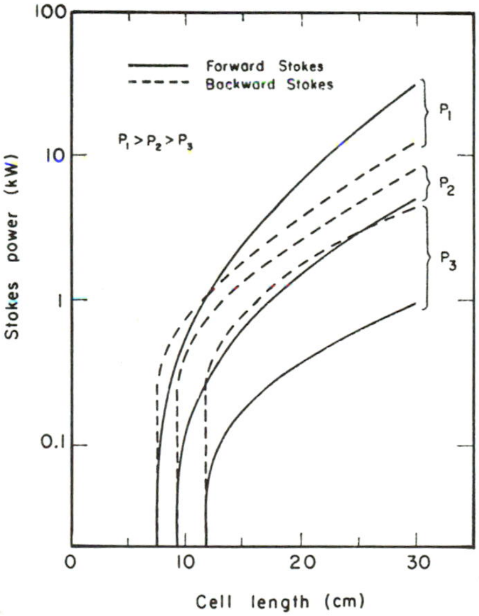

Fig. 7.7. First-order forward and backward Stokes power versus the toluene cell length at three laser powers $P_1 = 80$, $P_2 = 67$, and $P_3 = 53\ \mathrm{MW/cm^2}$. [After Y. R. SHEN and Y. J. SHAHAM: Phys. Rev. **163**, 224 (1967)]

active lengths, but the forward and the backward Raman intensities are supposed to be equal by symmetry. The experimental results in self-focusing liquids, however, show a clear forward-backward asymmetry, as seen in Fig. 7.7. This is again due to self-focusing. Since SRS is first initiated from the first focus at or near the end of the cell, the forward and backward Raman radiation should experience very different Raman amplification. Moreover, the fact that the backward Raman radiation always meets head-on with the undepleted incoming laser beam could lead to a very short sub-nanosecond Raman pulse [7.50].

Self-focusing also imposes a strong frequency modulation on the beam. This then explains the observed spectral broadening of both laser and Raman radiation in self-focusing liquids [7.51]. In non-self-focusing media, the spectral broadening may be due to successive beats of laser and Raman radiation to many orders [7.52, 53].

First-order anti-Stokes radiation should appear around the directions defined by $\boldsymbol{k}_a = 2\boldsymbol{k}_1 - \boldsymbol{k}_s$. However, in self-focusing liquids, another cone of anti-Stokes radiation at somewhat larger angle from the axis can often be observed [7.45]. This is presumably due to anti-Stokes emitted in the filamentary focal region, because the condition of phase matching along the surface of the filament is now important [7.54].

7.3.2 Raman Oscillation in Non-Self-Focusing Media

Even in non-self-focusing media, study of SRS using the set-up in Fig. 7.6 shows a sharp threshold in the growth of Stokes intensity versus input laser power. An example is shown in Fig. 7.8, which gives the Stokes output versus the laser input in liquid nitrogen [7.55]. Self-focusing was not observed in this case. As the laser power increases, the Stokes output first increases linearly as a result of spontaneous scattering and then grows quasi-exponentially. At a certain input power I_{Li}, the Stokes output shows a sudden rise. Finally, it levels off because of depletion of laser power. This sharp threshold was first believed to be due to feedback from Rayleigh scattering [7.55]. Recently, SPARKS [7.25] has pointed out that this is in fact an intrinsic behavior of Raman oscillation resulting from parametric instability. Using the full quantum description [Eq. (7.3a) with $W_{fi}\varrho_i \propto n_1(n_2+1)(n_\psi+1)$, $W_{if}\,\varrho_f \propto (n_1+1)n_2 n_\psi$, and $(\partial/\partial t + 1/T_1)(n_\psi - \overline{n}_\psi) = W_{fi}\varrho_i - W_{if}\varrho_f$, where $\overline{n}_\psi$ is the number of thermally excited phonons], he found a solution of the coupled equations. His calculated results for SRS in liquid N_2 agree very well with experimental data, as show in Fig. 7.8. However, it should be pointed out that for liquid N_2, it is probably more appropriate to use the localized molecular vibration model rather than the harmonic phonon model, i.e., we should use ϱ_i, ϱ_j, and (7.24) instead of n_f and the differential equation for n_f.

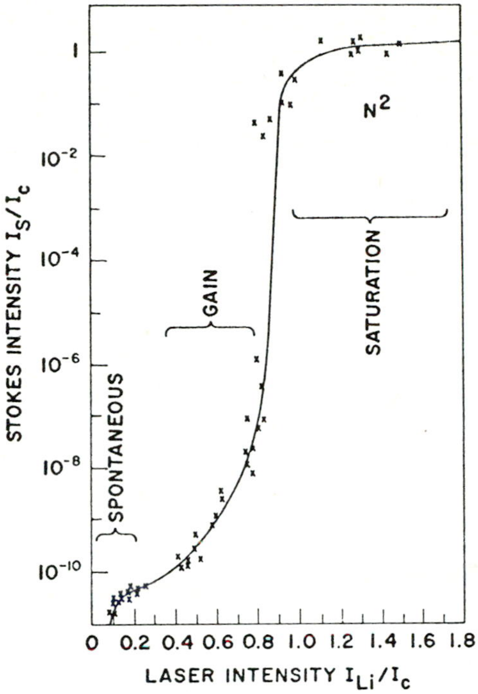

Fig. 7.8. Comparison of experimental data [7.55] and theoretical curve [7.25] of first-order Stokes power as a function of incident laser power in liquid nitrogen

Very recently, HAIDACHER and MAIER [7.128] have shown that the sharp Raman threshold anomaly is not due to a parametric instability, but due to diffuse reflection from cell windows.

7.3.3. Raman Gain Measurements

The set-up in Fig. 7.6 is, of course, only good for studying Raman oscillation. In order to study stimulated Raman amplification, one must not use a single oscillator system, but should use a combined oscillator-amplifier system [7.57]. A typical set-up is shown in Fig. 7.9. The Raman gain can be obtained by measuring the ratio of the Stokes input to the Stokes output of the amplifier. In Fig. 7.10, the results of LALLEMAND et al. [7.57] on hydrogen gas are shown to be in good agreement with the theoretical curve. Such a Raman gain measurement is, however, not very successful in self-focusing liquids because self-focusing would occur in

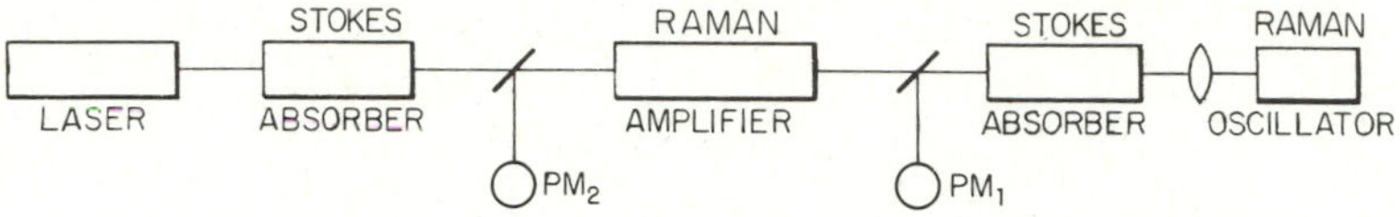

Fig. 7.9. Experimental set-up for measuring backward Stokes gain. (After [7.57])

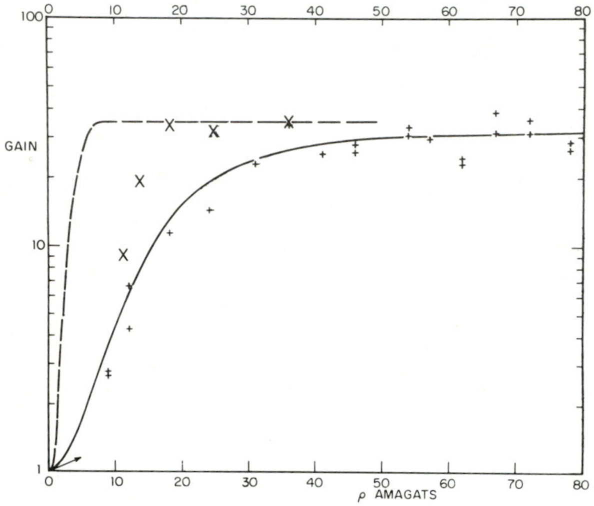

Fig. 7.10. Stokes Raman gain in H_2 gas as a function of pressure. The data points X for forward gain should be compared with the dashed theoretical curve (cell length, 80 cm; peak input intensity: 20 MW/cm^2). The data points + for backward gain should be compared with the solid theoretical curve (cell length: 30 cm; peak input intensity: 60 MW/cm^2). (After [7.57])

the amplifier before the amplification factor differs appreciably from 1 [7.58].

7.3.4. Anti-Stokes and Higher-Order Raman Radiation

Stimulated anti-Stokes radiation was first observed by TERHUNE [7.9]. As shown in Fig. 7.11, it appears in the form of bright multi-colored rings. Different rings often correspond to different orders of anti-Stokes. In liquids, anti-Stokes rings up to the 4th order can be easily photographed on color film. CHIAO and STOICHEFF [7.10b] showed in calcite

Fig. 7.11. Multicolor anti-Stokes rings created in benzene by a ruby laser beam. [After A. YARIV: *Quantum Electronics* (John Wiley, Inc., New York, 1967)]. (Picture taken by R. W. TERHUNE)

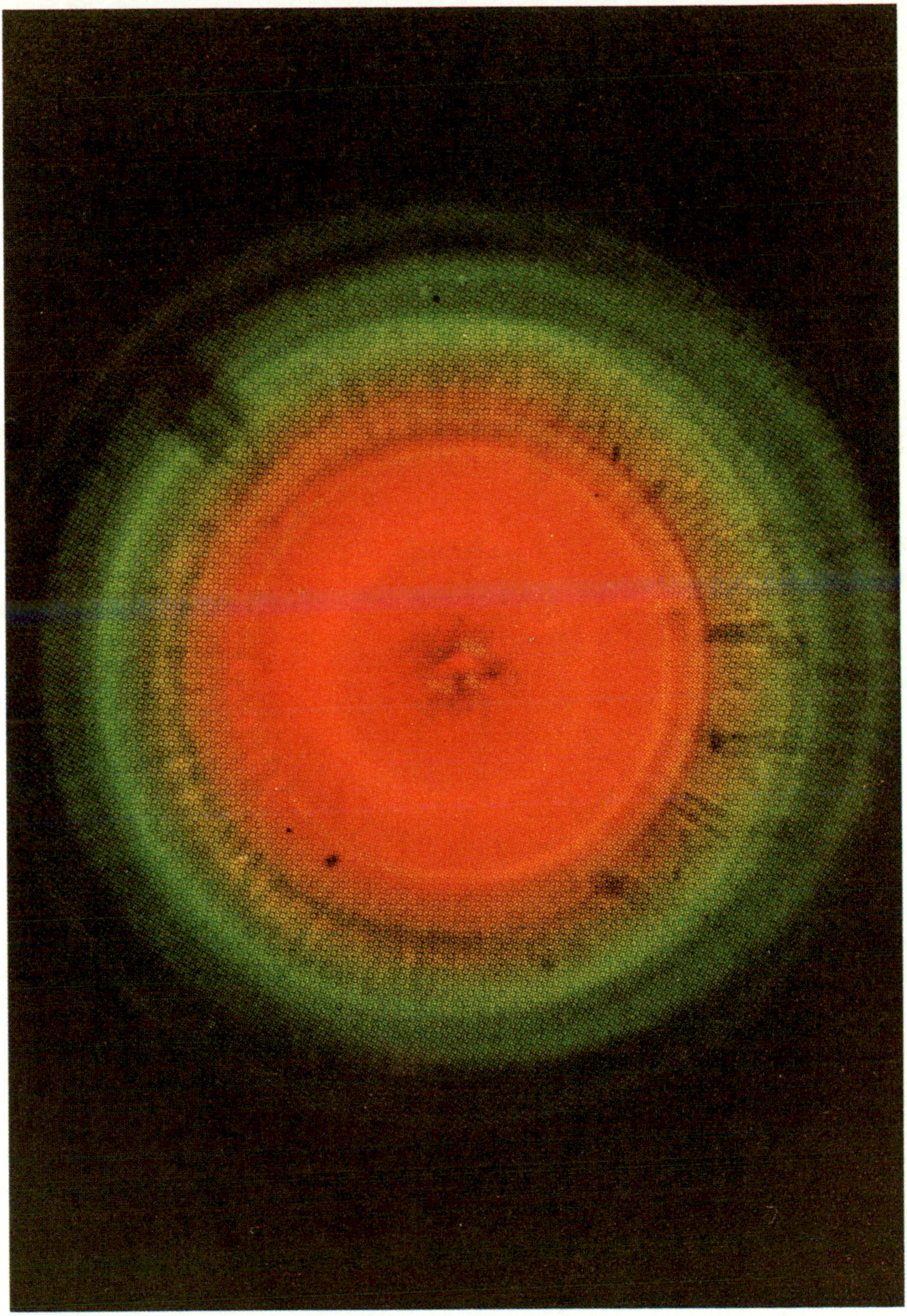

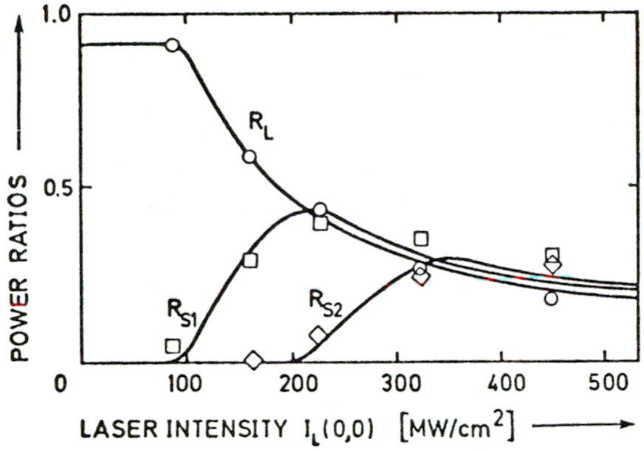

Fig. 7.12. Normalized transmitted laser (R_L), first (R_{S1}) and second Stokes (R_{S2}) power as a function of the incident laser intensity $I_L(0, 0)$. The experimental data of R_L, R_{S1}, and R_{S2} are represented by circles, rectangles, and diamonds respectively. The curves are calculated according to the theory in Subsection 7.2.5 with the finite beam cross-section taken into account. (After [7.42])

that the anti-Stokes is actually emitted with cone angles given by the phase-matching relation $\boldsymbol{k}_{a,n} = \boldsymbol{k}_{a,n-1} + \boldsymbol{k}_1 - \boldsymbol{k}_{s,n}$, where the integer n indicates the order. We may recall that the theory (Subsection 7.2.4) predicts for the first anti-Stokes a ring with a central dark band at the phase-matching directions. The reason the dark band was not observed is presumably due to the fact that a practical laser beam has a spread in $\boldsymbol{k}$-space which tends to smear the dark band in the anti-Stokes ring.

GARMIRE made a detailed study of the anti-Stokes rings produced in self-focusing liquids [7.45]. Two sets of cones were observed, Class I and II. Class I appeared at the normal phase-matching directions and Class II at some anomalous directions presumably determined by surface phase matching along the filaments created by self-focusing [7.45, 54].

Higher-order Stokes radiation has also been observed mostly along the axis in the forward and backward directions. Quantitative studies of generation of higher-order Stokes radiation are generally difficult because of the many competing nonlinear processes present. VON DER LINDE et al. [7.42] using subnanosecond laser pulses were, however, able to carry out a quantitative study in a special case. The short input pulse effectively suppresses both the stimulated Brillouin scattering and the backward Raman radiation. The effect of self-focusing can also be minimized by using high pulse intensity. Then, the forward SRS becomes the only effective nonlinear process in the medium. This is just the condition under which (7.36) is valid. Figure 7.12 shows that the experimental results agree very well with the theoretical curves calculated from (7.36), taking into account the real beam profile.

7.3.5. Stimulated Anti-Stokes Absorption

As seen in Subsection 7.2.4, the gain of the stimulated first-order anti-Stokes radiation is negative in directions where the Stokes—anti-Stokes coupling is weak. With both laser and anti-Stokes radiation present, the laser field will be amplified at the expense of the anti-Stokes. Thus, if the laser beam and a beam from a broadband source around ω_a propagate together in the medium, we expect to find a dark absorption band at ω_a in the broadband output spectrum. This was observed by JONES and STOICHEFF [7.59], and has been suggested as a useful spectroscopic technique to study molecular vibration.

7.3.6. Competition Between Different Raman Modes

Under normal conditions with nanosecond pulse excitation, only one Raman mode which has the maximum Raman gain participates in SRS. This is usually the Raman mode with both large cross-section and narrow linewidth. The effective depletion of laser power into this Raman mode forbids SRS to occur with other Raman modes. In transient SRS, it is, however, possible to have several Raman modes show up simultaneously, as we shall see later.

7.3.7. Competition Between Stimulated Raman Scattering and Other Non-Linear Optical Processes

We have already seen the effect of self-focusing on SRS. Other nonlinear processes, such as stimulated Brillouin scattering, two-photon absorption, etc., may also affect SRS. They may compete with SRS in depleting the laser power and consequently suppress SRS [7.50]. Quantitative studies of the influence of other nonlinear processes on SRS have not yet been done.

7.3.8. Stimulated Raman Scattering in Solids

We give in Table 7.2 a list of crystals in which SRS has been observed. The study of SRS in crystals has not been as extensive as in liquids mainly because a crystal is less flexible for investigation then a liquid and often much more expensive.

Table 7.2. Observed stimulated Raman lines in a number of crystals

Crystal	Stimulated Raman shift [cm^{-1}]	Ref.
$CaCO_3$	1086	a
Si	521	b
Diamond	1332	a
InSb	0—300	c
$LiNbO_3$	42—200	d
α-Sulfur	216, 470	a
$CaWO_4$	911	a
KDP	915, 93	e

[a] G. Eckardt: IEEE J. Quantum Electron. **2**, 1 (1966).
[b] J. M. Ralston, R. K. Chang: Phys. Rev. B**2**, 1858 (1970).
[c] C. K. N. Patel, E. D. Shaw: Phys. Rev. B**3**, 1279 (1971).
[d] J. Gelbwachs, R. H. Pantell, H. E. Puthoff, J. M. Yarborough: Appl. Phys. Lett. **14**, 258 (1969).
[e] M. K. Srivastava, R. W. Crow: Optics Commun. **8**, 82 (1973).

7.4. Stimulated Polarization Scattering

In more general cases, the material excitation ψ discussed in Section 7.2 can be excited not only by Raman process (or two-photon), but also by direct infrared (one-photon) absorption. In other words, the excitation is both infrared and Raman active. This happens, for example, with phonons in polar crystals. Coupling of lattice vibration and infrared wave was first studied by Huang [7.60]. In general, the mixed excitational wave resulting from direct coupling of photon and material excitation is known as polariton [7.61], and obeys the so-called polariton dispersion curve.

SRS in such a medium will excite the mixed infrared and material excitational wave or simply the polariton wave, and therefore is called stimulated polariton scattering (SPS). The theory of SPS has been discussed by Loudon [7.37], by Butcher et al. [7.37], by Shen [7.62], by Henry and Garrett [7.63], and by many others [7.64]. We shall present the coupled-wave approach here [7.14, 62].

Consider four waves interacting with one another in SPS: the laser E_1, the Stokes E_s, the infrared E_3, and the material excitation ψ. If we assume negligible depletion of the laser power, then the wave equations for the other three waves can be written as (see Section 7.2)

$$\begin{aligned}
&[\nabla^2+(\omega_s^2\varepsilon_s/c^2)]E_s=-(4\pi\omega_s^2/c^2)P^{NL}(\omega_s)\\
&[\nabla^2+(\omega_3^2\varepsilon_3/c^2)]E_3=-(4\pi\omega_3^2/c^2)[P^{(1)}(\omega_3)+P^{NL}(\omega_3)]\\
&\hbar(\omega_3-\omega_{fi}+i\Gamma)\psi=F^{(1)}+F^{NL},
\end{aligned}\tag{7.37}$$

where $\omega_3=\omega_1-\omega_s$.

We have assumed that ψ has a resonant frequency ω_{fi} independent of wave vector. The direct coupling between E_3 and ψ leads to an effective interaction energy

$$\langle \mathscr{H}_{\rm eff}^{(1)} \rangle = -NA_{fi}(\varrho_i - \varrho_f)\psi^* E_3 + {\rm c \cdot c.}, \tag{7.38}$$

where $A_{fi}E_3 = \langle f|\mathscr{H}'|i\rangle$ and $\mathscr{H}'$ is the usual interaction Hamiltonian between light and matter. We then obtain

$$\begin{aligned} P^{(1)}(\omega_3) &= -\partial\langle \mathscr{H}_{\rm eff}^{(1)}\rangle/\partial E_3^* = NA_{fi}^*(\varrho_i - \varrho_f)\psi \\ F^{(1)} &= +\partial\langle \mathscr{H}_{\rm eff}^{(1)}\rangle/\partial[\psi^* N(\varrho_i - \varrho_f)] = -A_{fi}E_3 \,. \end{aligned} \tag{7.39}$$

The nonlinear coupling between waves leads to

$$\langle \mathscr{H}_{\rm eff}^{\rm NL} \rangle = -NM_{fi}^{\rm s}(\varrho_i - \varrho_f)E_1 E_{\rm s}^* \psi^* - \chi^{(2)} E_1 E_{\rm s}^* E_3^* + {\rm c \cdot c.} \tag{7.40}$$

from which we find

$$\begin{aligned} P^{\rm NL}(\omega_{\rm s}) &= \chi^{(2)} E_1 E_3^* + NM_{fi}^{\rm s}(\varrho_i - \varrho_f)E_1 \psi^* \\ P^{\rm NL}(\omega_3) &= \chi^{(2)} E_1 E_{\rm s}^* \\ F^{\rm NL} &= -M_{fi}^{\rm s} E_1 E_{\rm s}^* \,. \end{aligned} \tag{7.41}$$

Elimination of ψ in (7.37) yields [7.14]

$$\begin{aligned} [\nabla^2 + (\omega_{\rm s}^2/c^2)(\varepsilon_{\rm s}^*)_{\rm eff}]E_{\rm s}^* &= -(4\pi\omega_{\rm s}^2/c^2)\chi_{\rm eff}^{(2)} E_1^* E_3 \\ [\nabla^2 + (\omega_3^2/c^2)(\varepsilon_3)_{\rm eff}]E_3 &= -(4\pi\omega_3^2/c^2)\chi_{\rm eff}^{(2)} E_1 E_{\rm s}^* \,, \end{aligned} \tag{7.42}$$

where

$$\begin{aligned} (\varepsilon_{\rm s}^*)_{\rm eff} &= \varepsilon_{\rm s}^* + 4\pi\chi_{\rm R}^{(3)}|E_1|^2 \\ \chi_{\rm R}^{(3)} &= -N|M_{fi}^{\rm s}|^2(\varrho_i - \varrho_f)/[\hbar(\omega_3 - \omega_{fi} + {\rm i}\Gamma)] \\ (\varepsilon_3)_{\rm eff} &= \varepsilon_3 - N|A_{fi}|^2(\varrho_i - \varrho_f)/[\hbar(\omega_3 - \omega_{fi} + {\rm i}\Gamma)] \\ \chi_{\rm eff}^{(2)} &= \chi^{(2)} - NA_{fi}^* M_{fi}^{\rm s}(\varrho_i - \varrho_f)/[\hbar(\omega_3 - \omega_{fi} + {\rm i}\Gamma)] \,. \end{aligned} \tag{7.43}$$

We have assumed for simplicity that $A_{fi}^* M_{fi}^{\rm s}$ is real. Note that $k_3 = (\omega_3/c) \cdot (\varepsilon_3)_{\rm eff}^{\frac{1}{2}}$ gives the polariton dispersion curve.

The equations in (7.42) are in the same form as the wave equations governing parametric amplification [7.65]. They are also similar to (7.32) for Stokes—anti-Stokes coupling. The solution of (7.42) is also similar to (7.34) and (7.35) and can be written down easily. For waves propagating into a medium with a plane boundary at $z=0$, we have [7.14]

$$\begin{aligned} E_s^* &= [\mathscr{E}_{s+}^* \exp(i\Delta K_+ z) + \mathscr{E}_{s-}^* \exp(i\Delta K_- z)] \exp(-i\boldsymbol{k}_s\cdot\boldsymbol{r}) \\ E_3 &= [\mathscr{E}_{3+} \exp(i\Delta K_+ z) + \mathscr{E}_{3-} \exp(i\Delta K_- z)] \exp(i\boldsymbol{k}_3\cdot\boldsymbol{r} + i\Delta k z)\,, \end{aligned} \tag{7.44}$$

where

$$\begin{aligned} k &= (\omega/c)\ (\varepsilon'_{\text{eff}})^{\frac{1}{2}} \\ \Delta k &= k_{1z} - k_{sz} - k_{3z}\,, \qquad k_z = \boldsymbol{k}\cdot\hat{z} \\ \Delta K_\pm &= \tfrac{1}{2}(\gamma_s - \gamma_3) \pm \tfrac{1}{2}[(\gamma_s+\gamma_3)^2 - 4\Lambda]^{\frac{1}{2}} \\ \gamma_s &= (k_s/2k_{sz})\,(i\alpha_s + 2k_R) \\ k_R &= (\omega_s^2/2k_{sz}c^2)4\pi\chi_R|E_1|^2 \\ \gamma_3 &= -\Delta k - i(k_3/2k_{3z})\alpha_3 \\ \Lambda &= (4\pi^2\omega_s^2\omega_3^2/c^2 k_{sz}k_{3z})\,(\chi_{\text{eff}}^{(2)})^2|E_1|^2 \\ |\mathscr{E}_3/\mathscr{E}_s^*| &= (\omega_3^2 k_{sz}/\omega_s^2 k_{3z})^{\frac{1}{2}}|\Lambda^{\frac{1}{2}}(\Delta K_\pm + \gamma_3)]\,. \end{aligned} \tag{7.45}$$

The gain corresponds to $G = -2\,\text{Im}\{\Delta K_\pm\}$. We consider only the mode with $G>0$.

Two special cases should be mentioned here. First, if there is neither linear nor nonlinear coupling between E_3 and the other waves, i.e., $A_{fi}=0$ and $\chi^{(2)}=0$, then the problem reduces to Stokes generation with $G=G_R$. Second, if the nonlinear coupling between ψ and E_s vanishes, i.e., $M_{fi}=0$, then the problem reduces to the simple case of parametric amplification [7.65]. The expressions of G and $|\mathscr{E}_3/\mathscr{E}_s^*|$ given here also reduce to those of HENRY and GARRETT [7.63] when ψ is replaced by $(2\omega_3/\hbar)^{\frac{1}{2}}Q$, the absorption coefficient α_s is neglected, and ε_3 is assumed to be real.

For given ω_3, the gain is a maximum at phase matching $\Delta k=0$ if the resonance of ω_{fi} is sufficiently narrow. HENRY and GARRETT [7.63] have calculated G_{max} and the corresponding $|\mathscr{E}_3/\mathscr{E}_s^*|$ for GaP at frequencies around the 366-cm^{-1} phonon mode. Their results are shown in Fig. 7.13. In this case, only a small section below $\omega_{fi}\equiv\omega_0$ can be phase matched. The two terms in $\chi_{\text{eff}}^{(2)}$ of (7.43) are of opposite signs for $\omega_3<\omega_0$ in GaP. As a result, G_{max} gradually reduces to zero as $\omega_3\to 250\ \text{cm}^{-1}$. SUSSMAN [7.66] has made similar calculations for the 248-cm^{-1} phonon mode of $LiNbO_3$. The results are shown in Fig. 7.14. The two terms in $\chi_{\text{eff}}^{(2)}$ are now of the same sign for $\omega_3<\omega_0$ and hence G_{max} increases gradually for $\omega_3<\omega_0$.

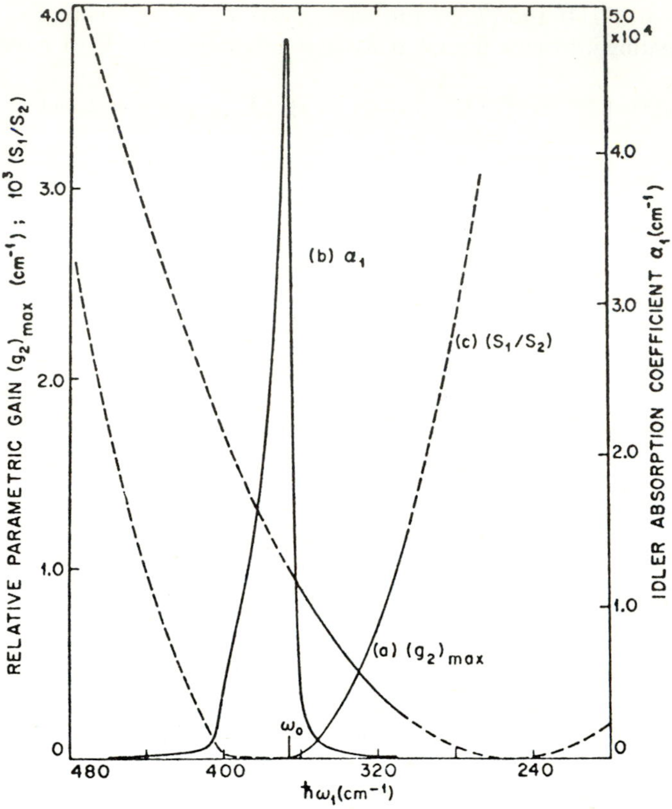

Fig. 7.13. (a) Relative parametric gain $(g_2)_{\max}$ versus the idler frequency ω_1 for gallium phosphide. Solid portion of the curve indicates the frequency range of the idler over which phase matching is possible. (b) Infrared absorption coefficient for gallium phosphide using $\Gamma = 4\,\mathrm{cm}^{-1}$. (c) Ratio of the idler flux density S_1 to the signal flux density S_2. (After [7.63])

Experimentally, SPS was first observed in $LiNbO_3$ by KURTZ et al. [7.11]. GELBWACHS et al. [7.12] demonstrated the fact that the Stokes frequency can be tuned over a narrow range by adjusting the relative angle between $\boldsymbol{k}_1$ and $\boldsymbol{k}_s$ to achieve phase matching with ω_3 sitting on the polation dispersion curve. In a resonator, up to 70% of the laser power can be converted into Stokes. Since the infrared E_3 should be generated simultaneously with the Stokes wave E_s, SPS can also be used to generate tunable coherent far-infrared radiation. Using a 1-MW Q-switched ruby laser beam of about 2 mm in diameter and a lens of 50 cm in focal length to focus the beam into a 3.3-cm a-axis $LiNbO_3$ crystal, YARBOROUGH et al. [7.67] have detected a far-infrared output

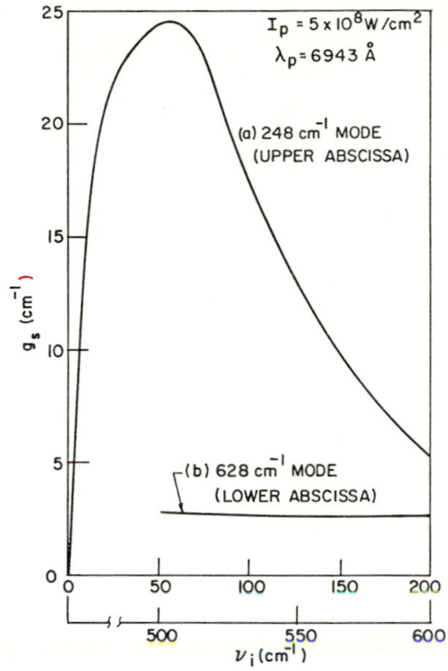

Fig. 7.14. Plots of stimulated gain coefficient (g_s) as a function of idler frequency for the 248 cm^{-1} mode (Curve *a*) and the 628 cm^{-1} mode (Curve *b*). (After [7.66])

with a peak power of 5 W and a tuning range from 50 to 238 cm^{-1}. Unfortunately, $LiNbO_3$ has a low damage threshold. A good single-mode laser should be used in order to avoid spatial inhomogeneity in the beam which increase the damage probability. Observation of far-infrared output by SPS in quartz has also been reported [7.68].

7.5. Stimulated Spin-Flip Raman Emission

We have emphasized in the previous sections that the material excitations involved in SRS can also be electronic excitations. A case of special interest is SRS in n-type InSb by electrons making spin-flip transition. The corresponding Raman process is shown schematically in Fig. 7.15.

Spontaneous spin-flip Raman scattering was first observed by Slusher et al. [7.69] following the theoretical predictions of Wolff [7.70] and Yafet [7.71]. Yafet showed that the Raman transition between the spin-up and spin-down states can be considered as a direct transition governed by an effective interaction Hamiltonian

$$\mathscr{H}''_{\mathrm{eff}} \cong \left(\frac{e^2}{m_s c^2}\right)\left[\frac{E_g \hbar\omega_1}{E_g^2 - (\hbar\omega_1)^2}\right] \boldsymbol{\sigma} \cdot (\boldsymbol{A}_1 \times \boldsymbol{A}_2^*)\,, \tag{7.46}$$

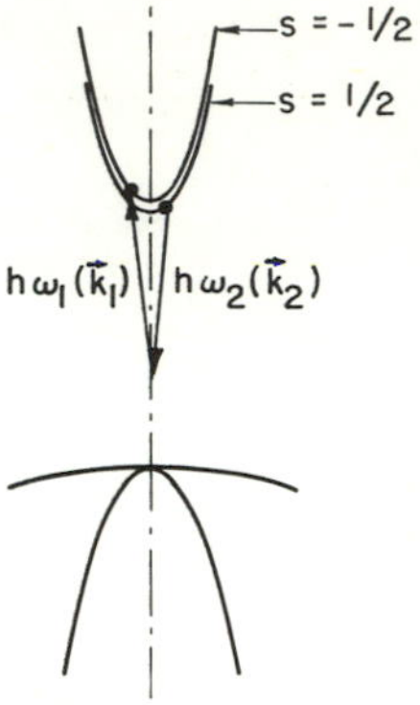

Fig. 7.15. Schematic of spin-flip Raman process in n-InSb

where m_s is the spin mass which is related to the free electron mass m and the spin g factor by $m_s = 2m/|g|$, E_g is the band gap, ω_1 is the incoming light frequency, $\boldsymbol{\sigma}$ is the spin operator, and $\boldsymbol{A}_1$ and $\boldsymbol{A}_2$ are the vector fields of the incoming and scattered radiation, respectively. The spin-flip frequency or the Raman frequency shift is

$$\Delta\omega = 2g\mu_B B\,, \tag{7.47}$$

where μ_B is the Bohr magneton and B is the applied magnetic field. From the first golden rule and (7.46), we find readily the spin-flip Raman cross-section as

$$(d\sigma/d\Omega)_{SF} \cong (e^2/m_s c^2)^2 (\omega_2/\omega_1)\,[E_g\hbar\omega_1/(E_g^2 - \hbar^2\omega_1^2)]^2 \tag{7.48}$$

for $\boldsymbol{A}_1$ perpendicular to $\boldsymbol{A}_2$.

In InSb, since $g \simeq 50$, we have $m_s \simeq 0.04$ m and if $E_g\hbar\omega_1 \approx (E_g^2 - \hbar^2\omega_1^2)$, $(d\sigma/d\Omega)_{SF}$ would be about 600 times larger than the Thompson scattering cross-section for free electron. For $\hbar\omega_1 \approx E_g$ $(d\sigma/d\Omega)_{SF}$ can be even much larger as a result of resonance enhancement.

Using a CO_2 laser ($\omega_1 = 940\ \text{cm}^{-1}$), SLUSHER et al. [7.69] found experimentally in InSb $(d\sigma/d\Omega)_{SF} \simeq 10^{-23}\ \text{cm}^2/\text{sr}$ which agrees fairly well with the theoretical prediction from (7.48). With a CO laser ($\omega_1 = 1880\ \text{cm}^{-1}$ while $E_g/\hbar = 1900\ \text{cm}^{-1}$), BRUECK and MOORADIAN [7.72] observed a strong resonant enhancement and found $(d\sigma/d\Omega)_{SF}$ can be 10^5 times larger than the Thomas cross-section (see Fig. 7.16). The spin-flip Raman linewidth is also extremely narrow at low temperatures. Depending on the carrier concentration and $\boldsymbol{k}_s \cdot \boldsymbol{B}$, where $\boldsymbol{k}_s$ is the scattering wave vector, the half width Γ can be as small as $0.15\ \text{cm}^{-1}$ at $n = 1 \times 10^{16}\ \text{cm}^{-3}$ [7.73].

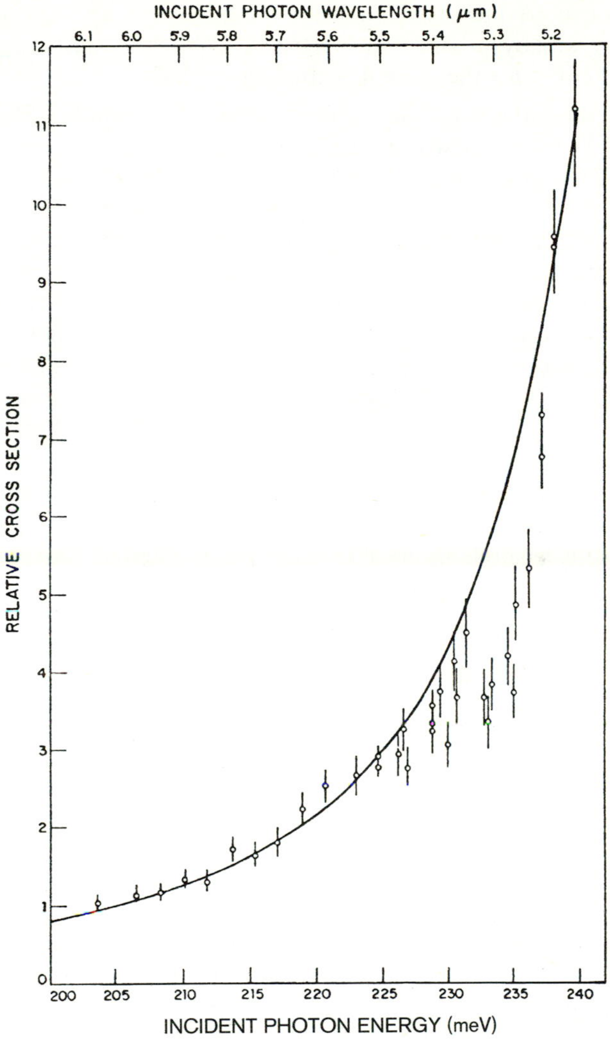

Fig. 7.16. Resonance enhancement of spontaneous spin-flip Raman scattering as a function of input photon energy ($n = 1 \times 10^{16}\ \mathrm{cm}^{-3}$, $H = 40$ kG, and $T \sim 30$ K). (After [7.72])

The above spontaneous scattering data clearly suggest that spin-flip SRS can easily be seen in InSb. By assuming a Lorentzian Raman line with a half width $\Gamma = 2\ \mathrm{cm}^{-1}$ and using (7.3) with $\varrho_i - \varrho_f = 1$, we find a spin-flip Raman gain $G_R = 1.7 \times 10^{-5}\, I\ \mathrm{cm}^{-1}$ in an n-type InSb with $n = N = 3 \times 10^{16}\ \mathrm{cm}^{-3}$, where I is the CO_2 laser intensity in $\mathrm{W/cm}^2$ [7.74]. As seen in Table 7.1, this is the largest known Raman gain for all materials.

The gain can be increased further by adjusting n appropriately with respect to Γ and by moving $\hbar\omega_1$ towards E_g. With a CO laser, G_R becomes $6 \times 10^{-4}\, I\, \mathrm{cm}^{-1}$ for the same n according to (7.48).

From the calculated gain, one expects that spin-flip SRS can be generated in InSb of few mm in length with a beam of $\sim 10^5$ W/cm^2 at 10.6 µm (CO_2 laser) or $\sim 10^3$ W/cm^2 at 5.3 µm (CO laser). PATEL and SHAW [7.13] first observed spin-flip SRS in InSb using a Q-switched CO_2 laser at 10.6 µm as the pump. With an input peak power of 1 KW focused into an area of 10^{-3} cm^2 in a sample of ~ 5 mm long and $n \approx 10^{16}$ cm^{-3} at $T \approx 18$ K, they obtained a Stokes peak power of 10 W. The Stokes frequency was tuned by an applied magnetic field according to (7.47). It was varied from 10.9 to 13.0 µm with B changing from 15 to 100 KG [7.74]. This then yields a practical tunable coherent source in the infrared. The Stokes output had a linewidth of less than 0.03 cm^{-1} [7.74].

BRUECK and MOORADIAN [7.75] found that spin-flip SRS in InSb could also be operated on the cw basis with a CO laser at 5.3 µm as the pump. Using a single-mode CO laser beam focused into an area of $\sim 5 \times 10^{-5}$ cm^2 in a 4.8-mm InSb with $n \approx 10^{16}$ cm^{-3} at $T \approx 30$ K, they obtained an SRS threshold of less than 50 mW, a power conversion efficiency of 50%, and an output power in excess of 1 W. DE SILETS and PATEL [7.76] have achieved a conversion efficiency of 80% with samples in a low magnetic field. The cw Stokes output can have a linewidth less than 1 KHz [7.77].

Anti-Stokes radiation and Stokes radiation up to the 4th order have been observed in the spin-flip Raman oscillator [7.75, 76, 78, 79]. The anti-Stokes is generated through coupling to the Stokes (Subsection 7.2.4) and the nth-order Stokes is generated by the $(n-1)$th-order Stokes (Subsection 7.2.5).

Very recently, stimulated spin-flip Raman scattering has also been observed in InAs by ENG et al. [7.79] using an HF laser pump. The threshold power was about 15 W and a conversion efficiency of 20% was achieved. This may become an important tunable source in the 3–5 µm region. SATTLER et al. [7.79] have observed tunable spin-flip Raman scattering in $Hg_{0.77}Cd_{0.23}Te$ pumped by a CO_2TEA laser. The output power could reach 1 W. It is believed that this tunable output is probably also the result of stimulated scattering.

The spin-flip transition can also be excited directly by an infrared wave through magnetic-dipole interaction. The interaction Hamiltonian for this direct process is

$$\mathscr{H}' = -g\mu_B \boldsymbol{\sigma} \cdot \boldsymbol{B}\,. \tag{7.49}$$

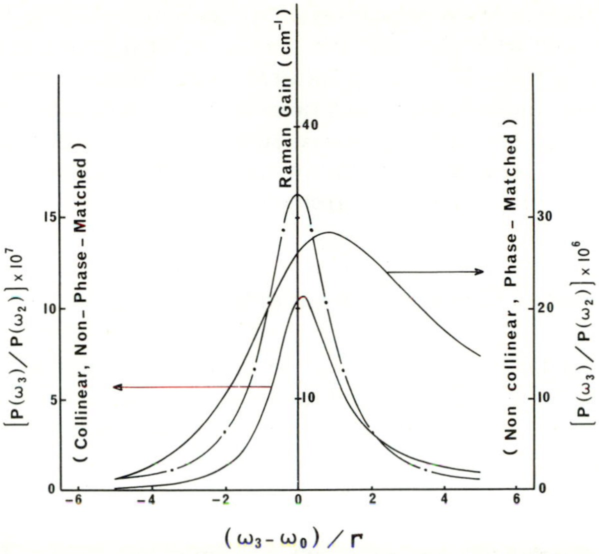

Fig. 7.17. Theoretical curves of the Raman gain g, and the ratios of the far-infrared output $P(\omega_3)$ to the Raman output $P(\omega_2)$ for the collinear phase-mismatched case and for the noncollinear phase-matched case. (See [7.16])

Since the spin-flip excitation can be excited by both the Raman process and direct absorption, we should in fact use the theory of stimulated polariton scattering in Section 7.4 to describe the spin-flip SRS. In this case, the effective interaction energies $\langle \mathscr{H}_{\text{eff}}^{(1)} \rangle$ and $\langle \mathscr{H}_{\text{eff}}^{\text{NL}} \rangle$ in (7.38) and (7.40) can be easily obtained from (7.49) and (7.46), and hence A_{fi} and M_{fi} are readily known [7.14]. The solution in (7.44) and (7.45) is then directly applicable to spin-flip SRS.

In the present case, we always have $(\gamma_s + \gamma_3)^2 \gg \Lambda$ because the background absorption at ω_3 is strong. As a result, the stimulated polariton gain is almost exactly equal to the stimulated Raman gain assuming $\Lambda = 0$, as shown in Fig. 7.17. We should, however, note that in stimulated polariton scattering, we expect to have the far-infrared generated together with the Stokes. In Fig. 7.17, we have also shown the ratio of the far-infrared output to the Stokes output for the collinear phase-mismatching case and for the non-collinear phase-matching case. The non-collinear phase-matching case appears to have a stronger far-infrared output.

No observation of direct far-infrared output from a spin-flip Raman oscillator has been reported yet, and only the collinear phase-nonmatching case has been tried [7.80]. However, by feeding both the laser and the Stokes into an InSb crystal, pulsed far-infrared output has been detected. Its maximum appears right at resonance [7.81]. The results agree very well with what the theory of stimulated polariton scattering would predict [7.14, 82]. The far-infrared output from InSb can of course be tuned over the same frequency range as the Stokes [7.83]. This therefor gives us a potential far-infrared source which is intense, coherent, and tunable. It is possible that the cw spin-flip SRS can also lead to a cw tunable far-infrared source of extremely narrow linewidth.

7.6. Transient Stimulated Raman Scattering

We have so far considered only the steady-state case of stimulated Raman emission. This is manifested by the time-independent wave equations we used in the theory. In the experiments, however, laser pulses are often used. Therefore, in general, we cannot neglect the time-dependence in the wave equations.

We can, however, still assume slow amplitude variation of the fields. The time-dependent wave equations corresponding to (7.5), (7.29), and (7.32) for infinite plane waves propagating in the forward direction along $\hat{z}$ in an isotropic medium are

$$\begin{aligned}
&\left(\frac{\partial}{\partial z}+\frac{1}{v_1}\frac{\partial}{\partial t}\right)\mathcal{E}_1(z,t)=\mathrm{i}\left(\frac{2\pi\omega_1^2}{c^2k_1}\right)N(M_{fi}^{\mathrm{s}})^*(\varrho_i-\varrho_f)\mathcal{E}_2A\\
&\left(\frac{\partial}{\partial z}+\frac{1}{v_{\mathrm{s}}}\frac{\partial}{\partial t}\right)\mathcal{E}_{\mathrm{s}}(z,t)=\mathrm{i}\left(\frac{2\pi\omega_{\mathrm{s}}^2}{c^2k_{\mathrm{s}}}\right)NM_{fi}^{\mathrm{s}}(\varrho_i-\varrho_f)\mathcal{E}_1A^*\\
&\hbar(\partial/\partial t+\Gamma)A(z,t)=\mathrm{i}M_{fi}^{\mathrm{s}}\mathcal{E}_1\mathcal{E}_{\mathrm{s}}^*\,,
\end{aligned}\tag{7.50}$$

where $\mathcal{E}_1$ and $\mathcal{E}_2$ were defined in (7.4), A is defined through the relation $\psi=A\exp[\mathrm{i}(k_1-k_{\mathrm{s}})z-\mathrm{i}\omega_{fi}t]$, $\omega_{\mathrm{s}}=\omega_1-\omega_{fi}$, v_1 and v_{s} are the group velocities at ω_1 and ω_{s},respectively.

Consider first the case where the amplitude variations of E_1 and E_{s} are sufficiently slow, so that $|\partial A/\partial t|$ is negligible compared with $|\Gamma A|$. Then, the material excitation $A(z,t)=\mathrm{i}M_{fi}^{\mathrm{s}}\mathcal{E}_1\mathcal{E}_{\mathrm{s}}^*/\hbar\Gamma$ follows almost instantaneously the time variation of $\mathcal{E}_1\mathcal{E}_{\mathrm{s}}^*$. If the dispersion of the medium is also negligible, then $v_1=v_{\mathrm{s}}$. By transformation of variables $z'=z$ and $t'=t-z/v$, (7.50) reduces to

$$\begin{aligned}
&\partial\mathcal{E}_1/\partial z'=\mathrm{i}(2\pi\omega_1^2/c^2k_1)\chi_{\mathrm{R}}^{(3)*}|\mathcal{E}_2|^2\mathcal{E}_1\\
&\partial\mathcal{E}_{\mathrm{s}}/\partial z'=\mathrm{i}(2\pi\omega_{\mathrm{s}}^2)/c^2k_{\mathrm{s}})\chi_{\mathrm{R}}^{(3)}|\mathcal{E}_1|^2\mathcal{E}_{\mathrm{s}}\\
&\quad\chi_{\mathrm{R}}^{(3)}=N|M_{fi}^{\mathrm{s}}|^2(\varrho_i-\varrho_f)\mathrm{i}\hbar\Gamma\,.
\end{aligned}\tag{7.51}$$

These equations are identical to (7.7) except that $\mathscr{E}_1$ and $\mathscr{E}_s$ are now functions of z and $t-z/v$. In other words, $\mathscr{E}_1$ and $\mathscr{E}_s$ follow the steady-state variation in the retarded time coordinate. This is the quasi-steady-state case. Using the retarded time, all our previous theoretical discussions are valid for this case.

For backward SRS, however, we should replace v_s in (7.50) by $-v_s$. The steady-state solution is no longer applicable unless the amplitude variations of the input pulses are negligible in the time which takes light to traverse the entire length of the medium. MAIER et al. [7.50] have found the following general solution for this case (assuming $|v_1|=|v_s|$)

$$|\mathscr{E}_s|^2\left(t+\frac{z}{v},t-\frac{z}{v}\right)=\frac{|\mathscr{E}_s|^2(t+z/v,0)}{F_s(t+z/v)+\exp[-F_1(t-z/v)]},$$

where

$$\begin{aligned}F_s(r+z/v)&=\int_0^{t+z/v} g|\mathscr{E}_s(y,0)|^2\,dy\\ F_1(t-z/v)&=\int_0^{t-z/v} g|\mathscr{E}_1(0,y)|^2\,dy\\ g&=-(4\pi\omega_s^2/c^2k_s)\,\mathrm{Im}\{\chi_R^{(3)}\}\,.\end{aligned}\tag{7.52}$$

For given initial conditions at $z=0$, one can then calculate the output Stokes intensity. An example is shown in Fig. 7.18. It is seen that with a sufficiently long medium, the backward Stokes pulse is sharpened through amplification drastically. Physically, pulse sharpening occurs because the wavefront of the backward Stokes pulse continuously sees the fresh undepleted incoming laser beam and gets full amplification while the lagging part of the pulse does not. This phenomenon was actually observed in Kerr liquids where the initial Stokes pulse was generated by self-focusing at the end of the cell [7.50].

Consider next the case where $\mathscr{E}_1\mathscr{E}_s^*$ varies rapidly such that in (7.50) $|\partial A/\partial t|$ is no longer negligible compared with $|\Gamma A|$. This means that even the material excitation cannot reach its steady-state value during the pulse. Consequently, the forward Stokes generation also shows a transient behavior. This type of transient SRS occurs when the laser pulsewidth T_P is smaller than, or comparable with the de-phasing time $T_2=1/\Gamma$ of the material excitation, or more rigorously, when $T_P<G_{Rm}lT_2$ where G_{Rm} is the steady-state Raman gain in (7.9) at the peak of the input pulse and l is the length of the medium [7.84, 85]. In gases, T_2 is of the order of 10^{-9} sec or longer. In liquids, T_2 is usually in the range of picoseconds. One must then use picosecond mode-locked pulses in general to study

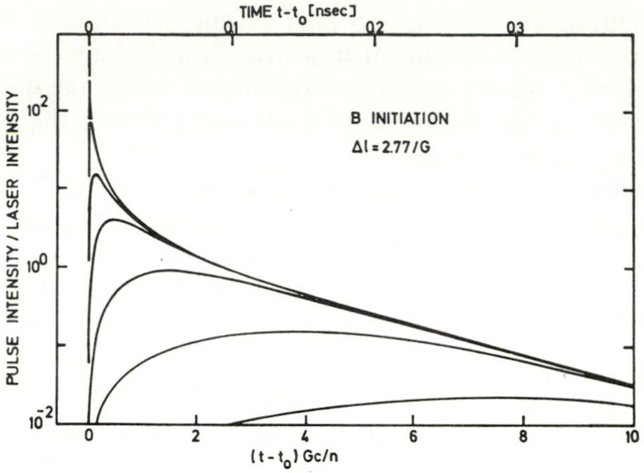

Fig. 7.18. Calculated normalized Raman pulse intensity as a function of time for an initial condition $|E_s| = |E_{s0}|(t-t_0)^3$ for $t > t_0$. The curves show the pulse development at length intervals of $\Delta l = 2.77/\mathrm{G}$. G is the Raman gain and was determined to be $0.7\,\mathrm{cm}^{-1}$ in CS_2. Lower scale is in dimensionless units; upper scale describes the experimental conditions. (After [7.50])

transient SRS, although Q-switched pulses may be used in some gas media [7.86]. With picosecond pulses, the backward Raman pulse intensity is hardly detectable because of the very limited interaction length with the incoming laser pulse. We can therefore limit our discussion to forward SRS only.

The theory of transient SRS has been discussed by many authors [7.85, 87–91]. It closely resembles the theory of transient stimulated Brillouin scattering [7.84]. Assume that both the depletion of laser power and the induced change of the population difference $(\varrho_i - \varrho_f)$ are negligible. Equation (7.50) then reduces to (assuming $v_1 = v_s$)

$$\begin{aligned} \left(\frac{\partial}{\partial z} + \frac{1}{v}\frac{\partial}{\partial t}\right)\mathscr{E}_s &= \mathrm{i}\eta_1 \mathscr{E}_1(t - z/v)A^* \\ \left(\frac{\partial}{\partial t} + \Gamma\right)A^* &= -\mathrm{i}\eta_2 \mathscr{E}_1^*(t - z/v)\mathscr{E}_s\,, \end{aligned} \tag{7.53}$$

where

$$\begin{aligned} \eta_1 &= (2\pi\omega_s^2/c^2 k_s)\, N M_{fi}^s(\varrho_i - \varrho_f) \\ \eta_2 &= (M_{fi}^s)^*/\hbar\,, \end{aligned}$$

and $\mathscr{E}_1(t - z/v)$ is prescribed by the initial condition.

Combining the two equations in (7.53) and using the new variables $z' = z$ and $t' = t - z/v$, we obtain a second-order partial differential equation [7.91]

$$[\partial^2/\partial t'\partial z' - \eta_1\eta_2|\mathscr{E}_1(t')|^2]U = 0\,, \tag{7.54}$$

where $U = F\exp(\Gamma t')$ and F stands for either A^* or $\mathscr{E}_s$. By defining $\tau = \int_{-\infty}^{t'} |\mathscr{E}_1(t'')|^2 dt''$ as the integrated energy in the laser pulse up to t', the equation is further reduced to the standard form of a hyperbolic equation

$$(\partial^2/\partial\tau\partial z' - \eta_1\eta_2)U = 0$$

which can now be solved with arbitrary initial conditions. The solution takes the form [7.91],

$$\begin{aligned}\mathscr{E}_s(z',t') &= \mathscr{E}_s(0,t') + (\eta_1\eta_2 z)^{\frac{1}{2}}\mathscr{E}_1(t')\int_{-\infty}^{t'} e^{-\Gamma(t'-t'')}\\ &\quad\cdot\{\mathscr{E}_1^*(t'')\mathscr{E}_s(0,t'')[\tau(t')-\tau(t'')]^{-\frac{1}{2}}\mathrm{I}_1(2[\eta_1\eta_2(\tau(t')-\tau(t''))z']^{\frac{1}{2}})\}dt''\\ A^*(z',t') &= \mathrm{i}\eta_1\int_{-\infty}^{t'} e^{-\Gamma(t'-t'')}\\ &\quad\cdot\mathscr{E}_1^*(t'')\mathscr{E}_s(0,t'')\mathrm{I}_0(2[\eta_1\eta_2(\tau(t')-\tau(t''))z']^{\frac{1}{2}})\}dt''\,,\end{aligned} \tag{7.55}$$

where the input conditions are $A^*(z') = 0$ at $t' \to -\infty$ and $\mathscr{E}_s(z',t') = \mathscr{E}_s(0,t')$ at $z = z' = 0$, and I_i is the ith order Bessel function of imaginary argument.

The Bessel functions have the asymptotic limits that $\mathrm{I}_0(x) \cong 1$ and $\mathrm{I}_1(x) \cong x$ for $x \ll 1$, and $\mathrm{I}_i(x) \cong (2\pi x)^{\frac{1}{2}}\exp(x)$ for $x \gg 1$. The Stokes amplitude therefore first increases linearly with z. Then, in the limit of large amplification, it varies exponentially in the form

$$\begin{aligned}\mathscr{E}_s(z',t') &\propto \mathscr{E}_1(t')\int_{-\infty}^{t'} \mathscr{E}_1^*(t'')\mathscr{E}_s(0,t'')[\tau(t')-\tau(t'')]^{-1}\\ &\quad\cdot\exp\{-\Gamma(t'-t'') + 2[\eta_1\eta_2(\tau(t')-\tau(t''))z']^{\frac{1}{2}}\}\,.\end{aligned} \tag{7.56}$$

For a rectangular laser pulse, it is easy to show from (7.56) that if the pulse is sufficiently long, then $\mathscr{E}_s$ takes on a steady-state exponential gain when $(t - t_0) > G_R z T_2$ where t_0 is the starting time of the pulse. For this reason, T_P (pulsewidth) $< G_{Rm} z T_2$ is used as the condition for the occurrence of transient SRS as we mentioned before.

If $T_P < T_2$, thr factor $\exp[-\Gamma(t'-t'')]$ can be neglected during the laser pulse. We can see from (7.55) and (7.56) that the Stokes generated does not grow appreciably at the leading edge of the laser pulse, but increases rapidly towards the middle part. It finally drops off following the laser pulse shape at the tail. Therefore, the Stokes peak always appears after the laser peak and the Stokes pulse should be narrower than the laser pulse. The material excitation behaves in a similar manner, but towards the tail it decays exponentially as $\exp(-\Gamma t)$ even after the laser intensity has dropped to almost zero. In the limit of large amplification, we have from (7.56) [7.91]

$$(\mathscr{E}_s)_{\max} \propto \exp(G_T z/2)\,, \tag{7.57}$$

where the transient gain G_T is given by

$$G_T = 4[\eta_1 \eta_2 \langle |\mathscr{E}_1|^2 \rangle T_P/z]^{\frac{1}{2}}$$

$$\langle |\mathscr{E}_1|^2 \rangle T_P \equiv \int_{-\infty}^{\infty} |\mathscr{E}_1(t)|^2 dt\,.$$

This transient gain is then independent of the laser pulse shape. For a pulse of the form $\mathscr{E}_1(t') = \mathscr{E}_{1m} \exp(-|t'/T|^n)$, it has been shown [7.91] that the peak of the Stokes pulse is delayed from the laser peak by a time $t_D = T(\frac{1}{2} \log G_{Rm} z)^{1/n}$.

CARMAN et al. [7.91] have carried out numerical calculations of transient SRS for various pulse shapes. Their results of $G_T z$ for Gaussian laser input pulses of different pulsewidths versus the maximum steady-state gain $G_{mR} z$ are presented in Fig. 7.19. It shows that when $T_P > G_{Rm} z T_2$, we have $G_T \simeq G_{Rm}$, and otherwise, $G_T < G_{Rm}$ as the transient gain should be. At sufficiently large $G_{Rm} z$, the curves also show a $z^{\frac{1}{2}}$ dependence for $G_T z$ as predicted by (7.57). The variation of the time delay t_D with G_T for Gaussian laser pulses of different pulse widths is shown in Fig. 7.20. As expected, t_D increases with G_T, more rapidly at large values of G_T. In Fig. 7.20, sharpening of the Stokes pulse with increasing G_T is also shown. The effects of phase modulation and linear dispersion have also been studied [7.91].

Earlier experiments by HAGENLOCKER et al. [7.86] first suggested the transient behavior of SRS in gases with Q-switched laser pulses. Later, with picosecond mode-locked pulses, transient effects of SRS in liquids were observed [7.92–97]. Since the transient gain G_T depends only on the total Raman cross-section ($\propto \eta_1 \eta_2$) while the steady-state gain G_{Rm} is also inversely proportional to the linewidth Γ, it is possible to observe in transient SRS some Raman modes which are not observed in steady-

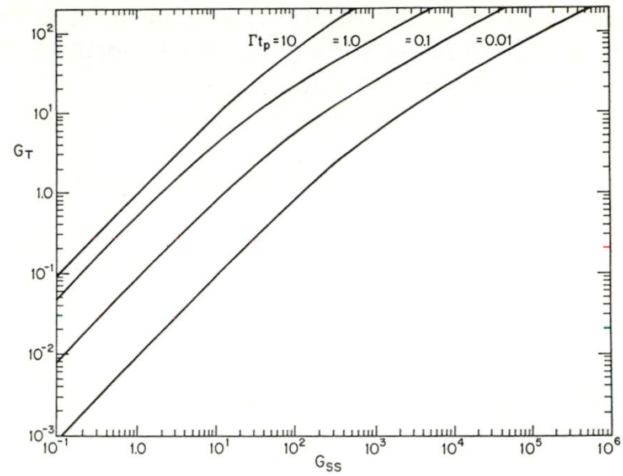

Fig. 7.19. The transient Raman gain coefficient for Gaussian laser input pulses with the same total energy, but different pulse widths. The steady state gain coefficient G_{SS} corresponds to a constant intensity laser output equal to the maximum laser pulse intensity. (After [7.91])

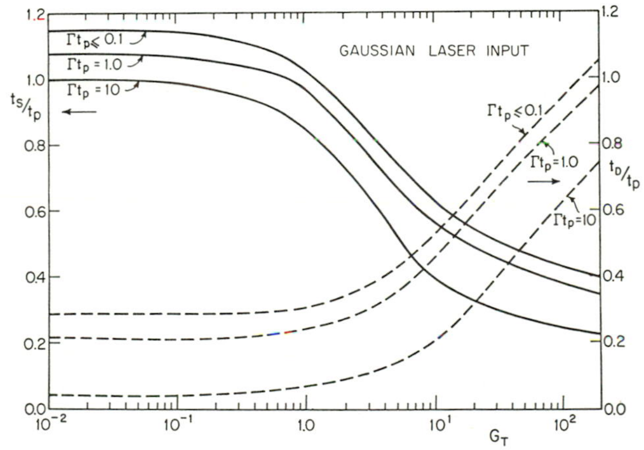

Fig. 7.20. The variation of Stokes pulse width t_S and delay t_D with transient gain coefficient, for Gaussian laser input pulses of various widths $t_\circ$ given in terms of the optical phonon dephasing time Γ^{-1}. (After [7.91])

state SRS [7.96]. More than one Raman mode can show up in transient SRS [7.96, 98]. CARMAN and MACK [7.99] have recently made quantitative measurements on transient SRS in SF_6 gas. They chose SF_6 because of its small linear dispersion and the absence of other nonlinear effects during SRS. Their results agree well with the predictions of the theory. A better experiment to test the theory of transient SRS is, however, to measure the temporal variation of Stokes amplification in an amplifier cell as we discussed earlier in Subsection 7.3.2. Such an experiment, however, has not yet been carried out.

7.7. Applications of Stimulated Raman Scattering

We have seen how SRS can be used to generate coherent light at new frequencies. There are a number of other applications of SRS. We shall discuss only a few of them in the following.

7.7.1. Measurements of Vibrational Lifetimes

In analogy to the magnetic resonance case, there are two characteristic lifetimes for each molecular or lattice vibration: the longitudinal lifetime T_1 and the transverse (or dephasing) lifetime T_2 [7.34]. These lifetimes cannot be obtained simply from linewidth measurements. Only in the limit of homogeneous broadening is T_2 equal to the inverse halfwidth, but even then T_2 can be very different from T_1. The two lifetimes can, however, be measured directly by watching the vibrational decay after a short pulse excitation. Both the pulse excitation and the monitoring of vibrational decay can be achieved through Raman transitions. In the case of condensed matter, the vibrational lifetimes are often in the picosecond range. Then, the SRS method appears to be the only way to measure the vibrational lifetimes directly.

As discussed in Section 7.2 and 7.6, the time-dependent excitation of the coherent vibrational field and the population of excited molecular vibration is governed by the following set of equations [from (7.50), (7.24), and (7.25)]

$$\begin{aligned}
\left(\frac{\partial}{\partial z}+\frac{1}{v_1}\frac{\partial}{\partial t}\right)\mathscr{E}_1(z,t) &= \mathrm{i}\left(\frac{2\pi\omega_1^2}{c^2k_1}\right)N\left(\frac{2\omega_0}{\hbar}\right)^{\frac{1}{2}}(M_{fi}^{\mathrm{s}})^*(\varrho_i-\varrho_f)\mathscr{E}_{\mathrm{s}}Q \\
\left(\frac{\partial}{\partial z}+\frac{1}{v_{\mathrm{s}}}\frac{\partial}{\partial t}\right)\mathscr{E}_{\mathrm{s}}(z,t) &= \mathrm{i}\left(\frac{2\pi\omega_{\mathrm{s}}^2}{c^2k_{\mathrm{s}}}\right)N\left(\frac{2\omega_0}{\hbar}\right)^{\frac{1}{2}}M_{fi}^{\mathrm{s}}(\varrho_i-\varrho_f)\mathscr{E}_1Q^* \\
\left(\frac{\partial}{\partial t}+\frac{1}{T_2}\right)Q(z,t) &= \mathrm{i}M_{fi}^{\mathrm{s}}(2\hbar\omega_0)^{-\frac{1}{2}}\mathscr{E}_1\mathscr{E}_{\mathrm{s}}^* \\
\left(\frac{\partial}{\partial t}+\frac{1}{T_1}\right)\Delta\varrho &= \frac{\mathrm{i}N}{2\hbar}\left(\frac{2\omega_0}{\hbar}\right)^{\frac{1}{2}} \\
&\quad\cdot[M_{fi}\mathscr{E}_1\mathscr{E}_{\mathrm{s}}^*Q^*-M_{fi}^*\mathscr{E}_1^*\mathscr{E}_2Q](\varrho_i-\varrho_f) \\
\varrho_i-\varrho_f &= \varrho_i^0-\varrho_f^0-2\Delta\varrho\,.
\end{aligned} \tag{7.58}$$

We assume that the experiment is designed to have negligible depletion of the $\mathscr{E}_1$ field and $\Delta\varrho \ll (\varrho_i^0 - \varrho_f^0)$. Thus, with $\mathscr{E}_1 = \text{constant}$ and $(\varrho_i - \varrho_f)$ replaced by $(\varrho_i^0 - \varrho_f^0)$, the solution for $\mathscr{E}_s(z, t)$ and $Q(z, t)$ is just the transient solution discussed in the previous section [7.91]. The solution for $\Delta\varrho(z, t)$ is then obtained by solving the equation of $\Delta\varrho$ with the calculated $\mathscr{E}_s(z, t)$ and $Q(z, t)$. The temporal behavior of both Q and $\Delta\varrho$ can be monitored experimentally by a probing beam. The coherent vibrational field Q should scatter the probing beam coherently in the phase-matched direction. The population change $\Delta\varrho$ gives rise to a change in the intensity of the incoherent Stokes and anti-Stokes scattering.

However, since T_1 and T_2 of condensed matter are in the picosecond range, it is difficult to monitor the continuous variation of scattering of the probing beam with time electronically. We must resort to optical means with the help of picosecond pulses. We can use a picosecond pulse at ω_1 to excite the vibration via Stokes transition and then another picosecond pulse at ω_2, time-delayed by t_D from the first pulse, to probe Q and $\Delta\varrho$ via anti-Stokes scattering. For the coherent anti-Stokes scattering in the phase-matching direction, the anti-Stokes field $\mathscr{E}_a(z, t)$ at frequency $\omega_2 + \omega_0$ obeys the equation (see Section 7.2)

$$\left(\frac{\partial}{\partial z} + \frac{1}{v_a}\frac{\partial}{\partial t}\right)\mathscr{E}_a(z, t) = \mathrm{i}\left(\frac{2\pi\omega_a^2}{c^2 k_a}\right) N \left(\frac{2\omega_0}{\hbar}\right)^{\frac{1}{2}} \cdot (M_{fi}^a)^* (\varrho_i^0 - \varrho_f^0)\mathscr{E}_2(z, t) Q(z, t + t_D)\,, \tag{7.59}$$

where $\mathscr{E}_2$ is the probing pulse field, and the non-resonant contribution to the nonlinear polarization is neglected. The anti-Stokes output from the medium as a function of the time delay t_D is given by

$$\begin{aligned} S^{\mathrm{coh}}(t_D) &\propto \int |\mathscr{E}_a(l, t' = t - z/v)|^2 dt' \\ &= (\text{constant}) \int dt' \left| \int dz \mathscr{E}_2(z, t') Q(z, t' + t_D)\right|^2 . \end{aligned} \tag{7.60}$$

The incoherent (spontaneous) anti-Stokes scattering gives a signal.

$$S^{\mathrm{inc}}(t_D) = (\text{constant}) \int dt\, dz |\mathscr{E}_2(z, t)|^2 \Delta\varrho(z, t + t_D)\,. \tag{7.61}$$

Knowing the exciting and probing pulses, and the solution for $Q(z, t')$ in (7.55) [replacing A by $(2\omega_0/\hbar)^{\frac{1}{2}} Q$], we can then calculate S^{coh} and S^{inc} from the above equations. It can be shown from (7.55) that if the exciting pulse has a pulsewidth T_P smaller than or comparable with T_2, then $Q(z, t)$ will show an exponential decay $\exp(-t/T_2)$ at large t, or $S^{\mathrm{coh}}(t_D) \propto \exp(-t_D/T_2)$ for $t_D \gg T_2$. Similarly, from the equation for $\Delta\varrho$ in (7.58),

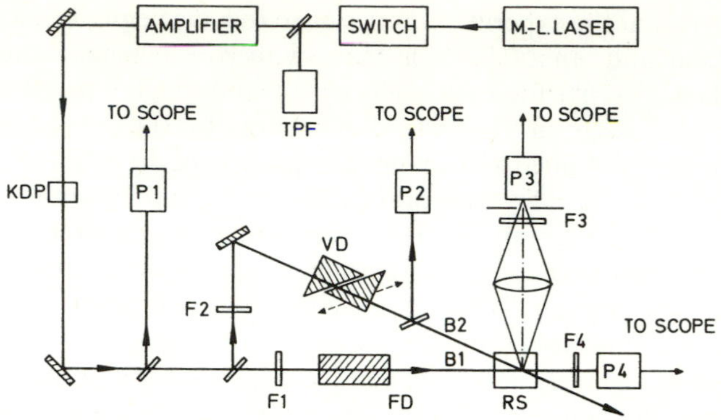

Fig. 7.21. Schematic of the experimental system for phonon lifetime measurement. The pump beam B1 at $\lambda \simeq 1.06$ μm and the probe beam B2 at $\lambda \simeq 0.53$ μm interact in the Raman sample RS. Glass rod for fixed optical delay: FD; glass prisms for variable delay: VD; filter: F; photodetector: P; two-photon-fluorescence system: TPF. (After [7.104])

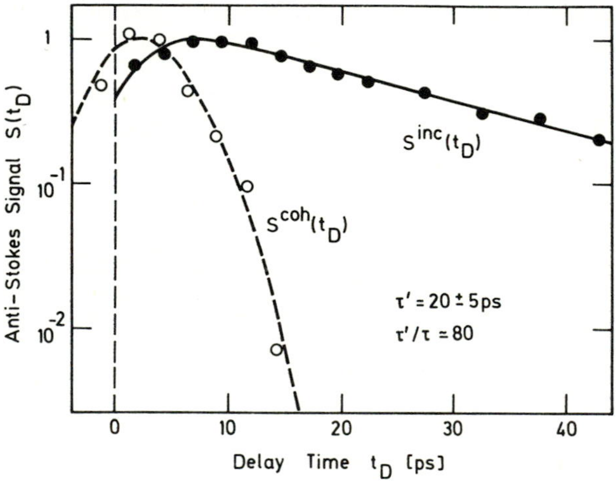

Fig. 7.22. Measured incoherent scattering $S^{\mathrm{inc}}(t_{\mathrm{D}})/S^{\mathrm{inc}}_{\mathrm{max}}$ (closed circles) and coherent scattering $S^{\mathrm{coh}}(t_{\mathrm{D}})/S^{\mathrm{coh}}_{\mathrm{max}}$ (open circles) versus delay time t_{D} for ethyl alcohol. The solid and dashed curves are calculated. (After [7.104])

we realize that $\Delta\varrho(z,t)$ or $S^{\mathrm{inc}}(t_{\mathrm{D}})$ will show an exponential decay tail with a time constant T_1. Thus, if $T_{\mathrm{P}} \lesssim T_1, T_2$, then we can obtain T_1 and T_2 directly by measuring the exponential decays of $S^{\mathrm{coh}}(t_{\mathrm{D}})$ and $S^{\mathrm{inc}}(t_{\mathrm{D}})$.

DEMARTINI and DUCUING [7.100] first used such a method to measure T_1 of the $4155\ \mathrm{cm}^{-1}$ vibrational excitation of gaseous H_2. In this case, T_1 is about 30 μsec at 0.03 atmospheric pressure. Ultrashort laser pulses are not needed for the measurements. Recently, ALFANO and SHAPIRO [7.101], and KAISER and his associates [7.102–105] have used picosecond mode-locked laser pulses to measure T_1 and T_2 of molecular or lattice vibration in liquids and solids. One of their experimental arrangements is shown in Fig. 7.21. The mode-locked pulse from a Nd glass laser is used to excite the vibration by SRS, and the second-harmonic of the mode-locked pulse is used to probe the vibrational excitation. An example of their results is shown in Fig. 7.22. The exponential tails of the $S^{\mathrm{inc}}(t_{\mathrm{D}})$ and $S^{\mathrm{coh}}(t_{\mathrm{D}})$ curves in the figure yield T_1 and T_2 readily. These were the first direct measurements of vibrational relaxation times in condensed matter. The same technique can of course be applied to the measurements of relaxation times of other types of excitations. By detecting the incoherent anti-Stokes signal at various frequencies ω_{a} as a function of t_{D}, one can also study the decay routes of a particular excitation [7.106, 107].

7.7.2. Measurements of Third-Order Nonlinear Susceptibilities

Consider now the steady-state case where the material excitation ψ with a resonant frequency ω_0 is driven by the incoming fields E_1 and E_{s} at ω_1 and ω_{s} respectively. A third incoming field E_2 at ω_2 is then used to probe ψ and the coherent anti-Stokes scattering at $\omega_{\mathrm{a}} = \omega_1 - \omega_{\mathrm{s}} + \omega_2$ is detected. From the discussion in Subsection 7.2.4, the anti-Stokes field E_{a} obeys the equation

$$\frac{\partial \mathscr{E}_{\mathrm{a}}}{\partial z} = \mathrm{i}\left(\frac{2\pi\omega_{\mathrm{a}}^2}{c^2 k_{\mathrm{a}z}}\right)\chi^{(3)}\mathscr{E}_2\mathscr{E}_1\mathscr{E}_{\mathrm{s}}^* \mathrm{e}^{\mathrm{i}\Delta k z}, \tag{7.62}$$

where

$$\Delta \boldsymbol{k} = \boldsymbol{k}_1 + \boldsymbol{k}_2 - \boldsymbol{k}_{\mathrm{s}}$$

$$\chi^{(3)} \equiv \hat{e}_{\mathrm{a}} \cdot \boldsymbol{\chi}^{(3)} : \hat{e}_2 \hat{e}_1 \hat{e}_{\mathrm{s}} = \chi_{\mathrm{R}}^{(3)} + \chi_{\mathrm{NR}}^{(3)}$$

$$\chi_{\mathrm{R}}^{(3)} = -N M_{fi}^{\mathrm{s}} (M_{fi}^{\mathrm{a}})^* (\varrho_i^0 - \varrho_f^0)/\hbar(\omega_1 - \omega_{\mathrm{s}} - \omega_0 + \mathrm{i}\Gamma),$$

and $\chi_{\mathrm{NR}}^{(3)}$ is the non-resonant contribution to $\chi^{(3)}$. At phase matching $\Delta k = 0$, the anti-Stokes output is given by

$$|\mathscr{E}_{\mathrm{a}}|^2 = \left(\frac{2\pi\omega_{\mathrm{a}}^2}{c^2 k_{\mathrm{a}z}}\right)^2 |\chi^{(3)}|^2 |\mathscr{E}_1|^2 |\mathscr{E}_2|^2 |\mathscr{E}_{\mathrm{s}}|^2 l^2. \tag{7.63}$$

With the incoming beam intensities kept constant, $|\mathscr{E}_a|^2$ as a function of $\omega_1 - \omega_s$ is a maximum when $|\chi^{(3)}|$ is a maximum.

If the ω_0 resonance is a narrow one, and the other resonances are far away, the dispersion of $\chi_{NR}^{(3)}$ is negligible when $\omega_1 - \omega_s$ varies around ω_0 over a few linewidths. Then, we can easily show from (7.62) that the maximum and minimum of $|\chi^{(3)}|$ occur at $(\omega_1 - \omega_s)_+$ and $(\omega_1 - \omega_s)_-$ respectively, with

$$(\omega_1 - \omega_s)_\pm = \omega_0 + \frac{1}{2}\left\{ -\frac{a}{\chi_{NR}^{(3)}} \pm \left[\left(\frac{a}{\chi_{NR}^{(3)}}\right)^2 + \Gamma^2\right]^{\frac{1}{2}} \right\}, \tag{7.64}$$

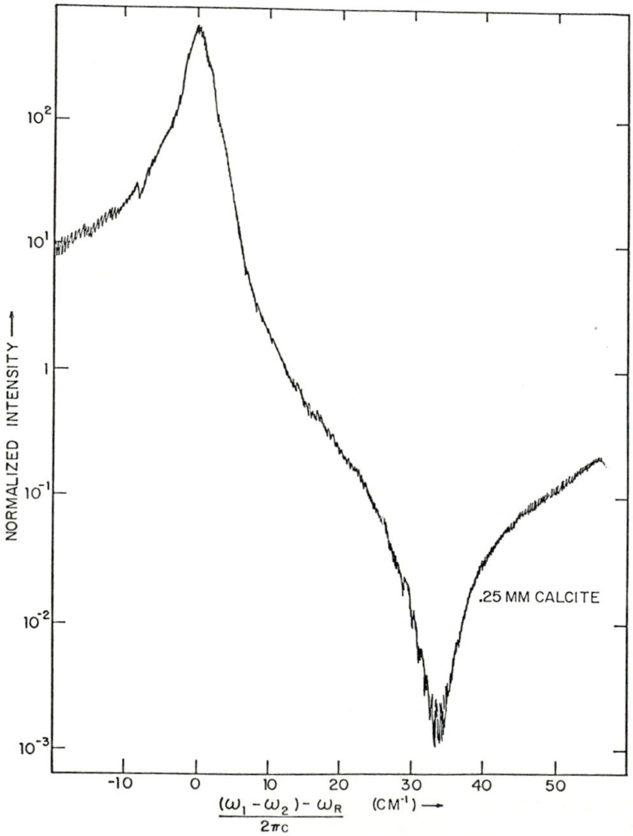

Fig. 7.23. Typical results for the anti-Stokes intensity at ω_3 in calcite. The polarizations were perpendicular to the optic axis of calcite. (After [7.112])

where

$$a = -NM^{s}_{fi}(M^{a}_{fi})^{*}(\varrho^{0}_{i} - \varrho^{0}_{f})/\hbar$$

such that $\chi^{(3)}_{R} = a/(\omega_1 - \omega_s - \omega_0 + i\Gamma)$. An example of $|\chi^{(3)}_{R}|$ versus $(\omega_1 - \omega_s)$ around ω_0 is shown in Fig. 7.23. From (7.64), we have

$$\begin{aligned} &(\omega_1 - \omega_s)_+ + (\omega_1 - \omega_s)_- = 2\omega_0 - a/\chi^{(3)}_{NR} \\ &[(\omega_1 - \omega_s)_+ - (\omega_1 - \omega_s)_-]^2 = (a/\chi^{(3)}_{NR})^2 + 4\Gamma^2 \,. \end{aligned} \quad (7.65)$$

In many cases where $|a/\chi^{(3)}_{NR}| \gg \Gamma$, the maximum of $|\chi^{(3)}|$ appears at $\omega_1 - \omega_s = \omega_0$, and then from (7.65) we can find $a/\chi^{(3)}_{NR}$ and Γ by measuring $(\omega_1 - \omega_s)_\pm$. We can, of course, determine ω_0 and Γ easily from spontaneous Raman scattering data. The interesting thing here is that we can now

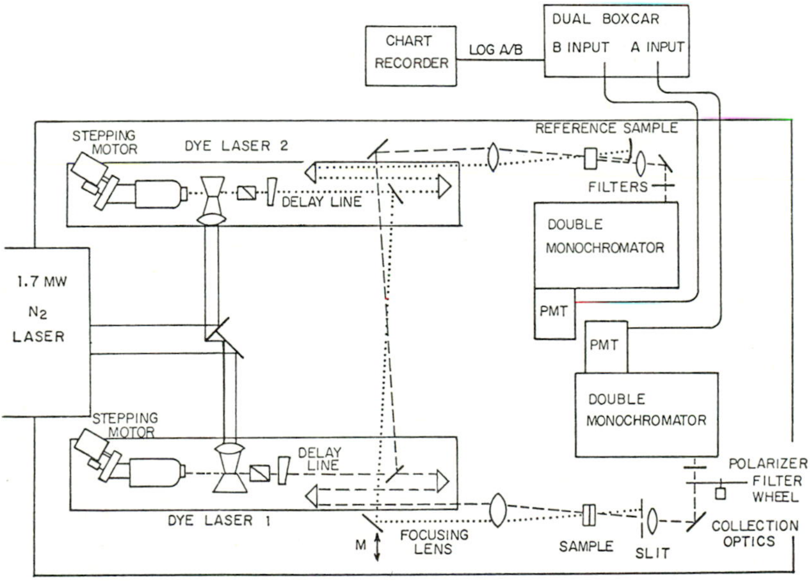

Fig. 7.24. The nonlinear spectroscopy system. The output of laser 1 is shown as a dashed line, that of laser 2 as a dotted line. Both laser beams are blocked after the samples, and the output due to frequency mixing is collected and directed into the monochromators. (After [7.112]). The signal from the reference arm and the ratioing system is used to compensate for laser fluctuations

obtain $a/\chi_{NR}^{(3)}$ directly from frequency measurements, and with the coefficient a deduced from the spontaneous Raman cross-section we can find $\chi_{NR}^{(3)}$. If the particular Raman mode is far away from other modes, then $\chi_{NR}^{(3)}$ is mainly due to electronic contribution. We can also obtain $\chi_{NR}^{(3)}$ by measuring the absolute intensity of the coherent anti-Stokes signal, but frequency measurements should be much more accurate than absolute intensity measurements.

This type of wave-mixing experiments has been carried out with $\omega_1-\omega_s$ around Raman-active vibrational modes in various materials by WYNNE [7.108, 109], and by BLOEMBERGEN and his associates [7.110–113a]. More recently, using this method, LEVENSON [7.112] and LEVENSON and BLOEMBERGEN [7.113] have measured $\chi_{NR}^{(3)}$ for a large number of liquids and solids. Their experimental set-up is shown in Fig. 7.24. Two N_2-laser-pumped dye lasers were used as the tunable sources. Their results for calcite are shown in Fig. 7.23. When there are several Raman-active modes close together, the analysis becomes somewhat more complicated, but it is clear that with the help of the spontaneous Raman scattering data, we can still deduce $\chi_{NR}^{(3)}$ from the wave-mixing results [7.113a]. In a crystal with no inversion symmetry, the modes can be both Raman active and infrared active. The coupled infrared and material excitational wave for the polariton wave discussed in Section 7.4 must now be used instead of ψ in the above formalism. In addition, the field at $\omega_a=\omega_1-\omega_s+\omega_2$ can also be generated by another two-step process in which the sum-frequency field $E(\omega_1+\omega_2)$ is first created and then mixed with $E^*(\omega_s)$ to produce $E(\omega_a)$. The analysis for this case is more complicated but quite straightforward [7.111, 114]. Interference between various contributions to E_a can be observed in the dispersion measurements [7.111]. We note again that the above analysis is applicable to any material excitation such as magnon, plasmon, etc., although experiments so far have only been performed on Raman-active vibrational modes.

7.7.3. Detection of Low-Concentration Substances

As is well known, Raman modes can be used to identify a substance. Likewise, the resonant curve of $|\chi^{(3)}|$ or $|E_a|^2$ in (7.63) can be used to identify a substance or to detect low-concentration substances in a mixture as long as $|\chi_R^{(3)}|_{max} \gtrsim |\chi_{NR}^{(3)}|$ in (7.62). This latter technique is most useful in gas mixtures where the Raman linewidths can be very narrow and consequently, $|\chi_R^{(3)}|_{max}$ can be quite large even for low concentrations [see (7.62)]. REGNIER and TARAN [7.115] have shown that with input laser powers of about 1 MW, they can detect a H_2 concentration of about 10 ppm in N_2 gas of 1 atm pressure. For comparison, the 100 ppm

H_2-in-N_2 mixture yields roughly 1 W of coherent anti-Stokes signal from 1 MW input laser pulses, while spontaneous scattering from the same focal volume with 1 MW of laser power would lead to an incoherent Stokes emission of 10^{-10} W per solid angle. The detection sensitivity can of course be greatly improved by letting ω_1 or ω_2 approach electronic resonances. Through the resonant enhancement in M_{fi} in (7.1), $|\chi_R^{(3)}|$ for a gas substance can increase by five to six orders of magnitude. This technique can be used to study the gas mixture in a flame, in a combustion engine, or in a supersonic jet flow [7.115]. Presumably, it can also be used to monitor the polluting substances in a smokestack and automobile exhaust.

7.7.4. Other Miscellaneous Applications

Plasma waves as a material excitation can be excited by SRS [7.36, 116]. The excited plasma waves will, of course, be damped with the energy turned into heat. This has been proposed as a possible means of heating a plasma for controlled fusion work [7.117]. However, no clean SRS experiment on a plasma has yet been reported.

The stimulated anti-Stokes absorption (Subsection 7.3.4) often known as the inverse Raman effect, leads to a new Raman spectroscopic technique [7.59, 118, 119]. By shining a laser pulse and a pulse of continuous spectrum simultaneously on a sample, the Raman spectrum can be obtained in one shot. More recently, the technique has been extended to the picosecond regime with the use of mode-locked pulses [7.120].

SRS is one of the effects which limit the propagation of high-intensity laser beams in materials. It is particularly important for high-intensity beam propagation in air for optical communications or other purposes. We have seen in (7.2, 3) that the Raman gain G_R is proportional to $\omega_2|M_{fi}|^2|E_1|^2$ where $|M_{fi}|^2$ usually increases as ω_1 increases. Consequently, SRS is often weaker at lower frequencies. Therefore, in order to transmit a high-power laser beam through a material, one should reduce G_R by using a larger beam size and a lower frequency.

7.8. Concluding Remarks

Because of limitation of space, we have omitted the discussion on a number of topics in SRS. We have not discussed the tensorial form of the Raman susceptibility $\overleftrightarrow{\chi}_R^{(3)}$ with respect to the symmetry of the medium and its tensorial relation with the spontaneous Raman cross-section

$d\vec{\sigma}/d\Omega$. We have not considered SRS in anisotropic crystals [7.121]. We have not mentioned SRS in optical waveguides [7.122]. We also have not discussed resonant SRS [7.123]. The latter case is particularly interesting in alkali vapor systems [7.124, 125]. Resonant SRS has recently been used to facilitate generation of tunable infrared [7.126] and tunable vacuum ultraviolet [7.127] by optical mixing in alkali vapor.

In conclusion, we believe the physics of SRS is basically understood. In actual experiments, the results of SRS may be complicated by other competing nonlinear optical effects. Influence of other nonlinear effects on SRS and vice versa are also qualitatively understood in most cases, but quantitative analysis is often quite difficult. SRS has already been used in a number of applications. With the recent advance of tunable lasers, it is expected that more research on resonant SRS will be carried out in the near future.

7.9. Addendum (1982)

There have been a series of important advances in the area of coherent Raman scattering in the past several years. First, on the theoretical understanding of stimulated Raman scattering, Ramer and Mostowski [7.129] have developed a unified quantum theory for the stimulated process with spontaneous emission and spatial dispersion. Then, on the application side, stimulated Raman scattering has been used to extend the tunable range of a coherent optical sources over a broad range in the infrared [7.130] (down to 257 μm) [7.131]. Even commercial Raman cells designed for this purpose are now available [7.132]. Anti-Stokes radiation from stimulated Raman scattering with dye lasers can also be used as a tunable UV source. More recently, Harris and coworkers have found that the spontaneous anti-Stokes scattering from a metastable state of an ion under tunable laser radiation can also be a useful tunable source in the XUV range [7.133]. Finally, on the spectroscopy applications of coherent Raman effects, the progress has been most remarkable. Coherent anti-Stokes Raman scattering (CARS) has already become an important tool for combustion studies and for studies in chemical and molecular physics [7.134]. It has also been used to obtain Doppler-free vibrational spectra of molecules in a molecular beam [7.135] to probe the internal energy distribution of dissociation fragments under the collisionless condition [7.136] and to study Raman transitions between unpopulated excited states [7.137]. Surface CARS with surface electromagnetic waves has been demonstrated to have the sensitivity to study molecular monolayers absorbed on a surface [7.138], and so does the stimulated Raman gain spectroscopy [7.139]. Both stimulated Raman gain spectroscopy and inverse Raman spectroscopy have been used to obtain high-resolution rotation-vibrational Raman spectra of molecules unsurpassed by any other method [7.140].

Acknowledgement

This paper was supported in part by the U.S. Atomic Energy Commission.

References

7.1. E. J. Woodbury, W. K. Ng: Proc. IRE **50**, 2347 (1962).
7.2. See E. J. Woodbury, G. M. Eckhardt: US Patent no. 3,371,265 (27 February 1968).
7.3. G. Eckhardt, R. W. Hellwarth, F. J. McClung, S. E. Schwartz, D. Weiner, E. J. Woodbury: Phys. Rev. Letters **9**, 455 (1962).
7.4. M. Geller, D. P. Bortfeld, W. R. Sooy: Appl. Phys. Letters **3**, 36 (1963).
7.5. B. P. Stoicheff: Phys. Letters **7**, 186 (1963).
7.6. G. Eckhardt, D. P. Bortfeld, M. Geller: Appl. Phys. Letters **3**, 137 (1963).
7.7. R. W. Minck, R. W. Terhune, W. G. Rado: Appl. Phys. Letters **3**, 181 (1963).
7.8. R. W. Hellwarth: Phys. Rev. **130**, 1850 (1963), Appl. Opt. **2**, 847 (1963).
7.9. R. W. Terhune: Solid State Design **4**, 38 (1963).
7.10. H. J. Zeiger, P. E. Tannenwald: *Proc. 3rd Intern. Conf. Quantum Electronics*, Paris (1963) ed. by P. Grivet and N. Bloombergen (Columbia University Press, New York, 1964), p. 1589.
R. Y. Chiao, B. P. Stoicheff: Phys. Rev. Letters **12**, 290 (1964).
7.11. S. K. Kurtz, J. A. Giordmaine: Phys. Rev. Letters **22**, 192 (1969).
7.12. J. Gelbwachs, R. H. Pantell, H. E. Puthoff, J. M. Yarborough: Appl. Phys. Letters **14**, 258 (1969).
7.13. C. K. N. Patel, E. D. Shaw: Phys. Rev. Letters **24**, 451 (1970).
7.14. Y. R. Shen: Appl. Phys. Letters **26**, 516 (1973).
7.15. S. L. Shapiro, J. A. Giordmaine, K. W. Wecht: Phys. Rev. Letters **19**, 1093 (1967).
7.16. See, for example, C. S. Wang: Phys. Rev. **182**, 482 (1969).
7.17. D. von der Linde, A. Laubereau, W. Kaiser: Phys. Rev. Letters **26**, 954 (1971).
7.18. R. R. Alfano, S. L. Shapiro: Phys. Rev. Letters **26**, 1247 (1971).
7.19. J. A. Armstrong, N. Bloembergen, J. Ducuing, P. S. Pershan: Phys. Rev. **127**, 198 (1962).
7.20. Y. R. Shen, N. Bloembergen: Phys. Rev. **137**, A1786 (1965).
7.21. N. Bloembergen: Am. J. Phys. **35**, 989 (1967). (Omission of [7.2] in this paper has led to an error in the description of the early history of stimulated Raman scattering).
7.22. W. Kaiser, M. Maier: In *Laser Handbook*, ed. by F. T. Arecchi and E. O. Schulz-Dubois (Nort-Holland Publishing Co., Amsterdam, 1972), p. 1077.
7.23. See, for example, W. Heitler: *The Quantum Theory of Radiation*, 3rd ed. (Cambridge University Press, New York, 1954), p. 192.
7.24. R. Glauber: Phys. Rev. **130**, 2529 (1963); **131**, 2766 (1963).
7.25. M. Sparks: Phys. Rev. Letters **32**, 450 (1974).
7.26. E. Garmire, E. Pandarese, C. H. Townes: Phys. Rev. Letters **11**, 160 (1963).
7.27. N. Bloembergen, Y. R. Shen: Phys. Rev. Letters **12**, 504 (1964).
7.28. V. T. Platonenko, R. V. Khokhlov: Sov. Phys. JETP **19**, 378 and 1435 (1964).

7.29. C. L. TANG, T. F. DEUTSCH: Phys. Rev. **138**, A1 (1965).
7.30. H. HAUS, P. L. KELLEY, H. ZEIGER: Phys. Rev. **138**, A690 (1965).
7.31. See, for example, N. BLOEMBERGEN, *Nonlinear Optics* (W. A. BENJAMIN, Inc., New York, 1965).
7.32. N. BLOEMBERGEN, Y. R. SHEN: Phys. Rev. A**133**, 37 (1964).
7.33. See, for example, [7.31], p. 27.
7.34. See, for example, A. ABRAGAM: *Principles of Nuclear Magnetism* (Oxford University Press, Oxford 1961);
P. S. HUBBARD: Rev. Mod. Phys. **33**, 249 (1961).
7.35. Y. R. SHEN, N. BLOEMBERGEN: Phys. Rev. **143**, 372 (1966).
7.36. N. BLOEMBERGEN, Y. R. SHEN: Phys. Rev. **141**, 298 (1966).
7.37. R. LOUDON: Proc. Phys. Soc. **82**, 393 (1963);
P. N. BUTCHER, R. LOUDON, T. P. MCLEAN: Proc. Phys. Soc. **85**, 565 (1965).
7.38. D. FRÖHLICH, E. MOHLER, P. WIESNER: Phys. Rev. Letters **26**, 554 (1971).
7.39. D. C. HAUEISEN, H. MAHR: Phys. Rev. Letters **26**, 838 (1971); Phys. Rev. B **8**, 2969 (1973).
7.40. D. BOGGETT, R. LOUDON: Phys. Rev. Letters **28**, 1051 (1972).
7.41. R. Y. CHIAO, C. H. TOWNES, B. P. STOICHEFF: Phys. Rev. Letters **12**, 592 (1964).
7.42. D. VON DER LINDE, M. MAIER, W. KAISER: Phys. Rev. **178**, 11 (1969).
7.43. F. J. MCCLUNG, W. G. WAGNER, D. WEINER: Phys. Rev. Letters **15**, 96 (1965);
G. BRET: Compt. Rend. Acad. Sci. **259**, 2991 (1964); **260**, 6323 (1965).
7.44. B. P. STOICHEFF: Phys. Letters **7**, 186 (1963).
7.45. R. W. TERHUNE: Solid State Design **4**, 38 (1964);
H. J. ZEIGER, P. E. TANNENWALD, S. KERN, R. BURENDEEN: Phys. Rev. Letters **11**, 419 (1963);
E. GARMIRE: In *Physics of Quantum Electronics*, ed. by P. L. KELLEY, B. LAX, and P. E. TANNENWALD (McGraw-Hill Book Co., New York, 1966), p. 167;
E. GARMIRE: Phys. Letters **17**, 251 (1965).
7.46. G. HAUCHECORNE, G. MAYER: Compt. Rend. Acad. Sci. **261**, 4014 (1965);
Y. R. SHEN, Y. J. SHAHAM: Phys. Rev. Letters **15**, 1008 (1965);
P. LALLEMAND, N. BLOEMBERGEN: Phys. Rev. Letters **15**, 1010 (1965).
7.47. P. L. KELLEY: Phys. Rev. Letters. **15**, 1005 (1965).
7.48. C. C. WANG: Phys. Rev. Letters **16**, 344 (1966).
7.49. M. M. T. LOY, Y. R. SHEN: Appl. Phys. Letters **19**, 285 (1971).
7.50. M. MAIER, W. KAISER, J. A. GIORDMAINE: Phys. Rev. **177**, 580 (1969); Phys. Rev. Letters **17**, 1275 (1966).
7.51. Y. R. SHEN, M. M. T. LOY: Phys. Rev. A **3**, 2099 (1971);
G. K. L. WONG, Y. R. SHEN: Appl. Phys. Letters **21**, 163 (1972).
7.52. N. BLOEMBERGEN, P. LALLEMAND: Phys. Rev. Letters **16**, 81 (1966).
7.53. A. PENZKOFER, A. LAUBEREAU, K. KAISER: Phys. Rev. Letters **31**, 863 (1973).
7.54. C. A. SACCHI, C. H. TOWNES, J. R. LIFSHITZ: Phys. Rev. **174**, 439 (1968).
7.55. J. B. GRUN, A. K. MCQUILLAN, B. P. STOICHEFF: Phys. Rev. **180**, 61 (1969).
7.56. J. H. DENNIS, P. E. TANNENWALD: Appl. Phys. Letters **5**, 58 (1964);
H. TAKUMA, D. A. JENNINGS: Appl. Phys. Letters **4**, 185; **5**, 239 (1964).
7.57. N. BLOEMBERGEN, G. BRET, P. LALLEMAND, A. PINE, P. SIMOVA: IEEE J. Quant. Electron. QE**3**, 197 (1967);
P. LALLEMAND, P. SIMOVA, G. BRET: Phys. Rev. Letters **17**, 1239 (1966).
7.58. N. BLOEMBERGEN, P. LALLEMAND: In *Physics of Quantum Electronics* ed. by P. L. KELLEY, B. LAX, and P. E. TANNENWALD, (McGraw-Hill Book Co., New York, 1966), p. 137.
7.59. W. J. JONES, B. P. STOICHEFF: Phys. Rev. Letters **13**, 657 (1964);
A. K. MCQUILLAN, B. P. STOICHEFF: In *Physics of Quantum Electronics*, ed. by

P. L. KELLEY, B. LAX, P. E. TANNENWALD (McGraw-Hill Book Co., New York, 1966), p. 192.
7.60. K. HUANG: Nature **167**, 779 (1951); Proc. Roy. Soc. (London), A**208**, 352 (1951); M. BORN, K. HUANG: *Dynamical Theory of Crystal Lattices* (Oxford University Press, London, 1954), Chapter II.
7.61. J. J. HOPFIELD: Phys. Rev. **112**, 1555 (1958).
7.62. Y. R. SHEN: Phys. Rev. **138**, A1741 (1965).
7.63. C. H. HENRY, C. G. B. GARRETT: Phys. Rev. **171**, 1058 (1968).
7.64. H. E. PUTHOFF, R. H. PANTELL, B. G. HUTH: J. Appl. Phys. **37**, 860 (1966); F. DE MARTINI: J. Appl. Phys. **37**, 4503 (1966); B. A. AKANEV, S. A. AKHMANOV, YU. G. KHRONOPULO: Sov. Phys. JETP **28**, 656 (1969).
7.65. See, for example, N. BLOEMBERGEN: *Nonlinear Optics* (W. A. BENJAMIN, Inc., New York, 1965), Section 4.4.
7.66. S. S. SUSSMAN: Microwave Lab. Report No. 1851, Stanford University (1970).
7.67. J. M. YARBOROUGH, S. S. SUSSMAN, H. E. PUTHOFF, R. H. PANTELL, B. C. JOHNSON: Appl. Phys. Letters **15**, 102 (1969).
7.68. S. BIRAUD-LAVAL, G. CHARTIER: Phys. Letters A**30**, 177 (1969).
7.69. R. E. SLUSHER, C. K. N. PATEL, P. A. FLEURY: Phys. Rev. Letters **18**, 77 (1967).
7.70. P. A. WOLFF: Phys. Rev. Letters **16**, 225 (1966).
7.71. Y. YAFET: Phys. Rev. **152**, 858 (1966).
7.72. S. R. J. BRUECK, A. MOORADIAN: Phys. Rev. Letters **28**, 161 (1972).
7.73. S. R. J. BRUECK, A. MOORADIAN: Phys. Rev. Letters **28**, 1458 (1972).
7.74. C. K. N. PATEL, E. D. SHAW: Phys. Rev. B**3**, 1279 (1971).
7.75. A. MOORADIAN, S. R. J. BRUECK, F. A. BLUM: Appl. Phys. Letters **17**, 481 (1970); S. R. J. BRUECK, A. MOORADIAN: Appl. Phys. Letters **18**, 229 (1971).
7.76. C. S. DESILETS, C. K. N. PATEL: Appl. Phys. Letters **22**, 543 (1973).
7.77. C. K. N. PATEL: Phys. Rev. Letters **28**, 649 (1972).
7.78. E. D. SHAW, C. K. N. PATEL: Appl. Phys. Letters **18**, 215 (1971).
7.79. R. S. ENG, A. MOORADIAN, H. R. FETTERMAN: Appl. Phys. Letters **25**, 453 (1974); J. R. SATTLER, B. A. WEBER, J. NEMARICH: Appl. Phys. Letters **25**, 451 (1974).
7.80. V. T. NGUYEN, T. J. BRIDGES: In *Proc. Laser Spectroscopy Conference*, Vail, Colo. 1973, ed. by R. G. BREWER and A. MOORADIAN (Plenum Press, New York 1974), p. 513.
7.81. V. T. NGUYEN, T. J. BRIDGES: Phys. Rev. Letters **29**, 359 (1972).
7.82. T. L. BROWN, P. A. WOLFF: Phys. Rev. Letters **29**, 362 (1972).
7.83. T. J. BRIDGES, V. T. NGUYEN: Appl. Phys. Letters **23**, 107 (1973).
7.84. N. M. KROLL: J. Appl. Phys. **36**, 34 (1965).
7.85. C. S. WANG: Phys. Rev. **182**, 482 (1969).
7.86. E. E. HAGENLOCKER, R. W. MINCK, W. G. RADO: Phys. Rev. **154**, 226 (1967).
7.87. S. A. AKHMANOV: Mat. Res. Bull. **4**, 455 (1969).
7.88. S. A. AKHMANOV, A. P. SAKHORUKOV, A. S. CHIRKIN: Zh ETF **55**, 143 (1968) [Translation: Sov. Phys. JETP **28**, 748 (1969)].
7.89. T. I. KUZNETSOVA: Zh ETF Pis'ma **10**, 153 (1969) [(Translation: Sov. Phys. JETP Letters **10**, 98 (1969)].
7.90. N. M. KROLL, P. L. KELLEY: Phys. Rev. A**4**, 763 (1971).
7.91. R. L. CARMAN, F. SHIMIZU, C. S. WANG, N. BLOEMBERGEN: Phys. Rev. A**2**, 60 (1970).
7.92. S. L. SHAPIRO, J. A. GIORDMAINE, K. W. WECHT: Phys. Rev. Letters **19**, 1093 (1967).
7.93. G. BRET, H. WEBER: IEEE J. Quant. Electron. QE**4**, 807 (1968).
7.94. M. J. COLLES: Opt. Commun. **1**, 169 (1969).
7.95. M. A. BOLSHOV, YU. I. GOLYAEV, V. S. DNEPROVSKII, I. I. NURMINSKII: Zh ETF **57**, 346 (1969) [Translation: Sov. Phys. JETP **30**, 190 (1970)].

7.96. R. L. CARMAN, M. E. MACK, F. SHIMIZU, N. BLOEMBERGEN: Phys. Rev. Letters **23**, 1327 (1969).
7.97. M. J. COLLES, G. E. WALRAFEN, K. W. WECHT: Chem. Phys. Letters **4**, 621 (1970).
7.98. M. E. MACK, R. L. CARMAN, J. REINTJES, N. BLOEMBERGEN: Appl. Phys. Letters **16**, 209 (1970).
7.99. R. L. CARMAN, M. E. MACK: Phys. Rev. A**5**, 341 (1972).
7.100. F. DE MARTINI, J. DUCUING: Phys. Rev. Letters **17**, 117 (1966).
7.101. R. R. ALFANO, S. L. SHAPIRO: Phys. Rev. Letters **26**, 1247 (1971).
7.102. D. VON DER LINDE, A. LAUBEREAU, W. KAISER: Phys. Rev. Letters **26**, 954 (1971).
7.103. A. LAUBEREAU, D. VON DER LINDE, W. KAISER: Phys. Rev. Letters **27**, 802 (1971).
7.104. A. LAUBEREAU, D. VON DER LINDE, W. KAISER: Phys. Rev. Letters **28**, 1162 (1972).
7.105. A. LAUBEREAU, D. VON DER LINDE, W. KAISER: Opt. Commun. **7**, 173 (1973).
7.106. R. R. ALFANO, S. L. SHAPIRO: Phys. Rev. Letters **29**, 1655 (1972).
7.107. A. LAUBEREAU, L. KIRCHNER, W. KAISER: Opt. Commun. **9**, 182 (1973).
A. LAUBEREAU, G. KEHL, W. KAISER: Opt. Commun. **11**, 74 (1974).
7.108. J. J. WYNNE: Phys. Rev. Letters **29**, 650 (1972).
7.109. J. J. WYNNE: Phys. Rev. B**6**, 534 (1972).
7.110. M. D. LEVENSON, C. FLYTZANIS, M. BLOEMBERGEN: Phys. Rev. B**6**, 3462 (1972).
7.111. E. YABLONOVITCH, C. FLYTZANIS, N. BLOEMBERGEN: Phys. Rev. Letters **29**, 865 (1972).
7.112. M. D. LEVENSON: IEEE J. Quant. Electron. QE-**10**, 110 (1974).
7.113. M. D. LEVENSON, N. BLOEMBERGEN: Phys. Rev. B**10**, 4447 (1974); J. Chem. Phys. **60**, 1323 (1974);
S. A. AKHMANOV, V. G. DMITRIEV, A. I. KOVERIGIN, N. I. KOROTEEV, V. G. TUNKIN, A. I. KHOLODNYKH: Zh ETP Pis'ma Redak **15**, 600 (1972) [JETP Letters **15**, 425 (1972)];
S. A. AKHMANOV, N. I. KOROTEEV, A. I. KHOLODNYKH: J. Raman Spectrosc. **2**, 239 (1974).
7.114. C. FLYTZANIS: Phys. Rev. B**6**, 1264 (1972).
7.115. P. R. REGNIER, J. P.-E. TARAN: Appl. Phys. Letters **23**, 240 (1973).
7.116. N. M. KROLL, A. RON, N. ROSTOKER: Phys. Rev. Letters **13**, 83 (1964);
G. C. COMISAR: Phys. Rev. **141**, 200 (1966).
7.117. B. I. COHEN, A. N. KAUFMAN, K. M. WATSON: Phys. Rev. Letters **29**, 581 (1973).
7.118. R. A. MCLAREN, B. P. STOICHEFF: Appl. Phys. Letters **16**, 140 (1970).
7.119. S. DUMARTIN, B. OKSENGORN, B. VODAR: Compt. Rend. Acad. Sci. **261**, 3767 (1965).
7.120. R. R. ALFANO, S. L. SHAPIRO: Chem. Phys. Letters **8**, 43 (1971).
7.121. See, for example, V. L. STRIZHEVSKII: Sov. Phys. JETP **32**, 914 (1971).
7.122. R. H. STOLEN, E. P. IPPEN, A. R. TYNES: Appl. Phys. Letters **20**, 62 (1972);
R. H. STOLEN, E. P. IPPEN: Appl. Phys. Letters **22**, 276 (1973).
7.123. See, for example, YA. S. BOBOVICH, A. V. BORTKEVICH: Sov. Phys. Uspekhi **14**, 1 (1971).
7.124. P. P. SOROKIN, N. S. SHIREN, J. R. LANKARD, E. C. HAMMOND, T. G. KAZYAKA: Appl. Phys. Letters **10**, 44 (1967).
7.125. M. ROKNI, S. YATSIV: Phys. Letters A**24**, 277 (1967); IEEE J. Quant. Electron. QE**3**, 329 (1967);
S. YATSIV, M. ROKNI, S. BARAK: IEEE J. Quant. Electron. QE**4**, 900 (1968);
S. BARAK, M. ROKNI, S. YATSIV: IEEE J. Quantum Electron. QE**5**, 448 (1969);
S. BARAK, S. YATSIV: Phys. Rev. A**3**, 382 (1971).
7.126. P. P. SOROKIN, J. J. WYNNE, J. R. LANKARD: Appl. Phys. Letters **22**, 342 (1973).
7.127. R. T. HODGSON, P. P. SOROKIN, J. J. WYNNE: Phys. Rev. Letters **32**, 343 (1974).
7.128. G. HADACHER, M. MAIER: *VIII Intern. Quant. Electron. Conf.*, San Francisco (1974), post-deadline paper Q. 7.

7.129. M.G.Raymer, J.Mostowski: Phys. Rev. A**24**, 1980 (1981).

7.130. D.C.Hanna, M.A.Yuratich, D.Cotter: *Nonlinear Optics of Free Atoms and Molecules*, Springer Series in Optical Sciences, Vol. 17 (Springer, Berlin, Heidelberg, New York 1979), Chap. 5.
V.Wilkie, W.Schmidt: Appl. Phys. **16**, 151 (1978).
T.R.Koree, R.C.Sze, D.L.Barker, P.B.Scott: IEEE J. QE-**15**, 337 (1979).

7.131. R.Frey, F.Pradere, J.Lukasik, J.Ducuing: Opt. Commun. **22**, 355 (1977).
R.Frey, F.Pradere, J.Ducuing: Opt. Commun. **23**, 65 (1977).
A.DeMartino, R.Frey, F.Pradere: Opt. Commun. **27**, 262 (1978).

7.132. Quanta Ray, Mountain View, CA, USA.

7.133. J.E.Rothenberg, J.F.Young, S.E.Harris: Opt. Lett. **6**, 363 (1981).
J.R.Wilson, R.W.Falcone, J.F.Young, S.E.Harris: Phys. Rev. Lett. **47**, 1827 (1981).

7.134. See, for example,
S.Druet, J.P.Taran: In *Chemical and Biochemical Applications of Nonlinear Optics*, ed. by A.B.Harvey (Academic Press, New York, 1980).
M.D.Levenson, J.L.Song: In *Coherent and Nonlinear Optics*, ed. by V.Letokhov, M.S.Feld, Topics in Current Phys., Vol. 21 (Springer, Berlin, Heidelberg, New York, 1980).
G.I.Eesley: *Coherent Raman Spectroscopy* (Pergamon Press, New York, 1981).

7.135. M.D.Duncan, P.Oesterlin, R.L.Byer: Opt. Lett. **6**, 90 (1981).

7.136. J.J.Valentine, D.S.Moore, D.S.Bomse: Chem. Phys. Lett. **83**, 217 (1981).

7.137. Y.Prior, A.R.Bogdan, M.Dagenais, N.Bloembergen: Phys. Rev. Lett. **46**, 111 (1981).

7.138. C.K.Chen, A.R.B.de Castro, Y.R.Shen, F.DeMartini: Phys. Rev. Lett. **43**, 946 (1979).

7.139. J.P.Heritage, D.L.Allara: Chem. Phys. Lett. **74**, 507 (1980).

7.140. A.Owyoung, P.Esherick: Opt. Lett. **5**, 421 (1980).

8. Overview

Light-scattering spectroscopy has recently become a standard, well established technique for the study of solids, with most of the necessary equipment commercially available. Highlights of this technique have extensively been discussed in [Rrf. 8.1, Chap. 1].

In 1975, the cw dye laser had just demonstrated its capability for the study of resonant light scattering phenomena. Hence, we devoted most of the introduction (Chap. 1) plus Chapter 3 and some of Chapter 2 to a discussion of the principles of resonant Raman scattering and the few available experimental results. In the meantime, this field has grown to maturity as a number of very diverse resonance phenomena have been observed, mainly in semiconductors, and theoretically interpreted. A detailed account of these phenomena has been given in [Ref. 8.1, Chap. 2] by M. CARDONA, incorporating a general introduction to light scattering in molecules and solids. A number of concepts used loosely in the literature, such as those of Raman tensor, Raman susceptibility and polarizability, scattering cross section, scattering efficiency, electron-phonon interaction constants, etc., are clarified. Tables of Raman tensors for the various crystallographic point groups and of the scattered intensities for several orientations in cubic crystals are given. Considerable efforts has been spent in the past few years to determine absolute scattering efficiencies in solid. In [Ref. 8.1, Chap. 2] the various methods used for measuring these efficiencies were discussed and a table of their values for many solids is given. Methods for theoretically calculating these efficiencies were also presented in [Ref. 8.1, Chap. 2].

The progress on hand would not have been achieved without a parallel development in the field of optical and laser instrumentation. One of the most challenging aspects of this development concerns optical multichannel detection systems. While to this date not yet in general use, they are ripe for widespread application, as outlined by R. K. CHANG and M. B. LONG in [Ref. 8.1, Chap. 3]. Modern instrumentation for Brillouin spectroscopy was discussed by C. WEISBUCH and R. G. ULBRICH [Ref. 8.2, Chap. 7].

Chapter 4 is devoted to electronic Raman scattering, a field where progress has also been considerable. In the past years we have witnessed the observation of scattering due to intervalley density fluctuations [8.3], acoustic plasmons [8.4], electron phonon coupled modes in non-polar materials [8.5], and light scattering by two-dimensional electron gases and by superlattices [8.6, 7]. We therefore devoted [Ref. 8.8, Chap. 2] to cover these advances.

Light scattering by amorphous semiconductors is the subject of Chapter 5. This topic has also become very important in the wake of the recent boom in amorphous semiconductors, especially hydrogenated amorphous silicon [8.9, 10]. Raman scattering techniques have become standard for the characterization of amorphous semiconductors. Some recent developments in the field are included in [Ref. 8.1, Chap. 2]. The related subject of disorder-induced Raman scattering in metals was discussed by M. V. Klein in [Ref. 8.2, Chap. 5].

Perhaps the most impressive growth has occurred in the field of Brillouin scattering, thanks mainly to the development of the multiple-pass Fabry-Perot interferometer [8.11, 12] and its tandem version [8.13]. These developments in instrumentation have made possible the study of strongly opaque materials such as metals (in particular, magnetic metals [8.11, 12]) and of surface excitations such as acoustic surface waves [8.16, 17]. In the course of these studies it has become clear that beside the standard photoelastic mechanism of Brillouin scattering a contribution due to *surface ripple* appears in highly absorbing materials [8.18, 19]. Of considerable interest are also the Brillouin studies of diffusive, very-low frequency excitations [8.15].

Equally spectacular is the development in the field of Brillouin scattering resonant at polaritons in semiconductors. The phenomenon had been predicted in 1972 (Sect. 6.4.3) [8.21], but was only observed in 1977 [8.14, 15]. It represents, in the editor's opinions, one of the rare and most beautiful examples of *a priori predictive* solid-state theory. Brillouin spectroscopy resonant at polaritons yields exciton-polariton dispersion relations and makes possible the study of additional boundary conditions (ABC's). The advances in the field since its discovery have been summarized in [Ref. 8.2, Chap. 7]. The reader will infer from [Ref. 8.2, Table 7.1] that the phenomenon has been observed and studied in at least 13 different materials.

Chapter 7 is devoted to stimulated Raman scattering and related phenomena. In the meantime these high-power pulsed laser spectroscopic techniques have also become standard analytical and research tools. These coherent scattering techniques gave rise to a puzzling number of acronyms such as CARS, RIKES, ARCS, etc. [Ref. 8.1, Fig. 4.1]. H. Vogt [Ref. 8.1, Chap. 4] has summarized these techniques and their *incoherent* high-power scattering counterparts such as the hyper-Raman effect [8.22]. An addendum to Chapter 7 also describes recent progress in coherent Raman spectroscopy. The general theory of CARS will be presented in a forthcoming monograph [8.23].

References

8.1. M. Cardona, G. Güntherodt (eds.): *Light Scattering in Solids II*, Topics in Applied Physics, Vol. 50 (Springer, Berlin, Heidelberg, New York, 1982). The contents are listed on p. 333.
8.2. M. Cardona, G. Güntherodt (eds.): *Light Scattering in Solids III*, Topics in Applied Physics, Vol. 51 (Springer, Berlin, Heidelberg, New York, 1982). The contents are listed on p. 337.
8.3. M. Chandrasekhar, E. O. Kane, M. Cardona: Phys. Rev. B **16**, 3579 (1977).
8.4. A. Pinczuk: Phys. Rev. Lett. **47**, 1487 (1981).
8.5. M. Cardona: J. Phys. Soc. Jpn. **49**, Suppl., 23 (1980).
8.6. G. Abstreiter, K. Ploog: Phys. Rev. Lett. **42**, 1308 (1979).
8.7. A. Pinczuk, H. L. Störmer, R. Dingle, J. M. Worlock, W. Wiegmann, A. C. Gossard: Solid State Commun. **32**, 1001 (1979).
8.8. M. Cardona, G. Güntherodt (eds.): *Light Scattering in Solids IV*, Topics in Applied Physics, Vol. 54 (Springer, Berlin, Heidelberg, New York, 1983) (in preparation).
8.9. M. Brodsky (ed.): *Amorphous Semiconductors*, Topics in Applied Physics, Vol. 36 (Springer, Berlin, Heidelberg, New York, 1979).
8.10. J. D. Joannopoulos, G. Lucovsky (eds.): *Amorphous Silicon*, Topics in Current Physics (Springer, Berlin, Heidelberg, New York, 1983) (in preparation).
8.11. J. R. Sandercock, W. Wettling: J. Appl. Phys. **50**, 7784 (1979).
8.12. M. Grimsditch, A. Malozemoff, A. Brunsch: Phys. Rev. Lett. **43**, 711 (1979).
8.13. J. R. Sandercock: US Patent 4,225,236 (1980).
8.14. R. G. Ulbrich, C. Weisbuch: Phys. Rev. Lett. **38**, 865 (1977).
8.15. G. Winterling, E. Koteles: Solid State Commun. **23**, 95 (1977).
8.16. A. A. Oliner (ed.): *Acoustic Surface Waves*, Topics in Applied Physics, Vol. 24 (Springer, Berlin, Heidelberg, New York, 1978).
8.17. V. Bortolani, F. Nizzoli, G. Santoro, A. Marvin, J. R. Sandercock: Phys. Rev. Lett. **43**, 224 (1979).
8.18. S. Mishra, R. Bray: Phys. Rev. Lett. **39**, 222 (1977).
8.19. R. Loudon, J. R. Sandercock: J. Phys. C **13**, 2609 (1980).
8.20. K. Lyons, P. A. Fleury: Phys. Rev. Lett. **37**, 161 (1976).
8.21. W. Brenig, R. Zeyhes, J. L. Birman: Phys. Rev. B **6**, 4617 (1972).
8.22. H. Vogt, G. Neumann: Phys. Status Solidi B **92**, 57 (1979).
8.23. R. L. Byer, M. A. Henesian: *High-Resolution Continuous-Wave Coherent Anti-Stokes Raman Spectroscopy*, Springer Series in Optical Sciences (Springer, Berlin, Heidelberg, New York, 1983) (in preparation).

Additional References with Titles

Proceedings of the Third International Conference on Light Scattering in Solids, Campinas, Brazil, 1975, ed. by M. Balkanski, R.C.C. Leite, and S.P.S. Porto (Flammarion, Paris, 1975) (References in pp. 1–6):

E. Anastassakis, H. Bilz, M. Cardona, P. Grünberg, W. Zinn: Acoustic and Optical One-Phonon Density of States in Antiferromagnetic GdS.

B. Bendow, Y. P. Tsay, S. S. Mitra: Statistical Interpretation of Raman Spectra of Amorphous Solids.

S. Buchner, E. Burstein, A. Pinczuk: Allowed, Field-Induced and Wave Vector-Dependent Resonance Raman Scattering at the E_1 Gap of InAs.

F. Cerdeira: Raman Scattering by Coupled Electron-Phonon Excitations in Heavily Doped Silicon.

Y. J. Chen, W. P. Chen, E. Burstein, D. L. Mills: Inelastic Light Scattering Using Surface Electromagnetic Waves.

A. Compaan, A. Genack, M. Washington, H. Z. Cummins: Experimental Tests of the Quadrupole-Dipole Raman Scattering Tensor in Cu_2O.

A. Compaan, J. R. Macdonald: Resonance Raman Study of Ion-Implantation Produced Damage in Cu_2O.

A. P. Defonzo, J. Tauc: Raman Scattering in Amorphous Systems.

J. Doehler, P. J. Colwell, S. A. Solin: Raman Studies of the Semiconductor to Metal Transition in Ge(As).

L. M. Falicov: Resonant Raman Scattering.

G. Favrot, R. L. Aggarwal, B. Lax: Stimulated Spin-Flip Raman Scattering in InSb at Magnetic Fields up to 180 KOe.

F. L. Galeener: Second Order Vibrational Spectra of Disordered Solids.

A. Z. Genack, H. Z. Cummins, M. A. Washington, A. Compaan: Symmetry-Forbidden Resonant Raman Scattering from polar phonons in Cu_2O.

S. Go, H. Bilz, M. Cardona: Bond Polarizabilities and Energy Bands of Covalent Semiconductors.

G. Hirlimann, R. Beserman, M. Balkanski, J. Chevalier: Raman Study of One and Two Phonons Coupling in Mixed $Ga_XIn_{1-X}P$.

K. P. Jain, M. Balkanski: Theory of Raman Scattering Line-Shapes in Semiconductors: Interference Effects.

K. Jain, M. V. Klein: Valley-Orbit Raman Scattering from Bound and Delocalized Electrons in n-Si.

M. Jouanne, R. Beserman, M. Balkanski: Electron-Phonon Interaction in Heavily Doped Silicon.

O. Keller: Theory of Brillouin Scattering.

W. Kiefer, B. D. McCombe, W. Richter, R. L. Schmidt, M. Cardona: Second Order Raman Scattering in Zincblende-Type Semiconductors.

P. J. Lin-Chung, K. L. Ngai: Intervalley Phonon Raman Scattering in Many-Valley Semiconductors.

J. MUZART, E. G. LLUESMA, C. A. ARGÜELLO, R. C. C. LEITE: Photon Induced Resonant Raman Scattering in CdS.

S. ONARI, E. ANASTASSAKIS, M. CARDONA: Resonant Raman Scattering in Mg_2Si and Mg_2Ge.

B. J. PARSONS, C. D. CLARK: Laser Raman Spectroscopy of Diamond at High Pressures.

Y. PETROFF, C. CARILLON, Y. R. SHEN: Some Aspects of Resonant Raman Scattering and Luminescence in Cu_2O.

A. S. PINE, G. DRESSELHAUS: Lineshape Asymmetries in Light Scattering from Opaque Materials.

L. A. RAHN, P. J. COLWELL, W. J. CHOYKE: Raman Scattering from Ion-Implanted Silicon Carbide.

J. REYDELLET, J. M. BESSON: Resonant Raman Scattering on Single Phonon Modes of GaSe.

W. RICHTER, R. ZEYHER, M. CARDONA: Resonant Raman Scattering under Uniaxial Stress: E_1—$E_1 + \Delta_1$ Gaps.

R. ROMESTAIN, S. GESCHWIND, G. E. DEVLIN: Determination of Spin-Flip Raman Scattering Cross Section from the Faraday Rotation.

J. R. SANDERCOCK: A High Resolution, High Contrast Interferometer, and Its Application to Brillouin Scattering from Magnetic Systems.

J. SHAH, J. C. V. MATTOS: Raman Scattering from Photoexcited Non-Equilibrium Excitations in Semiconductors.

S. A. SOLIN, M. GORMAN, J. DOEHLER: Uniaxial Stress Dependence of the Electronic Raman Spectra of Ge(As).

S. A. SOLIN, R. J. NEMANICH, G. LUCOVSKY: Vibrational Modes of Amorphous As_2S_3—GeS_2 and As_2Se_3—$GeSe_2$.

V. J. TEKIPPE, R. P. SILBERSTEIN, L. E. SCHMUTZ, M. S. DRESSELHAUS, R. L. AGGARWAL: Inelastic Light Scattering in Magnetic Semiconductors.

Y. F. TSAY, B. BENDOW, S. S. MITRA: Theory of Brillouin Effect: Calculation of Photoelastic Constants of Diamond and Zinblende Semiconductors.

R. TSU, L. ESAKI: Raman Scattering for LO Phonons via Quantum States.

R. TSU, R. RUTZ: Raman Scattering in AlN Single Crystals.

R. TUBINO, J. L. BIRMAN: Raman Intensities in Covalent Crystals: The Two Phonon Raman Spectrum of Diamond.

S. USHIODA, J. Y. PRIEUR: Raman Scattering from Surface Polaritons.

M. A. WASHINGTON, A. Z. GENACK, H. Z. CUMMINS, A. COMPAAN: First Order Resonant Raman Scattering in the Excited Yellow Exciton Series of Cu_2O.

G. WINTERLING: Very Low Frequency Raman Scattering from Glasses.

G. B. WRIGHT, H. TEMKIN: Electronic Raman Scattering from Shallow Acceptors.

P. Y. YU, N. M. AMER, Y. PETROFF, Y. R. SHEN: Resonant Raman Scattering in SbSI near the Absorption Edge.

R. ZEYHER: Calculation of Effective Electron-Two Phonon Deformation Potentials in Semiconductors.

M. ZIGONE, R. BESERMAN: Pressure Dependence of Impurity Induced Vibrational Modes in II—VI Compounds.

K. ANDO, C. HAMAGUCHI: Resonant Brillouin scattering by amplified phonons in CdS. Phys. Rev. **10**, 3876 (1975).

F. BECHSTEDT, R. ENDERLEIN, K. PEUKER: Theory of inter-valence-band electronic Raman scattering in cubic semiconductors without and with an external electric field. Phys. Stat. Sol. (b) **68**, 43 (1975).

G. BENZ, R. CONRADT: Anti-Stokes light scattering by a band-to-band transition in GaSb. Phys. Rev. Lett. **34**, 1551 (1975).

R. BERKOWICZ, B. P. KIETIS: Multiple Brillouin scattering from intense acoustic monochromatic flux in CdS. Phys. Stat. Sol. (a) **29**, 451 (1975).

W. BRAUN, J. S. LANNIN: Densities of valence states and Raman scattering in amorphous $CuGaSe_2$; in *Proc.* 12th *Intern. Conf. Physics of Semiconductors, Stuttgart* (1974), p. 1308.

S. V. GANTSEVICH, R. KATILYUS, N. G. USTINOV: Light scattering by nonequilibrium free carriers in a multivalley semiconductor. Fiz. Tverd. Tela **16**, 1106 (1974). [Sov. Phys. Solic State **16**, 711 (1974).]

S. V. GANTSEVICH, R. KATILYUS, N. G. USTINOV: Collision controlled light scattering in a multivalley semiconductor. Fiz. Tverd. Tela **16**, 1114 (1974). [Sov. Phys. Solid State **16**, 716 (1974).]

M. JOUANNE, R. BESERMAN, I. IPATOVA, A. SUBASHIEV: Electron-phonon coupling in highly doped n-type silicon. Solid State Commun. **16**, 1047 (1975).

R. C. KEEZER, G. LUCOVSKY, R. M. MARTIN: Infrared studies of the ring-chain concentration in amorphous S and Se. Bull. Am. Phys. Soc. II, **20**, 323 (1975).

O. KELLER: Brillouin-scattering cross section of off-axis phonons in CdS. Phys. Rev. **11**, 5059 (1975).

M. KRAUZMAN, R. M. PICK, H. POULET, G. HAMEL, B. PREVOT: Raman detection of one-phonon-two-phonon interactions in CuCl. Phys. Rev. Lett. **33**, 528 (1974).

I. B. LEVINSON, E. I. RASHBA: Threshold phenomena and bound states in the polaron problem. Usp. Fiz. Nauk **111**, 683 (1973). [Sov. Phys. Uspekhi **16**, 892 (1974).]

I. B. LEVINSON, E. I. RASHBA: Nonlinear polarization interaction of electrons with short-wave phonons. JETP Lett. **20**, 27 (1974).

J. F. MORHANGE, R. BESERMAN, M. BALKANSKI: Raman study of the vibrational properties of implanted silicon. Phys. Stat. Sol. (a) **23**, 383 (1974).

K. MURASE, S. KATAYAMA, Y. ANDO, H. KAWAMURA: Observation of a coupled phonon—damped-plasmon mode in n-GaAs by Raman scattering. Phys. Rev. Lett. **33**, 1481 (1974).

S. NAKASHIMA: Raman study of polytypism in vapor-grown PbI_2. Solid State Commun. **16**, 1059 (1975).

S. NAKASHIMA, H. KOJIMA: Raman scattering from plasmon—LO phonon coupled modes in ZnTe. Solid State Commun. **15**, 1699 (1974).

G. N. PAPATHEODOROU, S. A. SOLIN: Vibrational modes of vitreous As_2O_3. Solid State Commun. **16**, 5 (1975).

F. DAVID PEAT: Anisotropic electron-phonon coupling. Phys. Rev. B**3**, 3149 (1971).

RASHBA, E. I.: Theory of bound states of phonons with impurity centers and excitons. ZhETF Pis. Red. **15**, 577 (1972). [JETP Letters **15**, 411 (1972).]

C. RAZZETTI, M. P. FONTANA: Resonant Raman scattering in amorphous As_2S_3. Phys. Stat. Sol. (in press).

J. B. RENUCCI, R. N. TYTE, M. CARDONA: Resonant Raman scattering in silicon. Phys. Rev. **16**, 3885 (1975).

J. REYDELLET, J. M. BESSON: Double resonance in Raman scattering on single phonon modes of GaSe. Solid State Commun. **17**, 23 (1975).

W. RICHTER, K. PLOOG: Raman active phonons in α-boron. Phys. Stat. Sol. (b) **68**, 201, (1975).

W. SCHEUERMANN: Electronic Raman scattering by germanium p-acceptors and luminescence in GaAs. J. Raman Spectrosc. **3**, 101 (1975).

V. L. STRIZHEVSKII, YU. N. YASHKIR: The theory of Raman scattering by surface polaritons. Phys. Stat. Sol. (b) **69**, 175 (1975).

V. L. STRIZHEVSKII, YU. N. YASHKIR, H. E. PONAT: The influence of polariton motion on the polariton Raman scattering spectra. Phys. Stat. Sol. (b) **69**, 673 (1975).

V. STROM, J. R. HENDRICKSON, R. J. WAGNER, P. C. TAYLOR: Disorder-induced far infrared absorption in amorphous materials. Solid State Commun. **15**, 1871 (1974).
M. TANAKA, M. YAMADA, C. HAMAGUCHI: Brillouin scattering in GaSe. J. Phys. Soc. Japan **38**, 1708 (1975).
P. TRONC, M. BENSOUSSAN, A. BRENAC, G. ERRANDONEA, C. SEBENNE: Optical absorption edge and Raman scattering in Ge_xSe_{1-x} glasses for $1/2 > x > 1/3$". J. Physique **38**, 1498 (1977).
R. TSU, H. KAWAMURA, L. ESAKI: Raman scattering in the depletion region of GaAs. Solid State Commun. **15**, 321 (1974).
R. TUBINO, L. PISERI: A bond-polarizability approach. Phys. Rev. **11**, 5145 (1975).
R. S. TURTELLI, A. R. B. DE CASTRO, R. C. C. LEITE: Single particle scattering from hot electrons in GaAs. Solid State Commun. **16**, 969 (1975).
V. I. ZEMSKI, E. L. IVCHENKO, D. N. MIRLIN, I. I. RESHINA: Dispersion of plasmon-phonon modes in semiconductors: Raman scattering and infrared spectra. Solid State Commun. **16**, 221 (1975).

General and Recent References (1982)

General

W. F. MURPHY (ed.): *Proceedings of the 7th International Conference on Raman Spectroscopy* (North-Holland, Amsterdam, 1980).
E. D. SCHMID, R. S. KRISHNAN, W. KIEFER, H. W. SCHRÖTTER (eds.): *Proceedings of the 6th International Conference on Raman Spectroscopy* (Heyden, London, 1978).
W. HAYES, R. LOUDON (eds.): *Scattering of Light by Crystals* (Wiley, New York, 1978).
A. WEBER (ed.): *Raman Spectroscopy of Gases and Liquids*, Topics in Current Physics, Vol. 11 (Springer, Berlin, Heidelberg, New York, 1979).
H. BILZ, W. KRESS: *Phonon Dispersion Relations in Insulators*, Springer Series in Solid-State Sciences, Vol. 10 (Springer, Berlin, Heidelberg, New York, 1974).
M. BLOEMBERGEN (ed.): *Non-Linear Spectroscopy* (North-Holland, Amsterdam, 1977).
M. BALKANSKI (ed.): *Lattice Dynamics* (Flammarion, Paris, 1978).
M. AGRANOVICH, J. L. BIRMAN (eds.): *The Theory of Light Scattering in Solids* (Nauka, Moskow, 1976).
J. L. BIRMAN, H. Z. CUMMINS, K. K. REBANE (eds.): *Light Scattering in Solids* (Plenum Press, New York, 1979).
W. RICHTER: In *Springer Tracts in Modern Physics*, Vol. 78 (Springer, Berlin, Heidelberg, New York, 1976), p. 121.
S.P.S. PORTO, Memorial Issue: J. Raman Spectrosc. **10** (1981).
K. CHO (ed.): *Excitons*, Topics in Current Physics, Vol. 14 (Springer, Berlin, Heidelberg, New York, 1979).
W. E. BRON (ed.): International Conference on Phonon Physics, J. Physique C **6**, (1981).
J. G. DIL: Brillouin scattering in condensed matter. Rep. Prog. Phys. **45**, 285 (1982).
T. R. GILSON, P. J. HENDRA: *Laser Raman Spectroscopy* (Wiley, London, 1970).
M. S. FELD, V. S. LETOKHOV (eds.): *Coherent Nonlinear Optics*, Topics in Current Physics, Vol. 21 (Springer, Berlin, Heidelberg, New York, 1980).
A. B. HARVEY (ed.): *Chemical Applications of Nonlinear Raman Spectroscopy* (Academic Press, New York) (in press).
J. F. SCOTT: Spin flip Raman scattering in *p*-type semiconductors. Rep. Prog. Phys. **43**, 61 (1980).
R. J. H. CLARK, R. E. HESTER (eds.): *Advances in Infrared and Raman Spectroscopy*, Vol. 1–5 (Heyden, London, 1975).

Brillouin Scattering

W.C.Chang, S.Mishra, R.Bray: "Surface Brillouin Scattering: Comparison of Resonant Elasto-Optic and Ripple Mechanisms", Proc. of the 15th International Conference on Physics of Semiconductors, Kyoto (1980).

K.Cho, M.Yamane: Intensities of resonant Brillouin scattering from multicomponent exciton-polaritons. Solid State Commun. **40**, 121 (1981).

A.Domarkas, I.L.Drichko, A.M.Dyakonov: Investigation of photoelasticity mechanism in diffraction of light by sound in InSb. Sov. Phys. Solid State **23**, 1862 (1981).

S.C.Rand, B.P.Stoicheff: Elastic and photoelastic constants of CH_4 and CD_4 obtained by Brillouin scattering. Can. J. Phys. **60**, 287 (1981).

G.A.Smolenskii, I.G.Sinii, S.D.Prokhorova, A.A.Godovikov, R.Laiko, T.Levola, E.Karaemyaki: Brillouin scattering in prustite. Sov. Phys. Solid State **23**, 1178 (1981).

D.P.Vu, Y.Oka, M.Cardona: Resonant Brillouin scattering in CuBr. Phys. Rev. B **24**, 765 (1981).

V.Bortolani, F.Nizzoli, G.Santoro, J.R.Sandercock: Strong interference effects in surface Brillouin scattering from a supported transparent film. Phys. Rev. B **25**, 3442 (1982).

Instrumentation and Techniques

M.D'Orazio, B.Schrader: Calibration of the absolute spectral response of Raman spectrometers, J. Raman Spectrosc. **2**, 585 (1974).

M.Bridoux, A.Deffontaine, M.Delhaye, B.Rose, E.DaSilva: Picosecond multichannel Raman spectroscopy. J. Raman Spectrosc. **11**, 515 (1981).

O.S.Mortensen: Raman dispersion spectroscopy (RADIS). J. Raman Spectrosc. **11**, 329 (1981).

H.Yoshida, Y.Nakajima, S.Kobinata, S.Maede: Determination of order parameters in liquid crystals by resonant Raman method: 5CB as probed by β-carotene. J. Phys. Soc. Jpn. **50**, 3525 (1981).

W.Hüffer, R.Schieder, H.Telle, R.Raue, W.Brinkwerth: New efficient laser dye for pulsed and cw operation in the UV. Opt. Commun. **33**, 85 (1980).

H.Yamada, Y.Yamamoto: Illumination of flat or unstable samples for Raman measurements using optical fibres. J. Raman Spectrosc. **9**, 401 (1980).

Disordered Materials

M.Balkanski, L.M.Falicov, C.Hirliman, K.P.Jain: Localized exciton-phonon complex as an intermediate state in light scattering. Solid State Commun. **25**, 261 (1978).

M.Chandrasekhar, H.R.Chandrasekhar, M.Grimsditch, M.Cardona: Localized vibrations of boron in heavily doped Si. Phys. Rev. B **22**, 4825 (1980).

B.H.Bayramov, W.Toporov, Sh.B.Ubaydullaev: Disorder-induced Raman scattering of $In_{1-x}Ga_xP$. Solid State Commun. **37**, 963 (1981).

R.J.Nemanich, S.A.Solin, R.Martin: Light scattering study of boron nitride microcrystals. Phys. Rev. B **23**, 6348 (1981).

H.Richter, Z.P.Wang, L.Ley: The one phonon Raman spectrum in microcrystalline silicon. Solid State Commun. **39**, 625 (1981).

N.Saint-Cricq, R.Carles, J.B.Renucci, A.Zwick, M.A.Renucci: Disorder activated Raman scattering in $Ga_{1-x}Al_xAs$ alloys. Solid State Commun. **39**, 1137 (1981).

H.Wipf, M.V.Klein, W.S.Williams: Vacancy-induced and two-phonon Raman scattering. Phys. Status Solidi B **108**, 489 (1981).

T.Ishidate, K.Inoue, K.Tsuji, S.Minomura: Raman scattering in hydrogenated amorphous silicon under high pressure. Solid State Commun. **42**, 197 (1982).

R.H.Stolen, M.A.Bösch: Low-frequency and low-temperature Raman scattering in silica fibers. Phys. Rev. Lett. **48**, 805 (1982).

Electronic Excitations

D. A. ANDREEV, I.P. IPATOVA, A. V. SUBASHIEV: Light scattering from zero-sound in semiconductors. Indian J. Pure Appl. Phys. **16**, 322 (1978).

M. CHANDRASEKHAR, U. RÖSSLER, M. CARDONA: Intra- and inter-valence-band Raman scattering by free carriers in heavily doped *p*-Si. Phys. Rev. B **22**, 761 (1980).

V. V. ARTAMANOV, M. YA. VALAKH, V. A. KORNEICHUK: Resonance interaction between plasmons and two-phonon states in ZnSe crystals. Solid State Commun. **39**, 703 (1981).

L. M. FALICOV, C. A. BALSEIRO: Soft phonons and superconducting quasiparticles: a new Raman-active hybrid mode. J. Raman Spectrosc. **10**, 251 (1981).

I.P. IPATOVA, A. V. SUBASHIEV, V. A. VOITENKO: Electron light scattering from doped silicon. Solid State Commun. **37**, 893 (1981).

U. NOWAK, W. RICHTER, G. SACHS: LO-phonon-plasmon dispersion in GaAs. Phys. Status. Solidi B **108**, 131 (1981).

D. OLEGO, M. CARDONA: Self-energy effects of the optical phonons in heavily doped *p*-GaAs and *p*-Ge. Phys. Rev. B **23**, 6592 (1981).

D. OLEGO, M. CARDONA: Raman scattering by coupled LO-phonon plasmon modes and forbidden TO-phonon scattering in heavily doped *p*-type GaAs. Phys. Rev. B **24**, 7217 (1981).

A. PINCZUK, J. SHAH, P. A. WOLFF: Collective modes of photoexcited electron-hole Plasmas in GaAs. Phys. Rev. Lett. **47**, 1487 (1981).

K. M. ROMANEK, H. NATHER, E. O. GÖBEL: Light Scattering from optically excited electron-hole plasmas in GaAs. Solid State Commun. **39**, 23 (1981).

H. UWE, K. OKA, H. UNOKI, T. SAKUDO: Raman scattering from conduction electrons in $KTaO_3$. J. Phys. Soc. Jpn. **49**, 577 (1980).

A. KASUYA, Y. NISHINA, T. GOTO: Resonant electronic Raman scattering and electronic structures of bound exciton complexes in ZnSe. J. Phys. Soc. Jpn. **51**, 922 (1982).

N. B. AN, N. V. HIEU, N. A. VIET: On the role of the coulomb interaction of charge carriers in the electronic Raman scattering on donor levels in direct band gap semiconductors. Phys. Status Solidi B **109**, 463 (1982).

D. OLEGO, A. PINCZUK, A. C. GOSSARD, W. WIEGMANN: Plasma dispersion in a layered electron gas: a determination in GaAs-(AlGa)As heterostructures. Phys. Rev. Lett. (in press).

W. RICHTER, A. KROST, U. NOWAK, E. ANASTASSAKIS: Anisotropy and dispersion of coupled plasmon-LO-phonon modes in Sb_2Te_3 (in preparation).

C. Y. CHEN: Anisotropic Raman excitation of the coupled longitudinal optical phonon-plasmon modes in *n*-GaAs near the $E_0 + \Delta_0$ energy gap. Phys. Rev. (in preparation).

Magnetic Materials

K. B. LYONS, P. A. FLEURY: Magnetic energy fluctuations: observations by light scattering. Phys. Rev. Lett. **48**, 202 (1982).

N. NAGAOSA, E. HANAMURA: Microscopic theory on the Raman spectra of transition metal dichalcogenides in CDW state. Solid State Commun. **41**, 809 (1982).

A. PETROU, D. L. PETERSON, S. VENUGOPALAN, R. R. GALAZKA, A. K. RAMDAS, S. RODRIGUEZ: Zeeman effect of the magnetic excitations in a diluted magnetic semiconductor: a Raman scattering study of $Cd_{1-x}Mn_xTe$. Phys. Rev. Lett. **48**, 1036 (1982).

A. K. RAMDAS, S. RODRIGUEZ: Inelastic light scattering in diluted magnetic semiconductors, in *Novel Materials and Techniques in Condensed Matter* (Elsevier North-Holland, Amsterdam) (in preparation).

R. SOORYAKUMAR, M. V. KLEIN: Raman scattering from superconducting gap excitations in the presence of a magnetic field. Phys. Rev. B **23**, 3213 (1981).

G. GÜNTHERODT, A. JAYARAMAN, E. ANASTASSAKIS, E. BUCHER, H. BACH: Effect of configuration crossover on the electronic Raman scattering by 4*f* multiplets. Phys. Rev. Lett. **46**, 855 (1981).

Materials Studies

YU.S. POSONOV: Raman scattering of light in a ruthenium single crystal. Sov. Phys. Solid State **23**, 861 (1981).

I.P. PASHUK, N.S. PIOLZYRAILO, M.G. MATSKO: Exciton absorption, luminescence, and resonance Raman scattering of light in $CsPbCl_3$, $CsPbBr_3$. Sov. Phys. Solid State **23**, 1263 (1981).

M. TOKUMOTO, N. KOSHIZUKA, H. ANZAI, T. ISHIGURO: X-ray photoelectron and Raman spectra of TMTTF-TCNQ: estimation of the valence fluctuation time and the degree of charge transfer. J. Phys. Soc. Jpn. **51**, 332 (1982).

E.I. KAMITSOS, C.H. TZINIS, W.M. RISEN, JR.: Raman study of the mechanism of electrical switching in Cu-TCNQ films. Solid State Commun. **42**, 561 (1982).

Piezo-Raman Spectroscopy

I.I. NOVAK, V.V. BAPTIZMANSKII, L.V. ZHOGA: Effect of 2-D stress on the Raman spectrum of silicon. Opt. Spectrosk. **43**, 252 (1977).

R. TROMMER, H. MÜLLER, M. CARDONA, P. VOGL: Dependence of the phonon spectrum of InP on hydrostatic pressure. Phys. Rev. B **21**, 4869 (1980).

J. PASCUAL, J. CAMASSEL, P. MERLE, B. GIL, H. MATHIEU: Uniaxial-stress dependence of the first-order Raman spectrum of Rutile-type crystals: MgF_2. Phys. Rev. B **24**, 2101 (1981).

A. POLIAN, J.M. BESSON, M. GRIMSDITCH, H. VOGT: Brillouin scattering from GaS under hydrostatic pressure up to 17.5 GPa. Appl. Phys. Lett. **38**, 334 (1981).

E.A. VINOGRADOV, G.N. ZHIZHIN, N.N. MEL'NIK, S.I. SUBBOTIN, V.V. PANFILOV, K.R. ALLAKHVARDIEV: Resonance Raman scattering in GaSe and $TiGaSe_2$ under pressure. Sov. Phys. Solid State **22**, 1305 (1980).

S.R.J. BRUECK, B.Y. TSAUR, J.C.C. FAN, D.V. MURPHY, T.F. DEUTSCH, D.J. SILVERSMITH: Raman measurements of stress in silicon-on-sapphire device structures. Appl. Phys. Lett. **40**, 895 (1982).

D. OLEGO, M. CARDONA: Pressure dependence of Raman phonons of Ge and 3C-SiC. Phys. Rev. B **25**, 1151 (1982).

D. OLEGO, M. CARDONA, P. VOGL: Pressure dependence of the optical phonons and transverse effective charges in 3C-SiC. Phys. Rev. B **25**, 3878 (1982).

Pulsed Laser Spectroscopy

A. LAUBEREAU, W. KAISER: Vibrational dynamics of liquids and solids investigated by picosecond light pulses. Rev. Mod. Phys. **50**, 607 (1978).

W. ZINTH, H.J. POLLAND, A. LAUBEREAU, W. KAISER: New results on ultrafast coherent excitation of molecular vibrations in liquids. Appl. Phys. B **26**, 77 (1981).

D. VON DER LINDE, J. KUHL, H. KLINGENBERG: Raman scattering from nonequilibrium LO phonons with picosecond resolution. Phys. Rev. Lett. **44**, 1505 (1980).

J. KUHL, D. VON DER LINDE: In *Picosecond Phenomena* III, ed. by K.B. EISENTHAL, R.M. HOCHSTRASSER, W. KAISER, A. LAUBEREAU, Springer Series in Chemical Physics, Vol. 23 (Springer, Berlin, Heidelberg, New York, 1982), pp. 201–204.

W.H. HESSELINK, B.H. HESP, D.A. WIERSMA: In *Picosecond Phenomena* II, ed. by R. HOCHSTRASSER, W. KAISER, C.V. SHANK, Springer Series Chemical Physics, Vol. 14 (Springer, Berlin, Heidelberg, New York, 1980).

D.A. WIERSMA: In *Picosecond Phenomena* III, ed. by K.B. EISENTHAL, R.M. HOCHSTRASSER, W. KAISER, A. LAUBEREAU, Springer Series Chemical Physics, Vol. 23 (Springer, Berlin, Heidelberg, New York, 1982), pp. 179–183.

H.-P. GRASSL, M. MAISER: Efficient stimulated Raman scattering in silicon. Opt. Commun. **30**, 253 (1979).

Resonant Scattering

M. CARDONA: "Resonant Raman Scattering in Tetrahedral Semiconductors", in *Proceedings of the 7th International Conference on Raman Spectroscopy* (North-Holland, Amsterdam, 1980).

H. Kurita, O. Sakai, A. Kotani: Resonant Raman scattering and luminescence of excitonic polaritons. Solid State Commun. **40**, 127 (1981).

D. Olego, M. Cardona: Raman scattering by two LO-phonons near Γ in GaAs. Solid State Commun. **39**, 1071 (1981).

E. A. Vinogradov, G. N. Zhizhin, N. N. Mel'nik, V. L. Grachev: Characteristics of resonance Raman scattering of light in thin films. Sov. Phys. Solid State **21**, 1577 (1979).

D. N. Batchelder, D. Bloor: An investigation of the electronic excited state of a polydiacetylene by resonance Raman spectroscopy. J. Phys. C **15**, 3005 (1982).

J. M. Calleja, H. Vogt, M. Cardona: Absolute Raman scattering efficiencies of some zincblende and fluorite-type materials. Philos. Mag. A **45**, 239 (1982).

F. Meseguer, J. C. Merle, M. Cardona: Optical deformation potentials of the copper halides. Solid State Commun. (in press).

Second-Order Spectra, Density of States

M. M. Sushchinsky, V. S. Gorelik, O. P. Maximov: Higher-order Raman spectra of GaP. J. Raman Spectrosc. **7**, 26 (1978).

T. Nanba, I. Kawashima, M. Ikezawa: Far infrared and Raman scattering spectra due to two phonons in thallous halides. J. Phys. Soc. Jpn. **50**, 3063 (1981).

R. Carles, A. Zwick, M. A. Renucci, J. B. Renucci: A new experimental method for the determination of the one phonon density of states in GaAs. Solid State Commun. **41**, 557 (1982).

K. T. Tsen, D. A. Abramsohn, R. Bray: Two-phonon Raman scattering probe of non-equilibrium, high frequency acoustic phonons: the TA phonon bottleneck in GaAs. Phys. Rev. Lett. (in preparation).

Surface Enhanced Raman Scattering

T. E. Furtak, J. Reyes: A critical analysis of theoretical models for the giant Raman effect from adsorbed molecules. Surf. Sci. **93**, 351 (1980).

D. N. Batchelder, N. J. Poole, D. Bloor: Surface–enhanced Raman spectroscopy of molecular oxygen adsorbed on polydiacetylene single crystals. Chem. Phys. Lett. **81**, 560 (1981).

T. F. Heinz, C. K. Chen, D. Ricard, Y. R. Shen: Spectroscopy of molecular monolayers by resonant second-harmonic generation. Phys. Rev. Lett. (in preparation).

Theory

E. Anastassakis, M. Cardona: Internal strains and raman-active optical phonons. Phys. Status Solidi B **104**, 589 (1981).

A. V. Gol'tsev: Interband Raman scattering of light in semiconductors. Sov. Phys. JETP **54**, 175 (1981).

I. S. Gorban, V. P. Grischuk, V. A. Gubanov, V. M. Moroz, V. F. Orlenko, A. V. Slobodyamyuk: Circular dichroism in Raman scattering by class 4 crystals. Sov. Phys. Solid State **23**, 866 (1981).

S. M. Kostritskii, A. E. Semenov, E. V. Cherkasov: Characteristics of the angular dependence of the angular dependence of the spontaneous Raman scattering in uniaxial crystals. Sov. Phys. Solid State **23**, 1219 (1981).

D. L. Tonks, J. B. Page: General theory of vibrational mode mixing and frequency shifts in resonance Raman scattering. Chem. Phys. Lett. **79**, 247 (1981).

D. Bermejo, S. Montero, M. Cardona, A. Muramatsu: Transferability of the bond polarizabilities: from saturated hydrocarbons to diamond. Solid State Commun. **42**, 153 (1982).

Subject Index

Contents of **Light Scattering in Solids II**

Basic Concepts and Instrumentation (Topics in Applied Physics, Vol. 50)

Contents of **Light Scattering in Solids III**

Recent Results (Topics in Applied Physics, Vol. 51)

Contents of **Light Scattering in Solids IV**

Electronic Scattering, Spin Effects, Sers, and Morphic Effects
(Topics in Applied Physics, Vol. 54)

morphous Semiconductors

ditor: M. H. Brodsky
79. 181 figures, 5 tables. XVI, 337 pages.
opics in Applied Physices. Volume 36)
BN 3-540-09496-2

ontents: *M. H. Brodsky:* Introduction. – *Kramer, D. Weaire:* Theory of Electronic ates in Amorphous Semiconductors. – *A. Davis:* States in the Gap and Defects in morphous Semiconductors. – *G. A. E. Conell:* ptical Properties of Amorphous Semiconuctors. – *P. Nagels:* Electronic Transport in morphous Semiconductors. – *R. Fischer:* uminescence in Amorphous Semiconducors. – *I. Solomon:* Spin Effects in Amorphous Semiconductors. – *G. Lucovsky, T. M. Hayes:* Short-Range Order in Amorphous Semiconductors. – *P. G. LeComber, W. E. Spear:* Doped Amorphous Semiconductors. – *D. E. Carlson, C. R. Wronski:* Amorphous Silicon Solar Cells.

Fundamental Physics of Amorphous Semiconductors

Proceedings of the Kyoto Summer Institute, Kyoto, Japan, September 8–11, 1980

Editor: **F. Yonezawa**
1981. 91 figures. VIII, 181 pages. (Springer Series in Solid-State Sciences, Volume 25)
ISBN 3-540-10634-0

Contents: What are Non-Crystalline Semiconductors. – Defects in Covalent Amorphous Semiconductors. – Surface Effects and Transport Properties in Thin Films of Hydrogenated Silicon. – The Past, Present and Future of Amorphous Silicon. – The Effect of Hydrogen and Other Additives on the Electronic Properties of Amorphous Silicon. – New Insights on Amorphous Semiconductors from Studies of Hydrogenated a-Ge, a-Si, a-$SI_{1-x}Ge_x$ and a-GaAs. – Chemical Bonding of Alloy Atoms in Amorphous Semiconductors. –Theory of Electronic Properties of Amorphous Semiconductors. – Some Problems of the Electron Theory of Disordered Semiconductors. –The Anderson Localisation Problem. – Summary Talk. – Seminars Given During the KSI '80. – Photographs of the Participants of the KSI '80. – List of Participants.

Photoemission in Solids I

General Principles

Editors: **M. Cardona, L. Ley**
1978. 90 figures, 17 tables. XI, 290 pages. (Topics in Applied Physics, Volume 26)
ISBN 3-540-08685-4

Contents: *M. Cardona, L. Ley:* Introduction. – *W. L. Schaich:* Theory of Photoemission: Independent Particle Model. – *S. T. Manson:* The Calculation of Photoionization Cross Sections: An Atomic View. – *D. A. Shirley:* Many-Electron and Final-State Effects: Beyond the One-Electron Picture. – *G. K. Wertheim, P. H. Citrin:* Fermi Surface Excitations in X-Ray Photoemission Line Shapes from Metals. – *N. V. Smith:* Angular Dependent Photoemission. – Appendix.

Photoemission in Solids II

Case Studies

Editors: **L. Ley, M. Cardona**
1979. 214 figures, 26 tables. XVIII, 401 pages. (Topics in Applied Physics, Volume 27).
ISBN 3-540-09202-1

Contents: *L. Ley, M. Cardona:* Introduction. – *L. Ley, M. Cardona, R. A. Pollak:* Photoemission in Semiconductors. – *S. Hüfner:* Unfilled Inner Shells: Transition Metals and Compounds. – *M. Campagna, G. K. Wertheim, Y. Baer:* Unfilled Inner Shells: Rare Earths and Their Compounds. – *W. D. Grobman, E. E. Koch:* Photoemission from Organic Molecular Crystals. – *C. Kunz:* Synchrotron Radiation: Overview. – *P. Steiner, H. Höchst, S. Hüfner:* Simple Metals. – Appendix: Table of Core-Level Binding Energies. – Additional References with Titles. – Subject Index.

Springer-Verlag Berlin Heidelberg New York